Green Energy and Technology

Climate change, environmental impact and the limited natural resources urge scientific research and novel technical solutions. The monograph series Green Energy and Technology serves as a publishing platform for scientific and technological approaches to "green"—i.e. environmentally friendly and sustainable—technologies. While a focus lies on energy and power supply, it also covers "green" solutions in industrial engineering and engineering design. Green Energy and Technology addresses researchers, advanced students, technical consultants as well as decision makers in industries and politics. Hence, the level of presentation spans from instructional to highly technical.

Indexed in Scopus.

Indexed in Ei Compendex.

More information about this series at https://link.springer.com/bookseries/8059

Venkata Rao Ravipudi · Hameer Singh Keesari

Design Optimization of Renewable Energy Systems Using Advanced Optimization Algorithms

Springer

Venkata Rao Ravipudi
Department of Mechanical Engineering
S. V. National Institute of Technology
Surat, Gujarat, India

Hameer Singh Keesari
Department of Mechanical Engineering
Sreenidhi Institute of Science
and Technology
Hyderabad, Telangana, India

ISSN 1865-3529 ISSN 1865-3537 (electronic)
Green Energy and Technology
ISBN 978-3-030-95588-5 ISBN 978-3-030-95589-2 (eBook)
https://doi.org/10.1007/978-3-030-95589-2

This Springer imprint is published by the registered company Springer Nature Switzerland AG
The registered company address is: Gewerbestrasse 11, 6330 Cham, Switzerland

Preface

Global warming is a significant concern that raises a need for cleaner production of energy. Renewable energy resources are significant sources of contribution to such clean energy demands. Taking advantage of these renewable energy sources provides significant opportunities for handling energy-related problems. In the last few decades, researchers have focused on renewable energy resources like solar energy, bioenergy, wave energy, ocean thermal energy, tidal energy, geothermal energy, and wind energy. This has resulted in the development of new techniques and tools that could harvest energy from renewable energy sources. However, to meet energy demands and reduce investment, a rigorous study of energy extraction systems is required. Identifying, analyzing, and optimizing the effect of various parameters of a renewable energy system contribute significantly to assessing the system performance. Furthermore, it is always not preferable to present the optimum system parameters considering only a single objective as these systems have multiple objectives such as power output, system efficiency, investment cost, and economic and ecological factors. Hence, researchers have developed various optimization models of these systems and presented optimum system parameters through single- and multi-objective optimization using advanced optimization algorithms.

All evolutionary computation and swarm intelligence-based optimization algorithms are population-based algorithms and have control parameters such as population size, crossover probability, mutation probability, scaling factor, inertia weight, social and cognitive parameters, among others. Appropriate adjustment of the control parameters dictates the algorithm convergence toward the global optimum. Inappropriate adjustment of the control parameters leads to premature convergence and increased computational efforts. Also, selecting an appropriate population size for different optimization applications is a tedious job. Additionally, in multi-objective optimization, selecting the most suitable solution from non-dominated solution set is difficult.

In this book, recently developed Jaya and Rao (Rao-1, Rao-2, and Rao-3) algorithms are described for single- and multi-objective optimization of selected renewable energy systems. In addition, variants of the Jaya and Rao algorithms are presented to show the improvement in performances. Furthermore, variants of the

Jaya algorithm, namely multi-team perturbation guiding Jaya (MTPG-Jaya) algorithm and adaptive multi-team perturbation guiding Jaya (AMTPG-Jaya) algorithm, and variants of the Rao algorithms, namely elitist Rao (ERao-1, ERao-2, and ERao-3) algorithms and self-adaptive population Rao (SAP-Rao) algorithm, are demonstrated for optimization of the selected renewable energy systems. These algorithms have no algorithm-specific parameters and require only the common control parameters. Additionally, the applicability of multi-attribute decision-making methods in multi-objective optimization problems is discussed, and a decision-making procedure is recommended based on average rank to identify the best solution in a Pareto-front.

The Jaya and Rao algorithms and their variants are developed by our team, and these are gaining wide acceptance in the optimization research community. After its introduction in 2016, the Jaya algorithm is finding many applications in different fields of science and engineering. The major applications, as of December 2021, are found in the fields of electrical engineering, mechanical design, thermal engineering, manufacturing engineering, civil engineering, structural engineering, computer engineering, electronics engineering, physics, chemistry, biotechnology, and economics. Many research papers have been published in various reputed international journals of Elsevier, Springer-Verlag, Taylor & Francis, and IEEE Transactions, in addition to those published in the proceedings of international conferences. The number of research papers is continuously increasing at a faster rate. The Jaya algorithm and its variants have carved a niche in the field of advanced optimization, and many more researchers may find this as a potential optimization algorithm. The Rao algorithms are developed by our team in 2020, and these are also gaining acceptance in the optimization research community.

This book presents a comprehensive review on latest research and development trends at international level for parameter optimization of various renewable energy systems. Using examples of various renewable energy systems, the possibilities for parameter optimization with Jaya and Rao algorithms including their variants are demonstrated. The book presents real case studies, results of applications of the basic Jaya and Rao algorithms, and their variants and comparison with other advanced optimization techniques and highlights the best optimization technique to achieve best performance. The book also includes the validation of different variants of the Jaya and Rao algorithms through application to complex single- and multi-objective unconstrained benchmark functions. The algorithms and computer codes of different versions of Jaya and Rao algorithms are included in the book that will be very much useful to the readers. This book is expected to become a valuable reference for those wishing to do research on the use of advanced optimization techniques for solving single-/multi-objective combinatorial optimization problems related to the renewable energy systems.

We are grateful to Dr. Anthony Doyle and his team of Springer for their support and help in producing this book. We wish to thank various researchers and the publishers of international journals for publishing the research works of our team. Our special thanks are due to the director and the colleagues at Sardar Vallabhbhai National Institute of Technology, Surat, India. While every attempt has been made to ensure that no errors (printing or otherwise) enter the book, the possibility of these creeping into the book is always there. We will be grateful to the readers if these errors are pointed out. Suggestions for further improvement of the book will be thankfully acknowledged.

Surat, India
Hyderabad, India
December 2021

Venkata Rao Ravipudi
Hameer Singh Keesari

Contents

1 Introduction to Renewable Energy Systems 1
1.1 Solar Energy Systems 1
1.2 Wind Energy Systems 2
1.3 Hydroenergy Systems 3
1.4 Ocean Thermal Energy Systems 4
1.5 Geothermal Energy Systems 4
1.6 Bioenergy Systems 5
1.7 Nuclear Energy Systems 6
1.8 Other Emerging Renewable Energy Technologies 7
References 8

2 Selected Renewable Energy Systems and Formulation of Their Problems 11
2.1 Wind Farm Layout 11
2.1.1 Wake Model 11
2.1.2 Power Generation Model 13
2.1.3 Cost Model 13
2.1.4 Objective Function of the Wind Farm Layout Optimization 14
2.2 Solar Assisted Energy Systems 14
2.2.1 Solar-Assisted Brayton Heat Engine System 14
2.2.2 Solar-Assisted Stirling Heat Engine System 17
2.2.3 Solar-Assisted Carnot-Like Heat Engine System 22
2.3 Bio-Energy Systems 23
2.3.1 Single-Cylinder Direct-Injection Diesel Engine 24
2.3.2 Turbocharged DI Diesel Engine 25
2.3.3 Compression Ignition Biodiesel Engine with an EGR System 26
2.3.4 Microalgae-Based Biomass Cultivation Process 28
2.4 Hydro Energy and Geothermal Energy Systems 29
2.4.1 Hydropower Generation and Reservoir Operation 29

2.4.2 Ground Source Heat Pump-Radiant Ceiling Air Conditioning System ... 29
References ... 30

3 Advanced Engineering Optimization Techniques and Their Role in Energy Systems Optimization ... 33
References ... 46

4 Working of Jaya and Rao Optimization Algorithms and Their Variants ... 53
4.1 Working of the Jaya Algorithm and Its Modified Versions ... 53
4.1.1 Jaya Algorithm ... 53
4.1.2 Multi-team Perturbation-Guiding Jaya (MTPG-Jaya) Algorithm ... 55
4.1.3 Adaptive Multi-team Perturbation-Guiding Jaya (AMTPG-Jaya) Algorithm ... 61
4.1.4 Multi-objective Jaya and Multi-objective AMTPG-Jaya Algorithms ... 66
4.2 Working of the Rao Algorithms and Their Modified Versions ... 69
4.2.1 Rao Algorithms ... 69
4.2.2 Multi-objective Rao Algorithms ... 70
4.2.3 Elitist Rao Algorithms ... 71
4.2.4 Multi-objective Elitist Rao Algorithms ... 74
4.2.5 Self-Adaptive Population Rao (SAP-Rao) Algorithm ... 74
4.2.6 Multi-objective SAP-Rao Algorithms ... 79
4.3 Performance Indicators ... 80
4.3.1 Coverage ... 80
4.3.2 Spacing ... 80
4.3.3 Hypervolume ... 81
4.3.4 Inverted Generational Distance (IGD) ... 81
4.4 Computational Results Analysis on Single-objective Optimization of Unconstrained Benchmark Problems ... 82
4.4.1 Computational Results Analysis on 30 Unconstrained Standard Benchmark Problems ... 82
4.4.2 Computational Results Analysis on Unconstrained Unimodal and Multimodal Standard Benchmark Problems ... 93
4.5 Computational Results Analysis on Multi-objective Optimization Benchmark Problems ... 114
References ... 127

5 Multi-attribute Decision-Making Methods and Their Implementation in Energy Systems ... 131
5.1 Simple Additive Weighing (SAW) ... 132
5.2 Weighted Product Method (WPM) ... 132

5.3 Preference Ranking Organization Method for Enrichment Evaluation (PROMETHEE) 132
5.4 Technique for Order Preference by Similarity to Ideal Solution (TOPSIS) 133
5.5 Modified TOPSIS (MTOPSIS) 134
5.6 Compromise Ranking Method (VIKOR) 135
5.7 Complex Proportional Assessment (COPRAS) 136
5.8 Gray Relational Analysis (GRA) 137
References 139

6 Optimization of Wind Farm Layouts 141
6.1 Problem Definition and Wind Scenarios of the Wind Farm Layout Optimization 141
6.2 Case-I: Fixed Wind Speed and Fixed Direction 142
6.3 Case-II: Fixed Wind Speed and Variable Wind Direction 145
6.4 Case-III: Fixed Wind Speed and Variable Wind Direction 147
References 156

7 Optimization of the Selected Solar-Assisted Energy Systems 159
7.1 Optimization of a Solar-Assisted Brayton Heat Engine System 159
7.2 Optimization of a Solar-Assisted Stirling Heat Engine System 176
7.2.1 Case Study-1 176
7.2.2 Case Study-2 204
7.2.3 Case Study-3 216
7.3 Optimization of a Solar-Assisted Carnot-like Heat Engine System 230
References 247

8 Optimization of the Selected Bio-Energy Systems 249
8.1 Design Optimization of the Single-Cylinder Direct-Injection Diesel Engine 249
8.2 Design Optimization of a Turbocharged DI Diesel Engine 273
8.3 Design Optimization of a Compression Ignition Biodiesel Engine with an EGR System 288
8.4 Process Optimization of a Microalgae-Based Biomass Cultivation Process 313
References 327

9 Optimization of Hydroenergy and Geothermal Energy Systems 329
9.1 Optimization of a Hydropower Generation System 329
9.2 Optimization of a Ground Source Heat Pump-Radiant Ceiling Air Conditioning System 331
References 343

Appendices 345

Index 389

Chapter 1
Introduction to Renewable Energy Systems

Abstract Global warming is a significant concern that raises a need for cleaner energy production. In the last few decades, researchers have focused on exploiting renewable energy resources to meet the clean energy demands. This chapter presents a brief introduction to different renewable energy generation systems and the associated problems.

1.1 Solar Energy Systems

Solar energy has the greatest potential of all the sources of renewable energy. Solar power is vastly available and is a low-grade renewable energy. However, the use of this energy on commercial basis is somewhat facing difficulties due to the following reasons:

- The solar energy is very dilute and spread out.
- The intensity of radiation is weather dependent and season dependent. The variation due to the weather and season affects the quantum of radiation received in a place.
- The solar radiation is available only in daytime. The day and night cycle interrupts the continuous flow of energy.

Solar thermal energy and solar electric energy are the two ways in which solar energy can be used. A solar thermal system produces hot water or hot air, cooks food, dries materials, etc. Solar photovoltaic uses solar radiation in a solar electric energy system to produce electricity for household appliances and commercial and industrial buildings. Using a solar-assisted heat engine is a best way of extracting solar energy. The solar concentrator focuses the sunlight at the concentrator's focus point. The receiver located at the focus point of the concentrator absorbs the heat energy and transfers it to the heat engine that produces the useful work.

Another best way to extract solar energy is through solar photovoltaic energy systems. Electrical power is generated by converting solar radiation into the flow of electrons by using photovoltaic (PV) to obtain direct current electricity using semiconductors that exhibit the photovoltaic effect. Many solar panels comprising

V. R. Ravipudi et al., *Design Optimization of Renewable Energy Systems Using Advanced Optimization Algorithms*, Green Energy and Technology,
https://doi.org/10.1007/978-3-030-95589-2_1

many cells containing a semiconducting material are needed to generate direct current (DC) electricity.

The successful applications of solar energy are: cooling and heating of residential buildings, solar cookers, solar drying of animal and agricultural products, solar water heating, evaporation of seawater to produce salt, solar refrigeration and air conditioning, food refrigeration, solar distillation on a small scale, solar engines for water pumping, solar television, radio and tube lights, solar calculators, solar furnaces, solar drying, photovoltaic conversion, solar thermal power station, etc.

Solar energy systems are safe and have a well-established technology now compared to the other technologies related to electricity generation. Solar energy systems generally have a complex structure, making it a significant task to identify all possible environmental impacts for life cycle analysis and assessment. Comparatively lower ecological risks are associated with photovoltaic cell technologies. Many characteristics of these systems have been studied by the researchers. The performance measures such as economic and ecological aspects, system efficiency, and output power have been considered and the optimal design and process parameters are presented (Santos et al., 2017; Kajela & Manshahia, 2017; Khare et al., 2019a; Al-Shahri et al., 2021).

1.2 Wind Energy Systems

Wind energy can be economically used for the electrical energy generation. The winds are caused by two factors:

- Heating and cooling of earth's atmosphere which generates conversion currents. Heating and cooling are caused by the day and night solar energy cycle. In fact, it is the sun's energy which causes the wind.
- The rotation of the earth and its motion around the sun. Due to this, seasonal changes take place which affects the wind.

The differences in heating of the ground surface by the sun cause the movement of large air. If the wind speed is between 5 and 25 m/s, then this can be used for electricity generation. Wind energy conversion systems do electricity generation through wind. Wind energy conversion systems convert the wind's kinetic energy into electricity or other forms of energy.

Nowadays, wind power is a thoroughly established and sustainable branch of electricity generation. The horizontal axis wind turbine and vertical axis wind turbine are the two types of wind turbines used in wind energy conversion systems. Many types of windmills have been designed and developed. However, only a few have been found to be practically suitable and useful. Some of them are horizontal axis propeller-type windmills, sail-type horizontal axis windmill, multi-blade type of horizontal axis windmill, Savonius-type vertical axis windmill, and Darrieus-type vertical windmill. Among the different systems proposed, the proven and established wind energy systems are: (i). simple on-site domestic unit with storage battery and (ii). simple system with mechanical and solar storage.

The wind speed distribution and the average wind speed affect the power produced by a wind turbine. The strength of the wind varies from zero to storm force. The important criteria of power generation are the design of wind turbines and rotor blades. Each wind turbine design is related to a particular wind velocity to generate electrical energy (Khare et al., 2019a).

The noise pollution from turbines is significantly reduced in the modern wind turbine designs. The noise reflects the loss of energy and decreases in efficiency. Recent advancements in wind turbine design locate the blades upwind in place of downwind to reduce the level of infrasound and noise pollution (Khare et al., 2019a).

1.3 Hydroenergy Systems

The hydel power being a renewable perennial source of energy plays a vital role in generation of power. The quantity of hydroelectricity depends on the volume of water flow and the head created by the water reservoir. Once a dam is constructed, the electricity can be produced at a constant rate for many decades. During the operation, greenhouse gases and the atmospheric pollution are not created. However, the natural environment may be destroyed due to flooding of large land areas. The hydroelectric dams must be built in areas with the suitable conditions. The major difficulty in the development of hydropower projects is the relatively longer time required for its hydrological, topological, and geological investigations. Hydropower projects generate power at low cost. It is renewable, easy to manage, and pollution free. However, the major drawback is that it is dependent on the mercy of the nature. These plants are very suitable for peak load operation along with thermal plants being operated at base load. Depending on the load, the hydropower stations can be low head (less than 30 m), medium head (between 30 and 300 m), and high head (more than 300 m). Depending upon the plant capacity, the hydropower stations can be microhydel (less than 5 MW), medium capacity (5–100 MW), high capacity (101–1000 MW), and super plants (greater than 1000 MW).

A hydroenergy system is considered a renewable as well as a non-renewable energy system. Small hydroprojects have a relatively low environmental impact compared to a large hydroproject because they frequently have small reservoirs and civil construction work (Khare et al., 2019a).

The hydropower scheduling is a complex optimization problem that includes nonlinear dynamical hydraulic heads, nonlinear flows, and the interactional relationships of nonlinear input and output variables. The objective is to obtain optimal hydroresources available for maximum hydroelectric generation given a set of starting conditions and many complex constraints. The researchers have discussed the operation of reservoir and interconnected systems and developed the mathematical models for the optimum operation of reservoir for the power generation. The parameters such as reservoir level, time of release water, and flow of the stream are generally included in the mathematical models.

1.4 Ocean Thermal Energy Systems

A large amount of solar energy is collected and stored in tropical oceans. The heat contained in the oceans is converted into electricity by utilizing the fact that the temperature difference between the warm surface waters of the oceans and the water in the depths is about 20-25 °C. The basis of ocean thermal energy conversion (OTEC) systems lies on proper utilization of this storage with its associated temperature difference and its conversion into work. The surface water which is at higher temperature could be used to heat some low boiling organic fluid, the vapors of which would run a heat engine. The exit vapor would be condensed by pumping cold water from the deeper regions. The amount of energy available for ocean thermal power generation is enormous and is replenished continuously.

An OTEC plant operates based on a Rankine cycle, often with a refrigerant as the working fluid. The required heat exchanger areas of the evaporator and condenser and the water mass flow rates are very large and reduce the overall efficiency of the OTEC systems. The disadvantages of OTEC include low thermodynamic efficiency due to very low temperature difference, the size of boilers and condensers becomes big, capital cost goes up, and onshore installation requires long and big piping which add up to the cost. The materials suggested for heat exchangers are titanium or alloy of copper and nickel which are resistant to corrosion, but then the cost adds up.

The efficiency of an OTEC plant can be increased by using multiple cascaded stages instead of only utilizing a single stage. Upshaw et al. (2011) provided a methodology for analyzing the major plant design parameters and determining the optimal design of OTEC facilities. To take advantage of the potential of the ocean as an unlimited energy source, many researchers are working on developing new and efficient solutions. The optimization of condensers, vaporizers, heat exchangers, and turbines is very important. The optimization methodology can be developed from thermodynamic theory, new materials, and engineering designs.

1.5 Geothermal Energy Systems

Energy present as heat in the earth's crust (about 10 km depth) is potentially useful. This heat is apparent from increase in temperature of the earth with increasing depth below the surface. Although high and low temperatures occur, the average temperature at a depth of 10 km is present underneath the earth's surface at depths greater than about 200 °C. Hot molten or partially molten rock, called magma, with temperature of about 3000 °C is present underneath the earth's surface at depths greater than 24–40 km. The crust ranging in thickness from 15 to 159 km insulates the surface of the earth and protects it. However, at certain locations, where thickness is small and the crust is cracked, the subsoil water comes in contact with the hot matter and gets converted into steam or hot water. In some places, the steam/hot water issues under pressure. The steam coming under high pressure or water at high temperature can be used in conventional steam turbines for power generation or for heating purposes. Four different geothermal systems have been developed/proposed, and they are: (i). dry steam system, (ii). wet steam system, (iii). hot dry rack system, and (iv). magmatic system. Some of the advantages of geothermal energy are: versatility in use, continuous availability, less pollution, and low capital and generation cost compared to nuclear and coal plants. Some of the disadvantages include overall low efficiency, noisy drilling operations, corrosion of components due to salt, thermal pollution due to effluent if not reinjected, nuisance due to gaseous effluents, less life span compared to nuclear and coal plants, and requirement of large areas for exploitation of geothermal energy.

The heat energy can be used either for heating or for making electricity, depending on the temperature of the hydrothermal resource. Low-temperature geothermal energy is used to heat and cool homes and public apartments, and grow crops and dry lumber, vegetables, and fruits. To generate electrical energy, hydrothermal resources at high temperatures can be used.

Various designs have been proposed to improve the energy conversion efficiency of geothermal energy systems and their optimization (Lee et al., 2019). The principles of basic and state-of-the-art technologies are essential for developing the advanced energy systems. A comprehensive review of the geothermal energy systems must be carried out from systems design, analysis, and optimization.

The geothermal energy systems can be expected to proliferate soon to meet the demand of increasing worldwide primary energy demands coupled with the need for fossil fuel replacement. Improving the conversion efficiency is an important issue of the geothermal energy systems for the optimum use of geothermal resources. For efficient geothermal energy systems development, a deep understanding of the conversion principles and configurations of the existing technologies is helpful.

1.6 Bioenergy Systems

Biomass is a biological material derived from living or recently lived organisms. Electricity generation through a biomass system is a renewable energy source. Biofuel is considered as an environmentally friendly and sustainable alternative to petroleum-based fuels. Biofuels are produced from sugar, starch, vegetable oil, animal fats, switchgrass, cereals, agricultural waste, and microalgae. Biodiesel engine system uses the blend of biofuel and petrodiesel as the fuel. The performance of the biodiesel engine has been considerably affected by the blending proportions. Using biodiesel as an alternative to diesel can reduce the wear of the engine parts and the emissions such as carbon oxides, hydrocarbons, and particulate matter. However, there is a loss of power, lower fuel efficiency, and increased nitrogen oxides emission with biodiesels.

Food sources such as animal fats, sugar, starch, and vegetable oil are the first-generation biofuels. The second-generation biofuels are derived from cellulosic biomass, such as agricultural waste, switchgrass, and cereals, using thermochemical processes. Producing biofuels from microalgae is gaining popularity nowadays (Agarwal et al., 2017; Banerjee et al., 2016; Dickinson et al., 2017). Reducing exhaust emissions and enhancing engine performance of diesel engines is a crucial area in biodiesel research (Damanik et al., 2018).

Biomass energy plants are like the conventional power plants because both involve the combustion of a feedstock to generate electrical energy. Biomass energy systems are like conventional power plants in groundwater use and air emissions. Valuable biomass resources include energy crops that do not compete with food crops for land, portions of crop residues such as wheat straw or corn stover, clean municipal and industrial wastes, and sustainably harvested wood and forest residues (Khare et al., 2019a).

1.7 Nuclear Energy Systems

The use of nuclear energy is considered as a source of power and as an appropriate alternative to meet the ever-increasing demand. Opposition to the installation of this type of plants is mainly due to the fear of radiation hazards. ${}_{92}U235$ is one of the elements whose nucleus easily fissions. Other naturally available elements are stable, and the nucleus of them is not split easily. When a neutron enters the nucleus of ${}_{92}U235$, the nucleus splits into two fragments and releases 2 to 3 neutrons per fission. The difference in binding energy between the products of fission and the original nucleus is involved during the fission reaction. This process is called as nuclear fission. The neutrons released during the fission cause further fission reactions of uranium nucleus, and a chain reaction is maintained.

The chain reaction under uncontrolled condition is known as the atomic explosion. The breaking of U^{235} nucleus by neutron absorption can take place in several ways. Each way of splitting U^{235} nucleus ejects different number of neutrons. On an average, 2.5 neutrons are ejected per neutron absorbed. Nearly 0.3 neutron is lost due to escape at the surface out of 2.5 neutrons, and the remaining 2.2 neutrons are allowed to continue chain reaction. Enormous amount of heat energy will be evolved as the reaction rate will increase exponentially. Such reaction is known as uncontrolled chain reaction. It is called controlled chain reaction when only one neutron after every fission is allowed to continue to cause fission reaction. This type of reaction is used in nuclear reactors, and the energy evolved remains at a constant level. Usually out of 2.5 neutrons per fission, about 0.8 neutron is absorbed by U^{238} converting into a fissionable material Pu^{239}, 0.5 neutron is absorbed by the control rod, 0.2 neutron is lost from the reaction, and the chain reaction is maintained by the remaining 1 neutron. If the number of neutrons for next reaction is less than 1 and absorption is more than above, then the fission reaction will die out very quickly (Dubey et al., 2013).

The nuclear fuels are mainly classified as fissile fuels such as $_{92}U235$, $_{92}U233$, and $_{94}Pu^{239}$, and fertile fuels such as $_{92}U^{238}$ and $_{90}Th^{232}$. The basic components of a nuclear reactor of a power plant include fuel element such as natural or enriched uranium, moderator such as ordinary water or heavy water or graphite or beryllium, reflector, coolant such as ordinary water or heavy water or carbon dioxide, control rods to start and maintain the nuclear chain reaction or to shut down the reactor automatically under emergency conditions, shielding, and a reactor vessel. Different types of reactors are in use such as ordinary water-cooled reactor (either boiling water reactor or pressurized water reactor) and heavy water-cooled and moderated reactor.

The advantages of nuclear energy systems include less pollution, no contribution to global warming, very low fuel costs, long lifetime of the plant, and generation of large amount of electricity with a very small amount of nuclear fuel. The disadvantages include disposal of radioactive waste (hazardous to humans and the environment), high costs of building and safe decommissioning, catastrophic accidents, and local thermal pollution from wastewater that may affect the marine life.

1.8 Other Emerging Renewable Energy Technologies

Wave energy is a renewable energy system that extracts energy directly from surface waves. There is a huge amount of energy in ocean waves to provide up to 3 TW of electricity. However, the big limitation is that the wave energy cannot be harnessed everywhere.

Wave energy can be measured as an intense form of solar and wind energy. The differences in heating of the earth by the sun produce winds, and when air passes over open waterbodies, the wind is converted into waves. Such a mechanism is used to produce electrical energy with the help of wave energy. The major advantage of wave power is that it is easily predictable and can be used to calculate the amount of power that it can produce. However, wave power is in the early stages of development and costs involved may be higher (Khare et al., 2019a).

Tidal power or tidal energy is a type of hydropower that changes the energy acquired from tides in valuable types of energy. The tides in the ocean are due to the gravitational effect of the sun and moon on the earth. The effect of this force is apparent in the motion of water which shows a periodic rise and fall in levels which is in rhythms with the daily cycle of rising and setting of sun and moon. This periodic rise and fall of the water level of sea is called tide. These tides can be used to produce electric power which is known as tidal power.

The use of tides for electric power generation is practical in a few favorable situated sites where the geography of an inlet or bay favors the construction of a large-scale hydroelectric plant. A dam is to be built across the mouth of the bay to harness the tides. Water streams from the ocean into the tidal bowl through the water turbine at high tide. The stature of the tide is over that of the tidal bowl. Consequently, the turbine unit works, and the generator coupled to the turbine produces the power (Khare et al., 2019b).

The tidal energy has significant potential to meet the future power requirements. Tidal energy offers many advantages such as availability throughout the year irrespective of weather conditions, clean energy without pollution, requirement of small land area, and meeting the power requirements of local coastal areas. However, tidal energy is also linked to some disadvantages such as highly variable tidal range, limited power availability, suitable only for selected locations, and corrosion of machinery due to seawater.

The ocean currents are highly predictable. This makes it easier to construct tidal energy systems with the correct dimensions as the kind of power the equipment will be exposed to its known. It is effective at low speeds because of the higher density of water. However, it still has some environmental effects, and the tidal energy is considered reliable only when adequate energy storage solutions are available.

In the future, the number of renewable sources will continually increase with the increase in the demand for more energy and that may reduce the price of the renewable energy.

References

Agarwal, A. K., Agarwal, R. A., Gupta, T., & Gurjar, B. R. (2017). *Introduction to Biofuels*. In Biofuels, Springer. https://doi.org/10.1007/978-981-10-3791-7_1

Al-Shahri, O. A., Ismail, F. B., Hannan, M. A., Lipu, M. S. H., Al-Shetwi, A. Q., Begum, R. A., Al-Muhsen, N. F. O., & Soujeri, E. (2021). Solar photovoltaic energy optimization methods, challenges and issues: A comprehensive review. *Journal of Cleaner Production, 284*, 125465. https://doi.org/10.1016/j.jclepro.2020.125465

Banerjee, A., Guria, C., & Maiti, S. K. (2016). Fertilizer assisted optimal cultivation of microalgae using response surface method and genetic algorithm for biofuel feedstock. *Energy, 115*, 1272–1290. https://doi.org/10.1016/j.energy.2016.09.066

Dickinson, S., Mientus, M., Frey, D., Amini-Hajibashi, A., Ozturk, S., Shaikh, F., & El-Halwagi, M. M. (2017). A review of biodiesel production from microalgae. *Clean Technologies and Environmental Policy, 19*(3), 637–668. https://doi.org/10.1007/s10098-016-1309-6

Damanik, N., Ong, H. C., Tong, C. W., Mahlia, T. M. I., & Silitonga, A. S. (2018). A review on the engine performance and exhaust emission characteristics of diesel engines fueled with biodiesel blends. *Environmental Science and Pollution Research, 25*(16), 15307–15325. https://doi.org/10.1007/s11356-018-2098-8

Dubey, S. P., Memon, A. A., & Bhatt, M. K. (2013). A basic course in mechanical engineering. *New Popular Prakashan*, Surat, India.

Kajela, D., & Manshahia, M. S. (2017). Optimization of renewable energy systems : A review. *International Journal of Scientific Research in Science and Technology, 3*(8), 769–795.

Khare, V., Khare, C., Nema, S., & Baredar, P. (2019a). Introduction to energy sources. Tidal Energy Systems. https://doi.org/10.1016/b978-0-12-814881-5.00001-6

Khare, V., Khare, C., Nema, S., & Baredar, P. (2019b). Introduction of tidal energy. Tidal Energy Systems.https://doi.org/10.1016/b978-0-12-814881-5.00002-8

Lee, I., Tester, J. W., & You, F. (2019). Systems analysis, design, and optimization of geothermal energy systems for power production and polygeneration: State-of-the-art and future challenges.

Renewable and Sustainable Energy Reviews, 109, 551–577. https://doi.org/10.1016/j.rser.2019.04.058

Santos, S. F., Fitiwi, D. Z., Shafie-khah, M., Bizuayehu, A. W., & Catalão, J. P. S. (2017). Introduction to renewable energy systems. *Optimization in Renewable Energy Systems: Recent Perspectives*, 1–26. https://doi.org/10.1016/B978-0-08-101041-9.00001-6

Upshaw, C. R., & Webber, M. E. (2011). Integrated thermal-fluids system modeling of an ocean thermal energy conversion power plant for analysis and optimization. In *ASME 2011 5th International Conference on Energy Sustainability, Parts A, B, and C*, (pp. 1255–1264). https://doi.org/10.1115/ES2011-54595

Chapter 2
Selected Renewable Energy Systems and Formulation of Their Problems

Abstract This chapter presents the details of the case studies of selected renewable energy systems along with their problem formulations. The selected renewable energy system case studies include wind farm layout optimization for different wind scenarios, a case study of solar-assisted Brayton heat engine system, three case studies of solar-assisted Stirling heat engine system, a case study of solar-assisted Carnot-like heat engine system, three case studies of biodiesel engine, a case study of the microalgae cultivation process, a case study of hydropower energy generation, and a case study of a ground source heat pump-radiant ceiling air conditioning system.

2.1 Wind Farm Layout

A wind farm is a large area where several wind turbines are installed as a cluster. The wind farm layout optimization problem (WFLO) aims to minimize the cost of energy. It can be accomplished by optimally placing the number of wind turbines available such that maximum possible energy can be extracted. In the process of extracting the kinetic energy of wind, wind turbines decrease the wind speed at the back of the rotor and swirl the airflow; this is called wind turbine wake effect. Consequently, wind turbines in the wake effect of another turbine receive a modified wind inflow in terms of mean velocity and turbulence, which causes lesser energy production.

2.1.1 Wake Model

The Jensen wake model is used to determine the growth of the wake effect created by an upstream turbine on a downstream turbine. The wind velocity in the wake is affected by the distance between the upstream turbines to the downstream turbine (Shakoor et al., 2016). The assumptions made by Mosetti et al. (1994) and Grady et al. (2005) are followed in this book. According to Jensen wake model, the incident wind velocity (u) of a downstream wind turbine at a distance x as shown in Fig. 2.1 is given by the following equations (Moorthy & Deshmukh, 2013):

V. R. Ravipudi et al., *Design Optimization of Renewable Energy Systems Using Advanced Optimization Algorithms*, Green Energy and Technology,
https://doi.org/10.1007/978-3-030-95589-2_2

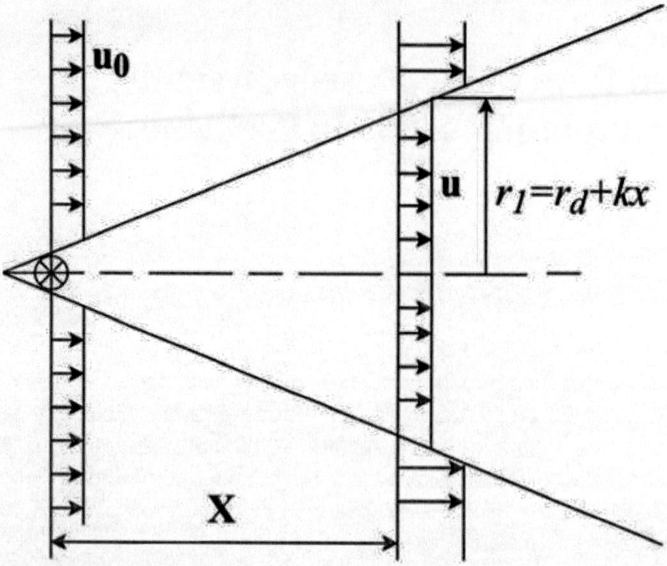

Fig. 2.1 Wake behind a wind turbine

$$u/u_0 = 1 - \left(2a/\left(1 + \frac{kx}{r_1}\right)^2\right) \tag{2.1}$$

$$r_1 = r_d\sqrt{\frac{1-a}{1-2a}} \tag{2.2}$$

$$C_T = 4a(1-a) \tag{2.3}$$

Where u_0 is the incident wind speed at the upstream turbine, a is axial induction factor, k is Entrainment constant, r_1 is downstream rotor radius, C_T is turbine thrust coefficient, and r_d is the rotor radius. Let z be the hub height of the turbine and z_0 be the surface roughness, then k is given by:

$$k = 0.5/\ln\frac{z}{z_0} \tag{2.4}$$

2.1.2 Power Generation Model

The algorithm proposed by Chowdhury et al. (2012) to calculate the total power produced by the wind farm is used. The power generated by turbine j for an incident velocity u is then given by:

$$P_j = \eta \frac{1}{2} \rho A u^3 \tag{2.5}$$

Taking efficiency of wind turbine (η) as 40%, air density (ρ) as 1.2 kg/m^3 and swept area of wind turbine,

$$P_j = 0.3u^3 \text{ kW} \tag{2.6}$$

The total power produced by the farm (P_{total}) consisting of N turbines is the summation of power produced by the individual wind turbines.

$$P_{\text{total}} = \sum_{\text{j=1}}^{\text{N}} \text{P}_\text{j} \tag{2.7}$$

Farm efficiency (η_{farm}) is given by Eq. 2.8. Let the power produced by a turbine when it has encountered wind with full velocity be P_{0j}.

$$\eta_{\text{farm}} = \frac{P_{\text{total}}}{\sum_{j=1}^{N} P_{0j}} \tag{2.8}$$

2.1.3 Cost Model

The total cost of the wind farm is the same as that used by Mosetti et al. (1994). For a wind farm consisting of N turbines, the total cost per year is given by the following equation:

$$Cost_{total} = N(\frac{2}{3} + \frac{1}{3}e^{-0.00174N^2}) \tag{2.9}$$

2.1.4 Objective Function of the Wind Farm Layout Optimization

The objective function of this WFLO problem has been chosen to minimize the cost of unit power generated, which is given by:

$$\text{Objective Function} = \text{Cost}_{\text{total}}/P_{\text{total}} \tag{2.10}$$

It can be observed from the objective of the WFLO model considered in this book, that the objective function is independent of the time variable.

2.2 Solar Assisted Energy Systems

Six multi-objective optimization case studies of solar assisted energy systems are considered for optimization to see if there can be any improvement in the performances of the selected systems. The subsequent section presents the solar-assisted Brayton heat engine system.

2.2.1 Solar-Assisted Brayton Heat Engine System

The schematic diagram of the solar-assisted regenerative Brayton heat engine system considered in this book is shown in Fig. 2.2. This system is driven by the combination of solar energy and fossil fuel. The heat absorbed at the receiver is used to heat the working fluid before entering the combustion chamber. Hence, less fuel is burned to heat the working fluid. The solar energy from the receiver is used as the supplement

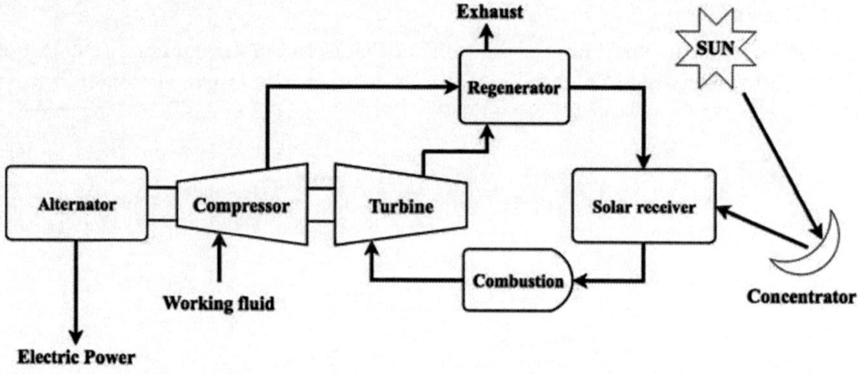

Fig. 2.2 Schematic diagram of a solar-assisted Brayton heat engine system

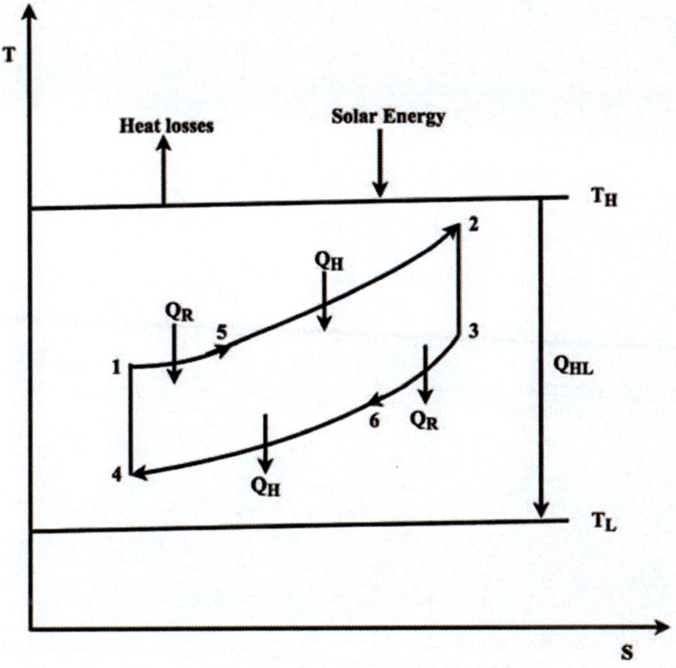

Fig. 2.3 T-S diagram of a solar-assisted Brayton heat engine system

to the fuel of a solar-assisted Brayton heat engine. Furthermore, the waste heat of working fluid exhausted from the gas turbine is used to preheat the working fluid coming out of the compressor. The working fluid from the combustion chamber expands rapidly and drives the turbine, which in turn drives the alternator Li et al. (2015).

The T-S diagram of the solar assisted regenerative Brayton engine is shown in Fig. 2.3. The Brayton cycle (with ideal regenerator) consists of four processes.

Process 1-2: an isobaric heat addition. During this process, the working fluid is heated by the regenerator (state 1-5) and then the heat source at T_H temperature (state 5-2).

Process 2-3: an isentropic expansion process, in which the hot working fluid drives a turbine.

Process 3-4: an isobaric heat rejection process in which heat released to the regenerator (state 3-6) and then to heat sink at T_L temperature (state 6-4).

Process 4-1 is an isentropic compression process. During this process, the working fluid at state 4 is compressed to state 1 by the compressor.

Let, T_1 be the temperature of the working fluid at state 1, T_0 be the ambient temperature, ε_H be the hot side heat exchange effectiveness, ε_L be the cold side heat exchange effectiveness, ε_R be the regenerator effectiveness, and C_{wf} be the working fluid heat capacity rate, then the power output is given by the following equation:

$$P = C_{wf}\varepsilon_H[T_H - (1-\varepsilon_R)T_1 - \varepsilon_R a_8] - C_{wf}\varepsilon_L[(1-\varepsilon_R)a_8 + \varepsilon_R T_1 - T_L] \tag{2.11}$$

where,

$$a_1 = (1-\varepsilon_H)(1-\varepsilon_L)(1-\varepsilon_R)\varepsilon_R \tag{2.12}$$

$$a_2 = a_4 T_1 + a_5 \tag{2.13}$$

$$a_3 = a_1 T_1^2 + a_6 T_1 + a_7 \tag{2.14}$$

$$a_4 = (1-\varepsilon_H)(1-\varepsilon_L)\varepsilon_R^2 + (1-\varepsilon_H)(1-\varepsilon_L)(1-\varepsilon_R)^2 - 1 \tag{2.15}$$

$$a_5 = \varepsilon_H(1-\varepsilon_R)(1-\varepsilon_L)T_H + (1-\varepsilon_H)\varepsilon_L\varepsilon_R T_L \tag{2.16}$$

$$a_6 = \varepsilon_H(1-\varepsilon_L)\varepsilon_R T_H + (1-\varepsilon_H)\varepsilon_L(1-\varepsilon_R)T_L \tag{2.17}$$

$$a_7 = \varepsilon_H\varepsilon_L T_H T_L \tag{2.18}$$

$$a_8 = \left[-a_2 - \sqrt{(a_2^2 - 4a_1a_3)}\right]/2a_1 \tag{2.19}$$

Now, the Brayton engine thermal efficiency (η_b) and solar concentrator efficiency (η_s) are given by:

$$\eta_b = \frac{\varepsilon_H[T_H - (1-\varepsilon_R)T_1 - \varepsilon_R a_8] - \varepsilon_L[(1-\varepsilon_R)a_8 + \varepsilon_R T_1 - T_L]}{\varepsilon_H[T_H - (1-\varepsilon_R)T_1 - \varepsilon_R a_8] + \xi(T_H - T_L)} \tag{2.20}$$

$$\eta_s = \eta_0 - \frac{1}{IR_C}[h_c(T_{H_{avg}} - T_0) + e_C\delta(T_{H_{avg}}^4 - T_0^4)] \tag{2.21}$$

where, $T_{\mathrm{H_{avg}}}$ is average absorber temperature, R_C is the collector concentrating ratio, e_C is the emissivity factor of the collector, h_c is the convection heat transfer coefficient, I is the solar irradiance, δ is the Stefan-Boltzmann constant, and η_0 is the collector optical efficiency. Now, the thermal efficiency of the system is given by:

$$\eta_m = \eta_s\eta_b \tag{2.22}$$

The non-dimensional thermo-economic performance function is given by:

$$F = \frac{\varepsilon_H[T_H - (1-\varepsilon_R)T_1 - \varepsilon_R a_8] - \varepsilon_L[(1-\varepsilon_R)a_8 + \varepsilon_R T_1 - T_L]}{\varepsilon_H[T_H - (1-\varepsilon_R)T_1 - \varepsilon_R a_8] - k\left[\frac{ln(1-\varepsilon_H)}{h_H} + \frac{ln(1-\varepsilon_L)}{h_L}\right]} \quad (2.23)$$

2.2.2 *Solar-Assisted Stirling Heat Engine System*

As shown in Fig. 2.4, the warm region of the Stirling engine is placed at the focal point of the concentrator. The concentrator reflects solar energy towards the focal point. The heat exchanger placed at the warm region of the Stirling engine transmits this reflected heat energy to the working fluid. The work output from the Stirling cycle is then utilized to drive a generator that produces electric power. Three case studies of the solar-assisted heat engine system are considered for optimization in this book. These case studies are presented in the subsequent sections.

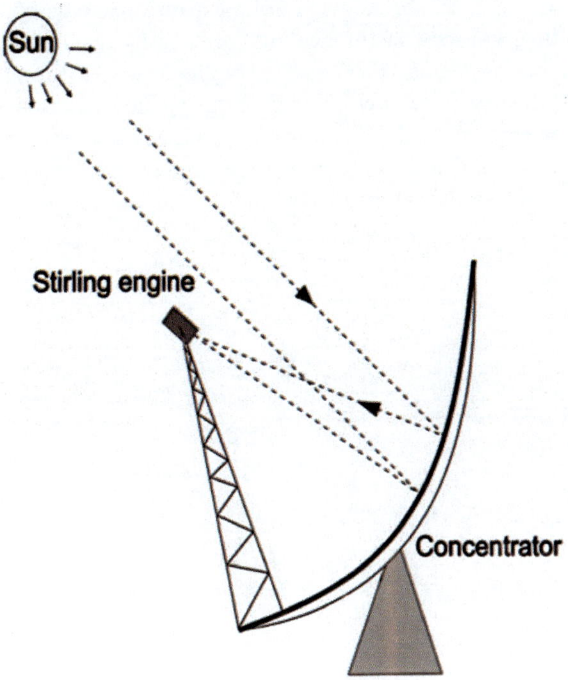

Fig. 2.4 Schematic diagram of the solar-assisted stirling heat engine

2.2.2.1 System Description of Case Study-1

Thermodynamic based optimization of the solar-assisted Stirling heat engine is considered in case study-1. The schematic diagram and TS diagram of the solar-assisted Stirling heat engine system considered in this book are shown in Fig. 2.5. The Stirling cycle (with ideal regenerator) consists of four processes. Process 1–2: an isothermal (at T_c temperature) compression involving heat rejection to the heat sink (at T_L temperature). Process 2–3: an isochoric (volume $V_2 = V_3$) heating process, in which the temperature of the working fluid rises to T_h by the regenerator. Process 3–4: an isothermal (at T_h temperature) expansion process, in which heat added to the working fluid from the heat source (at T_H temperature). Process 4–1: an isochoric (volume $V_4 = V_1$) process in which the regenerator absorbs heat from the working fluid; thus, the temperature falls from T_h to T_c. The detailed description and finite-time thermodynamic analysis of the considered system were presented by Ahmadi et al. (Ahmadi, Mohammadi, et al., 2013).

Let, n be the number of moles of the working fluid, C_v be the specific heat capacity of the working fluid during the regenerative process, R be the universal gas constant, λ be the ratio of volume during expansion and compression, M_1 be the regenerative time constant at the heating region, M_2 be the regenerative time constant at the cooling region, T_{H1} and T_{H2} be the heat source temperature before and after heat transferred to the working fluid during the isothermal expansion, and T_{L1} and T_{L2} be the heat sink temperature before and after heat transferred from working fluid during isothermal compression, then the power output of a solar-assisted Stirling system is given by the following equation:

$$P = \frac{nR(T_h - T_c)\ln\lambda}{\frac{nRT_h\ln\lambda + nC_v(1-\varepsilon_R)(T_h - T_c)}{C_H\varepsilon_R(T_{H1} - T_h) + \zeta C_H\varepsilon_H(T_{H1}^4 - T_h^4)} + \frac{nRT_c\ln\lambda + nC_v(1-\varepsilon_R)(T_h - T_c)}{C_L\varepsilon_L(T_c - T_{L1})} + \left(\frac{1}{M1} + \frac{1}{M2}\right)(T_h - T_c)} \tag{2.24}$$

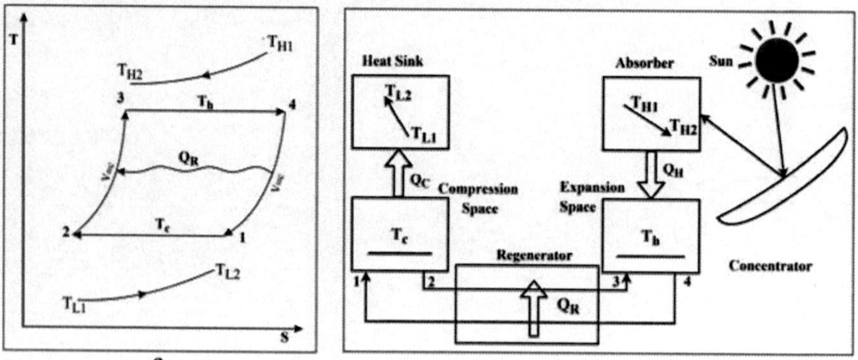

a b

Fig. 2.5 a T-S diagram and b schematic representation of the solar-assisted stirling heat engine

Now, Q_L be the net heat absorbed by the heat sink, Q_H be net heat released by the heat source, T_{Havg} and T_{Lavg} be the average temperatures of heat source and sink, η_t be the thermal efficiency of the Stirling engine, and η_s is the thermal efficiency of the dish collector, then the rate of entropy generation (σ) and thermal efficiency (η_m) of the system are given by the following equations respectively.

$$\sigma = \frac{1}{t}\left(\frac{Q_L}{T_{L_{avg}}} - \frac{Q_H}{T_{H_{avg}}}\right) \tag{2.25}$$

$$\eta_m = \eta_s \eta_t \tag{2.26}$$

$$\eta_t = \frac{nR(T_h - T_c)\ln\lambda}{nRT_h\ln\lambda + nC_v(1-\varepsilon_R)(T_h - T_c) + 0.5K_0\{(2-\varepsilon_H)T_{H1} - (2-\varepsilon_L)T_{L1} + (\varepsilon_H T_h - \varepsilon_L T_c)\}t} \tag{2.27}$$

$$\eta_s = \eta_0 - \frac{1}{IC}\left[h\left(T_{\text{H}_{\text{avg}}} - T_0\right) + \varepsilon\delta\left(T_{\text{H}_{\text{avg}}}^4 - T_0^4\right)\right] \tag{2.28}$$

where T_O is the ambient temperature, C is the collector concentrating ratio, ε is the emissivity factor of the collector, h is the convection heat transfer coefficient, I is the solar irradiance, δ is the Stefan-Boltzmann constant, and η_0 is the collector optical efficiency.

2.2.2.2 System Description of Case Study-2

Thermo-economic based optimization of the solar-assisted Stirling heat engine is considered in case study-2. The TS diagram of the solar-assisted Stirling heat engine system considered in this case study is shown in Fig. 2.5. The Stirling cycle consists of four processes. First: isothermal heat removal process, in which heat is rejected by the working fluid (at T_c temperature) to the heat sink (at T_L temperature). Second: an isochoric heat addition process, in which the temperature of the working fluid is increased to T_h by the regenerator. Third: isothermal heat addition process, in which heat added to the working fluid (at T_h temperature) from the heat source (at T_H temperature). Fourth: an isochoric heat removal process, in which the regenerator absorbs heat from the working fluid. In this case study, a MOO case is considered to find the optimal thermo-economic parameters to maximize dimensionless power and thermal efficiency.

In this MOO case, three objective functions are considered. Detailed thermodynamic and thermo-economic analysis of the solar-assisted Stirling heat engine is presented by Ahmadi et al. (2013). The objective functions of the optimization are thermal efficiency of solar-assisted Stirling system (η_{th}), the thermo-economic objective function (F), and the dimensionless power output (P). All objective functions are to be maximized. The decision variables of these objective functions are internal irreversibility parameter (ϕ), heat transfer area ratio (A_r), temperature ratio (χ), the temperature of the heat source (T_H), the temperature of the working fluid in the high-temperature isothermal process (T_h). The thermo-economic objective function is given by the following equation:

$$F = \frac{(1-\phi\chi)}{T_L\left[\frac{1+\left(\frac{1-f}{f}\right)A_r}{(T_H-T_h)} + \frac{\phi h_h \chi\left(1+\left(\frac{1-f}{f}\right)A_r\right)}{A_r h_c(\chi T_h - T_L)} + \psi h_h A_H(1-\chi)(1+\left(\frac{1-f}{f}\right)A_r)\right]} \tag{2.29}$$

$$\psi = \frac{1}{nRln\lambda}\left(\frac{1}{M_1} + \frac{1}{M_2}\right) \tag{2.30}$$

where, h_h is convection heat transfer coefficient for the high-temperature side, h_c is convection heat transfer coefficient for the low-temperature side, A_H is heat transfer area for the heat exchanger of the hot side, f is relative investment cost parameter of the heat exchanger hot side. Let n be the working fluid mole number, R is the universal gas constant, λ be the ratio of volume during the regenerative processes, M_1 be the regenerative time constant at the heating region, M_2 be the regenerative time constant at the cooling region. Now the dimensionless power output (P) and thermal efficiency (η_{th}) of the Stirling system are expressed using the following equations:

$$P = \frac{(1-\phi\chi)}{T_L\left[\frac{1}{(T_H-T_h)} + \frac{\phi h_h \chi}{A_r h_c(\chi T_h - T_L)} + \psi h_h A_H(1-\chi)\right]} \tag{2.31}$$

$$\eta_{th} = \eta_t \eta_s \tag{2.32}$$

where η_t is the thermal efficiency of the Stirling engine, and η_s is the thermal efficiency of the dish collector, expressed by

$$\eta_t = \frac{(1-\phi\chi)}{T_h + k_0(T_H - T_L)\left[\frac{1}{h_h A_H(T_H-T_h)} + \frac{\phi\chi T_h}{A_r A_H h_h(\chi T_h - T_L)} + \psi(1-\chi)\right]} \tag{2.33}$$

$$\eta_s = \eta_0 - \frac{1}{IC}[h(T_H - T_0) + \varepsilon\delta(T_H^4 - T_0^4)] \tag{2.34}$$

where C is the collector concentrating ratio, h is the convection heat transfer coefficient, T_0 is the ambient temperature, T_H is the absorber temperature, ε is the emissivity factor of the collector, δ is the Stefan-Boltzmann constant, I is the solar irradiance, and η_0 is the collector optical efficiency.

2.2.2.3 System Description of Case Study-3

The multi-objective optimization problem considered in this case study was formulated by Dai et al. (2018) for determining the optimal design parameters of a solar-assisted Stirling heat engine system considering conductive thermal bridging losses, finite heat transfer rate, and regenerative heat loss. The solar-assisted Stirling heat engine system considered in this case study consists of two isothermal and two isochoric processes, as shown in Fig. 2.5. From state-1 to state-2 is an isothermal compression process at constant temperature T_c. During this process, the working medium rejects heat energy (Q_L) to the heat sink at temperature T_L. From state-2 to state-3 is an isochoric heat addition process at constant volume V_1. During this process, the working medium absorbs heat energy from the regenerator to attain temperature T_h. From state-3 to state-4 is an isothermal expansion process at constant temperature T_h. During this process, the working medium absorbs heat energy (Q_H) from the heat source at temperature T_H. From state-4 to state-1 is an isochoric heat rejection process at constant volume V_2. During this process, the working medium rejects heat energy to the regenerator to attain temperature T_c.

The detailed thermodynamic analysis and description of the solar-assisted Stirling engine considered in this case study were given in Dai et al. (2018). By implementing finite-time thermodynamic analysis presented a theoretical model of this system to achieve optimum power output (*P*), system efficiency (η), and ecological coefficient of performance (COP_E). The proposed model considers finite-rate heat transfer, thermal bridging, and regenerative losses. Let C_v be the constant volume specific heat capacity of the working medium during the regenerative process; λ be the volumetric ratio during expansion and compression, *n* be the number of moles of the working medium, k_1 and k_2 be the rates of temperature change in hot heating and cooling regions respectively, *R* be the universal gas constant, $\mu_{2\text{-}3}$ and $\mu_{4\text{-}1}$ be the regeneration coefficients during heat energy addition and heat energy rejection process, *t* be the Stirling cycle time, α_h and α_c be the convective heat transfer coefficient during expansion and compression process. The power output (*P*) is given by the following equation:

$$P = \frac{nR(T_h - T_c)\ln\lambda}{t} \tag{2.35}$$

Let σ be the entropy generated, α_{leak} be the coefficient of heat leak, and then the following equations give the system efficiency (η) and COP_E:

$$\eta = \frac{nR(T_h - T_c)\ln\lambda}{\alpha_{leak}(T_H - T_L)t + \mu_{2-3}mC_v(T_h - T_c) + nRT_h\ln\lambda} \tag{2.36}$$

$$ECOP = \frac{nR(T_h - T_c)ln\lambda}{T_0\sigma} \tag{2.37}$$

$$t = \left(\frac{1-\mu_{2-3}}{k_1} + \frac{1-\mu_{4-1}}{k_2}\right)(T_h - T_c) + \frac{nRT_c\ln\lambda}{\alpha_c(T_l - T_L)}$$

$$+\frac{mC_v}{\alpha_h}\ln\frac{T_H-\mu_{2-3}T_c-(1-\mu_{2-3})T_h}{T_H-T_h}$$
$$+\frac{mC_v}{\alpha_c}\ln\frac{\mu_{4-1}T_h-T_L+(1-\mu_{4-1})T_c}{T_c-T_L}+\frac{nRT_h\ln\lambda}{\alpha_h(T_H-T_h)} \tag{2.38}$$

$$\sigma=\left(\frac{\alpha_{\mathrm{leak}}(\mathrm{T_H}-\mathrm{T_L})\mathrm{t}+\mu_{4-1}\mathrm{mC_v}(\mathrm{T_h}-\mathrm{T_c})+\mathrm{nRT_l}\ln\lambda}{T_L}\right.$$
$$\left.-\frac{\alpha_{\mathrm{leak}}(T_H-T_L)t+\mu_{2-3}mC_v(T_h-T_c)+n\mathrm{RT}_h\ln\lambda}{T_H}\right) \tag{2.39}$$

2.2.3 Solar-Assisted Carnot-Like Heat Engine System

The detailed description and thermodynamic analysis of the considered solar-driven heat engine system were presented by Sayyaadi et al. (2015). In this Carnot-like heat engine system, finite temperature difference for working and maximum ecological function conditions are considered to achieve optimum non-dimensional power output (W), non-dimensional ecological function (E), and thermal efficiency (η). The T-S diagram of the solar-assisted Carnot heat engine is shown in Fig. 2.6. From state-1 to state-2 is an isentropic heat addition process. During this process, working medium temperature rises to T_h from T_c. From state-2 to state-3 is an isothermal expansion process at constant temperature T_h. During this process, the working medium absorbs heat energy (Q_H) from the heat source at temperature T_H. From state-3 to state-4 is an isentropic heat rejection process. During this process, working medium temperature falls from T_h to T_c. From state-4 to state-1 is an isothermal compression process at constant temperature T_c. During this process, the working medium releases heat energy (Q_L) to the heat sink at temperature T_L.

Let α_{Hr} and α_{Hc} be the radiation and convection heat transfer coefficients respectively at the heat source, and α_{Lc} be the convection heat transfer coefficient at the heat sink, then the design variables of the system are as follows: the temperature ratio is $\gamma = T_h/T_H$, the convective coefficient ratio is $\phi = \alpha_{H,c}/\alpha_{L,c}$, the operating temperature ratio is $\tau = T_L/T_H$, and the heat source to the heat sink allocation parameter is $\beta = (\alpha_{H,r}/\alpha_{L,c})T_H^3$. Now, the non-dimensional power output (W), non-dimensional ecological function (E), and the thermal efficiency (η) are given by the following equations:

$$W=\eta\big[\beta\big(1-\gamma^4\big)-\phi(1-\gamma)\big] \tag{2.40}$$

$$E=\big[\beta\big(1-\gamma^4\big)+\phi(1-\gamma)\big]\times\left[\eta-\tau\frac{\beta\big(1-\gamma^4\big)+\phi(1-\gamma)-\gamma+1}{\gamma-\beta\big(1-\gamma^4\big)-\phi(1-\gamma)}\right] \tag{2.41}$$

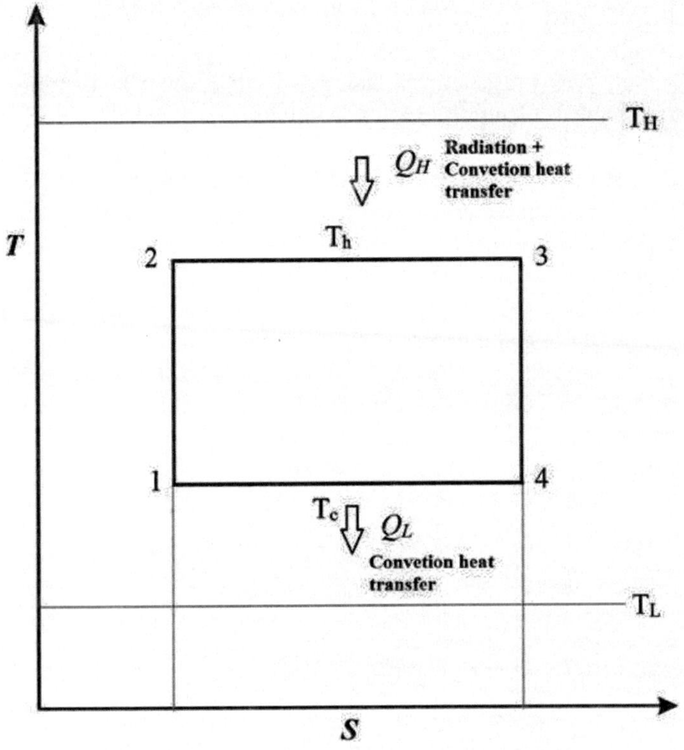

Fig. 2.6 T-S diagram of a solar-assisted Carnot-like heat engine system

$$\eta = 1 - \left[\tau/\left(\gamma - \beta\left(1 - \gamma^4\right) - \phi(1 - \gamma)\right)\right] \tag{2.42}$$

2.3 Bio-Energy Systems

Biofuel can be considered as a sustainable and environmentally friendly alternative to petroleum-based fuels. The performance of the biodiesel engine has considerably affected by the blending proportions. Using biodiesel as an alternative to diesel can reduce the wear of the engine parts and the emissions such as carbon oxides, hydrocarbons, and particulate matter. However, there is a loss of power, lower fuel efficiency, and increased nitrogen oxides emission with biodiesels.

Three multi-objective optimization case studies of biodiesel engine design and a multi-objective optimization case study of microalgae cultivation process optimization are considered for optimization to see if there can be any improvement in the performances of the selected systems. The subsequent section presents the design optimization of a biodiesel engine system.

2.3.1 Single-Cylinder Direct-Injection Diesel Engine

This case study was presented by Dhingra et al. (2014). In this case study, a single-cylinder direct-injection compression ignition engine that runs using Jatropha biodiesel blends was considered. The seven objectives of this case study are: minimizing the combustion parameters such as brake specific fuel consumption (*BSFC—kg/kWh*) and peak cylinder pressure (P_{max}*–bar*); maximizing the performance in terms of brake-thermal efficiency (*BTE–N-m*); and minimizing the emissions such as carbon mono oxide emission (*CO–%*), nitrogen oxides emission (NO_x*–ppm*), hydrocarbon emission (*HC–ppm*), and smoke emission opacity (S_m). The design variables are biodiesel blending ratio (X_1), load torque (X_2), and compression ratio (X_3) and their ranges are $11.25 \leq X_1 \leq 33.75$ (% *V/V*), $7.5 \leq X_2 \leq 12.5$ (*N-m*), and $13.5 \leq X_3 \leq 16.5$ (*V/V*). The regression models of this case study's objectives proposed by Dhingra et al. (2014) are as given below:

$$\begin{aligned} BSFC &= -46.68493 + 0.13685X_1 + 1.32378X_2 + 5.31712X_3 \\ &- 3.21186 \times 10^{-3}X_1^2 - 0.056480X_2^2 - 0.16911X_3^2 + 2.33817 \times 10^{-3}X_1X_2 \\ &- 1.13137 \times 10^{-3}X_1X_3 - 0.022062X_2X_3 \end{aligned} \tag{2.43}$$

$$\begin{aligned} BTE &= -2400.88522 + 10.28829X_1 \\ &+ 71.43483X_2 + 259.72937X_3 - 0.18203X_1^2 \\ &- 4.07476X_2^2 - 9.07056X_3^2 - 0.19143X_1X_2 \\ &- 4.19026 \times 10^{-3}X_1X_3 + 1.26091X_2X_3 \end{aligned} \tag{2.44}$$

$$\begin{aligned} P_{\max} &= -3669.50268 + 10.41183X_1 \\ &+ 129.7155X_2 + 396.37136X_3 - 0.28479X_1^2 \\ &- 5.68691X_2^2 - 12.90809X_3^2 - 0.037712X_1X_2 \\ &+ 0.18856X_1X_3 - 1.03709X_2X_3 \end{aligned} \tag{2.45}$$

$$\begin{aligned} \text{Sqrt}(CO) &= -102.47076 + 0.41566X_1 + 1.86746X_2 \\ &+ 11.94069X_3 - 7.55818 \times 10^{-3}X_1^2 - 0.11396X_2^2 \\ &- 0.40724X_3^2 - 3.75222 \times 10^{-3}X_1X_2 - 2.57927 \\ &\times 10^{-3}X_1X_3 + 0.035673X_2X_3 \end{aligned} \tag{2.46}$$

$$\begin{aligned} \text{Sqrt}(NO_x) &= -765.06345 + 3.33115X_1 + 12.59961X_2 \\ &+ 91.17741X_3 - 0.062945X_1^2 - 0.70774X_2^2 - 3.08084X_3^2 \\ &- 0.026616X_1X_2 - 0.01625X_1X_3 + 0.1666X_2X_3 \end{aligned} \tag{2.47}$$

$$\log_{10}(HC) = -182.12527 + 0.57186X_1 + 3.99162X_2 + 21.03397X_3 - 0.012787X_1^2 - 0.19348X_2^2 - 0.70153X_3^2 + 8.11844 \times 10^{-5}X_1X_2 + 2.71724 \times 10^{-4}X_1X_3 + 1.69324 \times 10^{-3}X_2X_3 \quad (2.48)$$

$$\text{Sqrt}(S_m) = 133.78384 - 0.22401X_1 - 3.48682X_2 - 15.04873X_3 + 0.010848X_1^2 + 0.18082X_2^2 + 0.49577X_3^2 + 3.59009 \times 10^{-3}X_1X_2 - 5.06832 \times 10^{-3}X_1X_3 + 0.041469X_2X_3 \quad (2.49)$$

2.3.2 *Turbocharged DI Diesel Engine*

This case study was presented by Shirneshan et al. (2016). In this case study, a turbocharged DI diesel engine using biodiesel and diesel blends was considered. The waste vegetable cooking oil was considered as the source of biodiesel. The specific fuel properties and engine specifications were presented in Shirneshan et al. (2016). By investigating the effect of design variables such as biodiesel blending ratio, engine speed, and engine load on the combustion, performance, and emissions of the considered biodiesel engine, Shirneshan et al. (2016) presented the regression models. The six objectives of this case study are: maximizing the brake power (*P*) and brake torque (*T*); minimizing the brake specific fuel consumption (*BSFC*) and emissions such as carbon mono oxide emission (*CO*), nitrogen oxides emission (NO_x), and hydrocarbon emission (*HC*). The design variables were the percentage of biodiesel in fuel (X_1), engine speed (X_2), and engine load (X_3) and their ranges were $0 \leq X_1 \leq 100$ (%), $1000 \leq X_2 \leq 2800(rpm)$, and $25 \leq X_3 \leq 100$ (%). The regression models of this case study's objectives are as follows:

$$P = -47.32 - 0.08X_1 + 0.056X_2 + 0.205X_3 + 0.0002X_1^2 - 1.4 \times 10^{-5}X_2^2 - 5.499 \times 10^{-4}X_3^2 + 0.0004X_1X_3 + 0.0002X_2X_3 \quad (2.50)$$

$$T = -299.277 - 0.524X_1 + 0.302X_2 + 4.654X_3 + 0.00362X_1^2 - 7.401 \times 10^{-5}X_2^2 - 0.00637X_3^2 - 3.771 \times 10^{-4}X_2X_3 \quad (2.51)$$

$$BSFC = 298.74 + 0.5X_1 - 0.088X_2 - 0.236X_3 + 0.0014X_1^2 + 2.67 \times 10^{-5}X_2^2 - 0.00018X_1X_2 - 0.00336X_1X_3 \quad (2.52)$$

$$\text{CO} = 0.109 - 3.43 \times 10^{-4}X_1 - 3.96 \times 10^{-5}X_2 - 7.53 \times 10^{-4}X_3 + 6.48 \times 10^{-7}X_1^2 + 6.32 \times 10^{-9}X_2^2 + 2.93 \times 10^{-6}X_3^2$$

$$+ 1.69 \times 10^{-6} X_1 X_3 + 7.33 \times 10^{-8} X_2 X_3 \quad (2.53)$$

$$NO_x = 216.71 - 0.264X_1 + 0.158X_2 + 0.755X_3 + 0.0114X_1^2 - 5.37 \times 10^{-5} X_2^2 + 0.0558X_3^2 - 0.00188X_2X_3 \quad (2.54)$$

$$HC = 56.38 - 0.028X_1 + 0.019X_2 - 0.554X_3 - 5.19 \times 10^{-4} X_1^2 - 2.1 \times 10^{-6} X_2^2 + 0.0026X_3^2 + 3.14 \times 10^{-5} X_1X_2 - 0.0011X_1X_3 + 8.38 \times 10^{-5} X_2X_3 \quad (2.55)$$

2.3.3 *Compression Ignition Biodiesel Engine with an EGR System*

This case study was presented by Jaliliantabar et al. (2019). In this case study, a compression ignition biodiesel engine with an exhaust gas recirculation system is considered for multi-objective optimization to see if there can be any improvement in the considered system performance. The design variables of this case study are the exhaust gas recirculation rate (*ER*), engine load percentage (*EL*), engine speed (*ES*) in rpm, and biodiesel percentage (*BP*). The ranges of design variables are as follows: $0 \le ER \le 30$, $25 \le EL \le 75$, $1800 \le ES \le 2400$, and $0 \le BP \le 15$. The objective functions of this case study are maximization of power output (*P*), and minimization of the brake specific fuel consumption (*BSFC*) and emissions such as carbon mono oxide emission (*CO*), nitrogen oxides emission (NO_x), hydrocarbon emission (*HC*), and smoke opacity (S_m). The regression models of the objectives of this case study are as follows:

$$\begin{aligned} \mathrm{P} = {} & 7.08 + (0.04 \times \mathrm{EL}) - \left(7.33 \times 10^{-3} \times \mathrm{ES}\right) \\ & - (0.01 \times \mathrm{ER}) + (0.11 \times \mathrm{BP}) \\ & + \left(4.73 \times 10^{-6} \times \mathrm{EL} \times \mathrm{ES}\right) + \left(3.7 \times 10^{-5} \times \mathrm{EL} \times \mathrm{ER}\right) \\ & - \left(1.13 \times 10^{-4} \times \mathrm{EL} \times \mathrm{BP}\right) + \left(5.87 \times 10^{-6} \times \mathrm{ES} \times \mathrm{ER}\right) \\ & - \left(4.77 \times 10^{-5} \times \mathrm{ES} \times \mathrm{BP}\right) + \left(6.69 \times 10^{-5} \times \mathrm{ER} \times \mathrm{BP}\right) \\ & - \left(7.88 \times 10^{-5} \times \mathrm{EL}^2\right) + \left(1.8 \times 10^{-6} \times \mathrm{ES}^2\right) \\ & + \left(5.9 \times 10^{-5} \times \mathrm{ER}^2\right) \\ & - \left(7.93 \times 10^{-5} \times \mathrm{BP}^2\right) \end{aligned} \quad (2.56)$$

$$\begin{aligned} \mathrm{CO} = {} & 5.55 - (0.08 \times \mathrm{EL}) - \left(4.3 \times 10^{-3} \times \mathrm{ES}\right) \\ & - (0.02 \times \mathrm{ER}) - \left(1.36 \times \mathrm{BP} \times 10^{-6}\right) \end{aligned}$$

$$+\left(2.59 \times 10^{-5} \times \mathrm{EL} \times \mathrm{ES}\right)+\left(5.48 \times 10^{-5} \times \mathrm{EL} \times \mathrm{ER}\right)$$
$$-\left(6.2 \times 10^{-4} \times \mathrm{EL} \times \mathrm{BP}\right)$$
$$+\left(9.2 \times 10^{-6} \times \mathrm{ES} \times \mathrm{ER}\right)+\left(3.66 \times 10^{-5} \times \mathrm{ES} \times \mathrm{BP}\right)$$
$$-\left(2.35 \times 10^{-4} \times \mathrm{ER} \times \mathrm{BP}\right)$$
$$+\left(5.2 \times 10^{-4} \times \mathrm{EL}^2\right)+\left(9.02 \times 10^{-7} \times \mathrm{ES}^2\right)$$
$$-\left(4.28 \times 10^{-5} \times \mathrm{ER}^2\right)$$
$$-\left(6.08 \times 10^{-4} \times \mathrm{BP}^2\right) \tag{2.57}$$

$$\begin{aligned}\mathrm{NO_x} = {} & 1514.26+(17.21 \times \mathrm{EL})-(1.61 \times \mathrm{ES}) \\ & +(0.51 \times \mathrm{ER}) \\ & +(3.73 \times \mathrm{BP})-\left(1.55 \times 10^{-3} \times \mathrm{EL} \times \mathrm{ES}\right) \\ & -(0.05 \times \mathrm{EL} \times \mathrm{ER}) \\ & -\left(8.8 \times 10^{-3} \times \mathrm{EL} \times \mathrm{BP}\right)+\left(4.73 \times 10^{-5} \times \mathrm{ES} \times \mathrm{ER}\right) \\ & -\left(1.48 \times 10^{-3} \times \mathrm{ES} \times \mathrm{BP}\right) \\ & -(0.07 \times \mathrm{ER} \times \mathrm{BP})-\left(0.1 \times \mathrm{EL}^2\right) \\ & +\left(3.86 \times 10^{-4} \times \mathrm{ES}^2\right) \\ & +\left(0.03 \times \mathrm{ER}^2\right)-\left(0.07 \times \mathrm{BP}^2\right)\end{aligned} \tag{2.58}$$

$$\begin{aligned}\mathrm{HC} = {} & 8.76-(7.49 \times \mathrm{EL})+(0.096 \times \mathrm{ES})-(0.17 \times \mathrm{ER}) \\ & -(0.18 \times \mathrm{BP})+\left(2.33 \times 10^{-3} \times \mathrm{EL} \times \mathrm{ES}\right)+(0.01 \times \mathrm{EL} \times \mathrm{ER}) \\ & -(0.06 \times \mathrm{EL} \times \mathrm{BP})+\left(4.69 \times 10^{-4} \times \mathrm{ES} \times \mathrm{ER}\right) \\ & -\left(1.61 \times 10^{-4} \times \mathrm{ES} \times \mathrm{BP}\right) \\ & -(0.03 \times \mathrm{ER} \times \mathrm{BP})+\left(0.05 \times \mathrm{EL}^2\right) \\ & -\left(2.8 \times 10^{-5} \times \mathrm{ES}^2\right) \\ & -\left(0.02 \times \mathrm{ER}^2\right)+\left(0.09 \times \mathrm{BP}^2\right)\end{aligned} \tag{2.59}$$

$$\begin{aligned}\mathrm{BSFC} = {} & -631.23-(28.02 \times \mathrm{EL}) \\ & +(1.5 \times \mathrm{ES})+(2.92 \times \mathrm{ER}) \\ & -(2.12 \times \mathrm{BP})+\left(2.5 \times 10^{-3} \times \mathrm{EL} \times \mathrm{ES}\right) \\ & +\left(7.07 \times 10^{-3} \times \mathrm{EL} \times \mathrm{ER}\right) \\ & -(0.02 \times \mathrm{EL} \times \mathrm{BP})-\left(2.02 \times 10^{-3} \times \mathrm{ES} \times \mathrm{ER}\right) \\ & -\left(1.6 \times 10^{-3} \times \mathrm{ES} \times \mathrm{BP}\right) \\ & -\left(4.17 \times 10^{-3} \times \mathrm{ER} \times \mathrm{BP}\right)+\left(0.19 \times \mathrm{EL}^2\right)\end{aligned}$$

$$
\begin{aligned}
&- \left(3.38 \times 10^{-4} \times ES^2\right)\\
&+ \left(0.03 \times ER^2\right) + \left(0.37 \times BP^2\right)
\end{aligned} \tag{2.60}
$$

$$
\begin{aligned}
S_m = {} & 15.09 - (0.36 \times EL)\\
& - \left(8.69 \times 10^{-3} \times ES\right) - (0.05 \times ER)\\
& + \left(6.72 \times 10^{-3} \times BP\right) + \left(8.06 \times 10^{-5} \times EL \times ES\right)\\
& - \left(1.45 \times 10^{-4} \times EL \times ER\right)\\
& - \left(3.8 \times 10^{-3} \times EL \times BP\right) + \left(5.57 \times 10^{-5} \times ES \times ER\right)\\
& - \left(2.74 \times 10^{-6} \times ES \times BP\right)\\
& - \left(5.06 \times 10^{-4} \times ER \times BP\right) + \left(4.17 \times 10^{-3} \times EL^2\right) - \left(1.9 \times 10^{-6} \times ES^2\right)\\
& - \left(1.04 \times 10^{-3} \times ER^2\right) + \left(5.2 \times 10^{-3} \times BP^2\right)
\end{aligned} \tag{2.61}
$$

2.3.4 *Microalgae-Based Biomass Cultivation Process*

Nowadays, producing biofuels from microalgae is gaining popularity. This case study's objectives include the maximization of biomass production, eicosapentaenoic acid (EPA), and lipid productions. Banerjee et al. (2016) presented the mathematical models of these objectives considering the cultivation light intensity (X_1 in $\mu mol/m^2/s$), temperature (X_2 in $°C$), and concentrations of NaCl (X_3 in M), $NaHCO_3$(X_4 in g/L) and NPK-10:26:26 fertilizer (X_5 in g/L) as the design variables. The regression models for the biomass-production (BMP), total lipid-production (TLP), and EPA generations in terms of the considered design variables are as follows:

$$
\begin{aligned}
\text{BMP (g/L)} = {} & 0.069 + 0.029X_1 - 0.0039X_2 - 0.0419X_3\\
& + 0.0014X_4 + 0.002X_5 + 0.0028X_1X_2 - 0.00037X_1X_3\\
& + 0.00068X_1X_4 + 0.0026X_1X_5 + 0.00037X_2X_3 - 0.0014X_2X_4\\
& + 0.00031X_2X_5 + 0.0015X_3X_4 + 0.0036X_3X_5 - 0.00068X_4X_5\\
& - 0.0046X_1^2 - 0.0022X_2^2 - 0.023X_3^2 - 0.012X_4^2 - 0.0072X_5^2
\end{aligned} \tag{2.62}
$$

$$
\begin{aligned}
\text{TLP (\%)} = {} & 40.87 + 5.07X_1 + 0.081X_2 - 3.49X_3 + 1.46X_4\\
& + 2.11X_5 - 0.26X_1X_2 - 0.67X_1X_3 - 0.31X_1X_4 - 0.029X_1X_5\\
& - 0.25X_2X_3 - 0.29X_2X_4 - 0.24X_2X_5 - 0.19X_3X_4 - 0.24X_3X_5\\
& - 0.28X_4X_5 - 1.61X_1^2 - 0.47X_2^2 - 2.39X_3^2 - 1.35X_4^2 - 1.45X_5^2
\end{aligned} \tag{2.63}
$$

$$
\text{EPA (\%)} = 19.66345 + 1.50675X_1 - 1.26475X_2 - 2.04X_3 + 0.11X_4
$$

$$+ 0.51X_5 + 0.081X_1X_2 - 0.009X_1X_3 + 0.0078X_1X_4 + 0.04X_1X_5$$
$$+ 0.14X_2X_3 - 0.096X_2X_4 - 0.066X_2X_5 + 0.0115X_3X_4 - 0.033X_3X_5$$
$$+ 0.089X_4X_5 - 1.42X_1^2 + 0.0219X_2^2 - 0.228X_3^2 - 0.059X_4^2 - 0.474X_5^2 \quad (2.64)$$

2.4 Hydro Energy and Geothermal Energy Systems

2.4.1 Hydropower Generation and Reservoir Operation

Optimization of the Nigerian Jebba hydropower plant performance characteristics to enhance its electricity generation case study is considered to demonstrate the application of advanced optimization algorithms. This case study was presented by Onokwai et al. (2020). This case study investigates the influence of pressure head and discharge with other constraints such as the temperature of the turbine, oil pressure for lubrication and cooling of the bearings, on estimated power generation (EP_w). By using the response surface methodology (RSM), the effect of the selected variables on the estimated power generation was measured, and presented regression models for the optimization of the plant EP_w. The detailed plant's description including technical specifications and validation of the modal was presented in Onokwai et al. (2020).

The input variables considered in this case study are the discharge (D), pressure drop between the head and tail-water (P_d), the stator temperature (T_{st}), the water pressure (P_{water}) for cooling the generator and the oil pressure (P_{oil}) for cooling and lubrication of the bearings. The objective of the RSM based optimization is to maximize the power generation (EP_w) of the power plant. The RSM based regression model of the objective is as follows (Onokwai et al. 2020):

$$EP_w = 5173 + 1.58D + 1636P_d - 1092T_{st} - 1033P_{\text{water}} + 884P_{\text{oil}}$$
$$- 0.000130D^2 - 42.7P_d^2 + 3.8T_{st}^2 - 1.3P_{\text{water}}^2 - 9.76P_{\text{oil}}^2 - 0.136P_dD$$
$$+ 0.0742T_{st}D + 0.0106P_{\text{water}}D - 0.0064P_{\text{oil}}D - 2.3P_dT_{st} + 31.6P_dP_{\text{water}}$$
$$+ 28.7P_dP_{\text{oil}} + 12.9P_{\text{water}}T_{st} - 8.1P_{\text{oil}}T_{st} - 16.1P_{\text{water}}P_{\text{oil}} \quad (2.65)$$

2.4.2 Ground Source Heat Pump-Radiant Ceiling Air Conditioning System

The ground source heat pump-radiant ceiling (GSHP-RC) air conditioning system case study was presented by Xie et al. (2020). This case study is to investigate the coupling mechanism between the system and building thermal environment and optimize the whole system globally. By using the response surface methodology (RSM), the effect of the controllable variables on the output variables was measured, and presented regression models for global optimization of the system. The detailed description including technical specifications and validation of the modal was presented in Xie et al. (2020).

The controllable variables considered in this case study are the water supply temperature of radiant ceiling (a), the indoor set temperature (b) and the water supply temperature of the heat pump (c). The objective functions of the system are seasonal performance factor (SPF), predicted mean value (PMV) and the operating cost (OC). The seasonal performance factor is the energy consumption evaluation index of the system, in which the power consumptions of ventilation units, heat pump units and pumps are all taken into account.

PMV is a thermal comfort evaluation index based on the human body heat balance equation and the subjective thermal sensation of the human body. The average value of PMV in the air conditioned rooms is used as an indicator for the thermal comfort of the system. In order to ensure the thermal comfort of the air-conditioned room, it is desirable that the average value of the PMV in the air-conditioned room varies between −0.5 and 0.5. It is better for the value close to 0. The system operating cost is used as the indicator of the system economy and it is the product of the electricity price and the total power consumption in the whole cooling season. The RSM based regression models of these objectives are as follows (Xie et al. 2020):

$$SPF = 2.422 - 0.01573a + 0.01933b + 0.2157c - 0.01102c^2 \tag{2.66}$$

$$PMV = -3.75 + 0.4103a - 0.303b + 0.01642b^2 - 0.01449\text{ab} \tag{2.67}$$

$$\begin{aligned} OC = {} & 77076 + 2412a - 5066b + 1364c + 47.7a^2 \\ & + 120.3b^2 - 106.5c^2 - 135ab + 0.2bc \end{aligned} \tag{2.68}$$

The following chapter presents a review on the application of advanced engineering optimization techniques to the various renewable energy systems.

References

Ahmadi, M. H., Mohammadi, A. H., Dehghani, S., & Barranco-Jiménez, M. A. (2013). Multi-objective thermodynamic-based optimization of output power of solar dish-Stirling engine by implementing an evolutionary algorithm. *Energy Conversion and Management, 75*, 438–445. https://doi.org/10.1016/j.enconman.2013.06.030

Ahmadi, M. H., Sayyaadi, H., Mohammadi, A. H., & Barranco-Jimenez, M. A. (2013). Thermo-economic multi-objective optimization of solar dish-Stirling engine by implementing evolutionary algorithm. *Energy Conversion and Management, 73*, 370–380. https://doi.org/10.1016/j.enconman.2013.05.031

Banerjee, A., Guria, C., & Maiti, S. K. (2016). Fertilizer assisted optimal cultivation of microalgae using response surface method and genetic algorithm for biofuel feedstock. *Energy, 115*, 1272–1290. https://doi.org/10.1016/j.energy.2016.09.066

Chowdhury, S., Zhang, J., Messac, A., & Castillo, L. (2012). Unrestricted wind farm layout optimization (UWFLO): Investigating key factors influencing the maximum power generation. *Renewable Energy, 38*(1), 16–30. https://doi.org/10.1016/j.renene.2011.06.033

Dai, D., Yuan, F., Long, R., Liu, Z., & Liu, W. (2018). Performance analysis and multi-objective optimization of a Stirling engine based on MOPSOCD. *International Journal of Thermal Sciences, 124*, 399–406. https://doi.org/10.1016/j.ijthermalsci.2017.10.030

Dhingra, S., Bhushan, G., & Dubey, K. K. (2014). Multi-objective optimization of combustion, performance and emission parameters in a jatropha biodiesel engine using non-dominated sorting genetic algorithm-II. *Frontiers of Mechanical Engineering, 9*(1), 81–94. https://doi.org/10.1007/s11465-014-0287-9

Grady, S. A., Hussaini, M. Y., & Abdullah, M. M. (2005). Placement of wind turbines using genetic algorithms. *Renewable Energy, 30*(2), 259–270. https://doi.org/10.1016/j.renene.2004.05.007

Jaliliantabar, F., Ghobadian, B., Najafi, G., Mamat, R., & Carlucci, A. P. (2019). Multi-objective NSGA-II optimization of a compression ignition engine parameters using biodiesel fuel and exhaust gas recirculation. *Energy, 187*, 115970. https://doi.org/10.1016/j.energy.2019.115970

Li, Y., Liao, S., & Liu, G. (2015). Thermo-economic multi-objective optimization for a solar-dish Brayton system using NSGA-II and decision making. *International Journal of Electrical Power and Energy Systems, 64*, 167–175. https://doi.org/10.1016/j.ijepes.2014.07.027

Mosetti, G., Poloni, C., & Diviacco, B. (1994). Optimization of wind turbine positioning in large windfarms by means of a genetic algorithm. *Journal of Wind Engineering and Industrial Aerodynamics, 51*(1), 105–116. https://doi.org/10.1016/0167-6105(94)90080-9

Moorthy, C. B., & Deshmukh, M. K. (2013). A new approach to optimise placement of wind turbines using particle swarm optimisation. *International Journal of Sustainable Energy, 34*(6), 396–405. https://doi.org/10.1080/14786451.2013.860140

Onokwai, A. O., Owamah, H. I., Ibiwoye, M. O., Ayuba, G. C., & Olayemi, O. A. (2020). Application of response surface methodology (RSM) for the optimization of energy generation from Jebba hydro-power plant, Nigeria. *ISH Journal of Hydraulic Engineering*. https://doi.org/10.1080/09715010.2020.1806120

Sayyaadi, H., Ahmadi, M. H., & Dehghani, S. (2015). Optimal design of a solar-driven heat engine based on thermal and ecological criteria. *Journal of Energy Engineering, 141*(3), 4014012. https://doi.org/10.1061/(asce)ey.1943-7897.0000191

Shakoor, R., Hassan, M. Y., Raheem, A., & Wu, Y. K. (2016). Wake effect modeling: A review of wind farm layout optimization using Jensen's model. *Renewable and Sustainable Energy Reviews, 58*, 1048–1059. https://doi.org/10.1016/j.rser.2015.12.229

Shirneshan, A., Samani, B. H., & Ghobadian, B. (2016). Optimization of biodiesel percentage in fuel mixture and engine operating conditions for diesel engine performance and emission characteristics by artificial bees colony algorithm. *Fuel, 184*, 518–526. https://doi.org/10.1016/j.fuel.2016.06.117

Xie, Y., Hu, P., Zhu, N., Lei, F., Xing, L., & Xu, L. (2020). Collaborative optimization of ground source heat pump-radiant ceiling air conditioning system based on response surface method and NSGA-II. *Renewable Energy, 147*, 249–264. https://doi.org/10.1016/j.renene.2019.08.109

Chapter 3
Advanced Engineering Optimization Techniques and Their Role in Energy Systems Optimization

Abstract This chapter presents a brief literature review on the optimization of the selected renewable energy systems and the role of advanced engineering optimization techniques in energy systems.

Global warming is a significant concern that raises a need for cleaner production of energy. In the last few decades, researchers have focused on renewable energy resources like solar energy, bio-energy, wave energy, ocean thermal energy, tidal energy, geothermal energy, and wind energy. This has resulted in the development of new techniques and tools that could harvest energy from renewable energy sources. Among all these sources, the solar, wind, and bio-energy sources have become major resources upon which the focus is made.

However, to meet energy demands and reduce investment, a rigorous study of energy extraction systems is required. Identifying, analyzing, and optimizing the effect of various parameters of a renewable energy system contribute significantly in assessing the system performance. Furthermore, it is always not preferable to present the optimum system parameters considering only a single objective as these systems have multiple objectives such as power output, system efficiency, investment cost, and economic and ecological factors. Hence, researchers have developed various optimization models of these systems and presented optimum system parameters through single- and multi-objective optimization using advanced optimization algorithms.

Wind farm layout optimization

Researchers have made several attempts for prediction and optimization of turbines location in the wind farm to maximize the net power generation. Mosetti et al. (1994) attempted the genetic algorithm (GA) to find the optimal location of turbines on a wind farm. Grady et al. (2005) improved the location of turbines by making proper improvements in the selection of parameters used in the GA. González et al. (2010) employed a variable-length genetic algorithm with novel crossover procedures to obtain the optimal positioning of wind turbines. Ituarte-Villarreal et al. (2011) determined the optimal positioning of turbines using viral-based optimization (VBO) algorithm. Chowdhury et al. (2012) presented the unrestricted wind farm

V. R. Ravipudi et al., *Design Optimization of Renewable Energy Systems Using Advanced Optimization Algorithms*, Green Energy and Technology,
https://doi.org/10.1007/978-3-030-95589-2_3

layout optimization (UWFLO) framework that determines the position of turbines in a wind farm and appropriate turbines installed at the same time. Also, using the constrained particle swarm optimization (PSO) algorithm, the UWFLO model was optimized. Chowdhury et al. (2013) enhanced the UWFLO methodology for designing commercial-scale wind farms and used an advanced mixed-discrete PSO algorithm to optimize the proposed model. Pookpunt and Ongsakul (2013) proposed a binary particle swarm optimization (BPSO) for determining the optimal placement of wind turbines.

Sessarego et al. (2014) used a hybrid non-dominated sorting genetic algorithm (NSGA-II) to optimize yearly energy production, flap wise root-bending moment, and wind turbine blade mass concurrently. Similarly, Turner et al. (2014) proposed mixed-integer linear and quadratic optimization formulations, Moorthy et al. (2014) proposed GA, and Moorthy and Deshmukh (2013) used PSO to optimize the placement of turbines in a wind farm. Shakoor et al. (2015) proposed modeling of WFLO to reduce wake effect losses and obtain optimized layout for wind farms. In finding the optimal arrangement of wind turbines under unidirectional and uniform wind speed scenario, Patel et al. (2015) have implemented the teaching–learning-based optimization (TLBO) algorithm, and Ogunjuyigbe et al. (2017) used GA.

DuPont et al. (2016) have presented a modeling system that includes precise cost and power modeling, partial wake interface, and unstable atmosphere effects. It is validated by employing within an extended pattern search (EPS) multi-agent system (MAS) optimization approach. Furthermore, DuPont and Cagan (2016) proposed hybrid EPS/GA (HEPS/GA) and a multi-objective extended pattern search for optimizing the multi-stage wind farms layout. Li et al. (2017) proposed the selection hyper-heuristics technique to solve a multi-objective WFLO problem. Patel et al. (2017) proposed enhanced versions of the basic teaching–learning-based optimization algorithm named PAL and AL to optimize the WFLO problem.

Biswas et al. (2018) proposed a decomposition-based multi-objective evolutionary algorithm for simultaneous optimization of the power output and farm efficiency of four case studies of the WFLO problem. Pillai et al. (2018) proposed a WFLO framework for the levelized cost of energy that includes the wind farm's electrical infrastructure, annual energy production, and cost as functions of the wind farm layout and also demonstrated the implementation of the proposed model to three different cases of the WFLO problems and optimized using the PSO algorithm.

Wang et al. (2019) proposed an optimization model for the WFLO problems based on mean wind farm power output and variability. Also, the proposed models are optimized through weighted optimization and confidence interval approaches using the GA. By making the wake effect on individual turbines uniform, Yang et al. (2019) proposed and tested a new objective function to increase the power output and reducing the wake losses. The simulated annealing (SA) algorithm was used for layout optimization, and its performance was validated for an actual wind farm.

Wu et al. (2020) presented a WFLO model considering wind turbine noise along with the power output and wake effect. Also, two strategies (strict noise control strategy and economically compensated control strategy) are presented for designing the wind farms, and by applying the PSO algorithm, the proposed models are optimized. Reddy (2020) proposed a framework that provides a large set of analytical wake models and wake superposition schemes, by considering terrain elevation and the ambient wind velocity profile. Also, an empirical relation for cost in terms of turbine rotor diameter and height was presented. So far, it can be observed that advanced optimization meta-heuristics like GA, enhanced GA, VBO, PSO, BPSO, hybrid NSGA-II, TLBO, EPS-MAS, HEPS/GA, BBO, SH, AL, and PAL algorithms have been implemented in finding optimal layout for WFLO.

Optimization of solar-assisted energy systems

Many researchers and scientists have focused upon developing and optimizing the solar aided heat engines in order to extract and utilize solar energy more efficiently. Ahmadi et al. (2013a) presented an optimization model of an irreversible solar-driven heat engine based on thermodynamic and thermo-economic criteria. Also, optimum designs were obtained using the NSGA-II algorithm to achieve maximum power and thermal efficiency of the system. Ahmadi et al. (2013b) investigated the Stirling engine's application powered by solar energy and proposed a mathematical model considering regenerative losses, thermal bridging losses, and finite rate heat transfer. Also, Ahmadi et al. (2013c) studied the effect of the various parameters such as regenerator effectiveness, volumetric ratio, absorber temperature, and temperature ratio on the performance of the solar-powered Stirling system. Furthermore, Pareto-optimal design points were obtained through the multi-objective optimization (MOO) using NSGA-II algorithm to achieve maximum power and thermal efficiency. Similarly, Ahmadi et al. (2013d) presented the optimum design of a solar-powered Stirling system through the MOO of the thermo-economic function, thermal efficiency, and dimensionless power output.

Li et al. (2015) presented optimum design parameters of an integrated solar-assisted irreversible open Brayton engine system with thermal bridging losses and an imperfect solar collector using the NSGA-II algorithm. Furthermore, the NSGA-II algorithm was employed by Sadatsakkak et al. (2015) for thermo-economic optimization of an irreversible closed regenerative Brayton cycle by taking the power output, ecological, and thermo-economic functions as MOO goals. Ahmadi et al. (2016d) presented a thermodynamic analysis of combined Stirling and Otto cycle for power production, and MOO was performed for optimal power output and thermal efficiency of the Stirling engine using the NSGA-II algorithm. Ahmadi et al. (2016c) optimized a finite speed thermodynamic analysis of Stirling heat engine considering internal irreversibilities such as fluid friction and imperfect regeneration.

Ahmadi et al. (2016a) presented a thermodynamic model of solar-driven regenerative irreversible Brayton cycle system and identified optimal process parameters of the system to achieve maximum power and thermal efficiency using the NSGA-II algorithm. Ahmadi et al. (2016b) presented the thermodynamic analysis of a solar-assisted Stirling system with high-temperature differential, and simultaneous optimization was carried out using the NSGA-II algorithm. Arora et al. () performed multi-objective optimization of a solar-driven Stirling system with regenerative losses by employing the NSGA-II algorithm. Bellos et al. (2017) investigated the solar energy supported gas turbine system and optimized to attain optimum collector area, electricity production, and fuel consumption.

Ahmadi et al. (2017) optimized a solar-driven small-scale transcritical CO_2 power cycle system with the liquid natural gas heat sink using the NSGA-II algorithm. Arora et al. (2017) performed multi-objective optimization using the NSGA-II algorithm to solar energy-driven Stirling heat engine system with regenerative heat losses, finite rate heat transfer, thermal bridging losses, and regeneration process time. Dai et al. (2018) proposed and optimized the finite-time thermodynamic model of Stirling heat engine by considering finite rate heat transfer, conductive thermal bridging losses, and regenerative heat losses.

Kim et al. (2018) presented a solar-assisted heat pump system for hot water supply. The proposed system uses a hybrid solar collector. The hybrid solar collector uses a solar thermal receiver and air source–receiver to collect solar energy. Furthermore, presented a thermodynamic model of the proposed system and performed a parametric analysis of the system's critical parameters. Sanaye and Taheri (2018) proposed a hybrid liquid desiccant heat pump system and considering energy, exergy, environmental, and economic aspects optimized using the MOGA. Total annual cost and exergy efficiency were considered as the objectives to identify the optimum parameters of the system.

Furthermore, solar-assisted systems applications related to several integrated energy systems have been researched. Jing et al. (2018) presented a combined MOO framework and multi-criteria decision-making for optimal designing and planning of a solar-assisted solid oxide fuel cell distributed power system. Rashidi and Khorshidi (2018) presented a mathematical model for a solar-based multi-generation system consisting of desalination unit, water heater, organic Rankine cycle evaporator, PV solar collectors, and a single effect absorption chiller unit by performing energy and exergy analysis of the system. Also, taking total cost rate and energy efficiency as the objectives performed MOO using multi-objective differential evolution (MODE) algorithm. Habibollahzade et al. (2018) presented an integrated power generation system which consists of a biomass-based solid oxide fuel cell (BSOFC), a Stirling engine, and an electrolyzer. In this system, the waste heat of the BSOFC unit was utilized to run the Stirling engine, and the excess power generated by the Stirling engine was then used for hydrogen production by proton exchange membrane electrolyzer.

Behzadi et al. (2019) proposed a solar-based integrated energy system and presented optimal design parameters which maximize exergy efficiency and hydrogen production rate and minimize total cost rate, through MOO of the proposed system using the genetic algorithm. Khanmohammadi et al. (2020) proposed a solar-assisted integrated power and refrigeration system with the thermoelectric generator (TEG) and investigated the effect of various parameters on the system's performance through energy, exergy, and exergo-economic analyses. Song et al. (2020) presented optimal parameters of a solar hybrid combined cooling, heating and power system using the NSGA-II algorithm.

It can be observed that the effect of various parameters of a solar-driven energy system had been researched in assessing the performance of the system. The performances of these systems had been studied and optimized simultaneously on thermodynamic, thermo-economic, ecological, and economic aspects. The multi-objective optimization considering these aspects was performed by advanced optimization algorithms such as NSGA-II, MOPSO, and MODE. For a MOO problem, these algorithms suggest a set of non-dominated Pareto-optimal solutions.

Optimization of bio-energy systems

Biofuel can be considered as a sustainable and environmentally friendly alternative to petroleum-based fuels. Biofuels can be produced from sources such as sugar, starch, vegetable oil, animal fats, agricultural waste, switchgrass, cereals, microalgae. Biodiesel engine system uses the blend of the biofuel and petrodiesel as the fuel. The performance of the biodiesel engine has considerably affected by the blending proportions. Using biodiesel as an alternative to diesel can reduce the wear of the engine parts and the emissions such as carbon oxides, hydrocarbons, and particulate matter. However, there is a loss of power, lower fuel efficiency, and increased nitrogen oxides emission with biodiesels.

First-generation biofuels are derived from food sources such as sugar, starch, vegetable oil, and animal fats. The second-generation biofuels are derived from cellulosic biomass, such as agricultural waste, switchgrass, and cereals, using thermochemical processes. Nowadays, producing biofuels from microalgae is gaining popularity. The microalgae combine the CO_2 with water using solar energy and create biomass more rapidly (Agarwal et al., 2017).

Microalgae consist of lipids, carbohydrates, and proteins and can be used to produce biofuels like biodiesel, biogas, and bioethanol. Microalgae consist of high oil and lipid content and have a faster growth rate. Nannochloropsis species is a popularly used microalgae for biofuel production. This species, in addition to generating relatively high-density cells and lipids, it contains the eicosapentaenoic acid (EPA), which is used as an additive in human diets. Nannochloropsis species is cultivated in artificial seawater (ASW) containing concentrated *f/2* medium and vitamins (Banerjee et al., 2016).

Kumari et al. (2014) proposed a culture medium using NPK-10:26:26 fertilizer and flue gas for cultivating the Spirulina *Platensis*. In another work, Kumari et al. (2015) investigated the effect of different light sources on the growth of the *Spirulina Platensis* using NPK-10:26:26 fertilizer assisted cultivation medium. Kumar et al. (2015) presented optimal nutrients and environmental parameters of NPK-10:26:26 fertilizer-based cultivation medium for the cultivation of *Dunaliella tertiolecta* species. The response surface methodology (RSM), GA, and NSGA-II algorithms were employed to identify the optimal parameters for the maximum production of biomass and lipid.

Dharma et al. (2016) presented the optimal parameters of biodiesel production from *Jatropha curcus* and *Ceiba pentandra* feedstocks using RSM. Rahman et al. (2017) proposed a two-step process and presented optimal operating conditions for maximizing biodiesel production from *Spirulina maxima* microalgae. Nayak et al.

(2018) proposed an optimal production process and presented optimum process parameters for maximum biomass production using *Scenedesmus* species, using flue gas as the source of CO_2 and domestic wastewater as the culture medium. The optimum process parameters were identified using an artificial neural network (ANN)-assisted genetic algorithm. Yusuff et al. (2019) investigated *Leucaena leucocephala* seed oil as a feedstock for biodiesel generation. Santya et al. (2019) investigated the catfish oil, waste cooking oil, and chicken eggshells as feedstock for biodiesel production. Also, process-optimization was performed on the sensitivity of the molar ratio of oil to methanol and catalyst.

Enhancing engine performance and reducing exhaust emissions of diesel engines is another crucial area in biodiesel research. The major biodiesel engine performance characteristics are engine torque, brake specific fuel consumption (BSFC), exhaust gas temperature, brake-thermal efficiency, power output, and brake power. The exhaust gas emissions are carbon monoxide (CO), carbon dioxide (CO_2), unburned hydrocarbons (HC), nitrogen oxides (NO_x), smoke opacity, and particulate matter (PM). Enhancing the performance and reducing the emissions can be achieved by modifying combustion chamber geometry, the timing of fuel injection, and compression ratio, installing exhaust gas recirculation (EGR) systems, and formulating biodiesel blends that have appropriate physicochemical properties (Damanik et al., 2018).

Ong et al. (2014) investigated *Calophyllum inophyllum* oil as the feedstock for biodiesel production. Wong et al. (2015) presented the process parameters that have improved the biodiesel engine performance using the cuckoo search (CS) algorithm. Shirneshan et al. (2016) determined optimal engine speed, engine load, and biodiesel ratio using the ABC algorithm to achieve optimum engine performance and emission characteristics. Bharadwaz et al. (2016) presented optimum parameters of a variable compression ratio biodiesel engine to improve the engine performance. Using the DoE based on RSM experiments, the effect of the compression ratio, fuel blend, and load on engine performance and emissions are modeled. Finally, optimal parameters are identified using the derringers desirability approach.

Khoobbakht et al. (2016) investigated the effect of engine load, speed, and fuel blend levels on the performance of the biodiesel direct-injection diesel engine. Also, optimum input parameters were determined using RSM, which has resulted in the decrement of CO, HC, NO_x, and smoke opacity. Hasni et al. (2017) presented the optimum process parameters to produce biodiesel from Brucea javanica seeds using the RSM.

Yatish et al. (2018) presented the optimal biodiesel production process parameters using RSM based on CCD for extracting biodiesel from bauhinia variegata oil. Jaliliantabar et al. (2019) identified optimum operating parameters of a biodiesel CI engine with an EGR system through MOO using the NSGA-II algorithm. Dey et al. (2020a) investigated the performance and emission characteristics of a signal cylinder, four-stroke direct-injection CI engine considering different diesel, ethanol, and palm oil biodiesel blends. Dey et al. (2020b) investigated the application of artificial neural network (ANN) model for predicting the brake specific energy consumption (BSEC) and emissions such as NO_x, HC, and CO_2 of a single-cylinder diesel engine which operates with diesel-palm biodiesel-ethanol blends. Also, using a fuzzy interface system, optimum engine operating parameters were identified from the experimental and ANN predicted data. Hirkude and Belokar (2020) investigated the effect of compression ratio, fuel blend, and engine load on the performance of single-cylinder water-cooled naturally aspirated direct-injection diesel engine. Also, predicted the BSEC, BTE, exhaust gas temperature, and smoke opacity were predicted using the RSM.

From the above literature, it can be observed that researchers had focused on the optimum nutrient concentration and environmental factors, which result in an increase in the productivity of lipid, biomass, and EPA and reduce the cost of cultivation. Furthermore, optimum process parameters have been identified using response surface methodology and various optimization algorithms. Furthermore, researchers are focused on finding the optimum operating parameters of biodiesel engine systems, which would improve engine performance and reduce emissions. Also, optimum system parameters had been identified using various methods such as RSM, GRA, and Taguchi and various optimization algorithms such as SA, PSO, CS, ABC, GA, and NSGA-II.

Hydropower generation and reservoir operation

Hydroelectricity is a clean and renewable energy whose quantity depends on the volume of water flow and the amount of head created by the water reservoir. In cascade reservoirs, the reservoir water level of a downstream plant is influenced by the generation of the upstream plant. Therefore, hydropower scheduling is a complicated nonlinear dynamical optimization problem that includes nonlinear flows, nonlinear dynamical hydraulic heads, and the interactional relationships of nonlinear input and output variables. The objective is to obtain the optimal utilization of the hydro resources available for maximum hydroelectric generation given a set of starting conditions and many complex constraints in the hydropower system.

Optimizing hydropower reservoirs are complex because of their nonlinear objective function and constraints. Sharma et al. (2015) developed a decision support system (DSS) for the operation scheduling and optimization of Sewa Hydro Electric Plant to improve the flexibility of decision support system for reservoir operation under availability-based tariff (ABT) regime. Upon implementation on the actual project site, the developed DSS demonstrated to be stable and managed to give reliable decisions.

Li and Qiu (2016) proposed a long-term multi-objective optimization model for integrated hydro/PV power system considering the smoothness of power output process and the total amount of annual power generation of the system simultaneously. The PV power output was firstly calculated by hourly solar radiation and temperature data, which was then taken as the boundary condition for reservoir optimization. For hydropower, due to its great adjustable capability, a month was taken as the time step to balance the simulation cost. This multi-objective optimization problem was then solved by using the modified version of NSGA-II algorithm.

Shang et al. (2017) presented a practical genetic algorithm based solution for solving the economic load dispatch problem (ELDP) and compared the performance with that of dynamic programming (DP). Specifically, their performance was comprehensively evaluated in terms of addressing the ELDP through a case study of 26 turbines in the Three Gorges Hydropower Plant with a focus on calculation accuracy, calculation time, and algorithm stability. Kumar and Yadav (2018) proposed the TLBO and Jaya algorithms for solving the discrete time four-reservoir operation (DFRO), the continuous time four-reservoir operation (CFRO), and the ten-reservoir operation (TRO). The results obtained by the TLBO and JA were compared with different approaches from the literature. It was found that both Jaya algorithm and TLBO algorithm provided a satisfactory solution. Niu et al. (2018) presented a parallel multi-objective PSO to resolve the cascade hydropower reservoirs operation balancing benefit and firm output of Lancang cascade hydropower system in southwest China. The results indicated that PMOPSO can provide satisfying scheduling results in different cases.

Wang et al. (2019) proposed a double-layer model for coordinating the operations of cascaded hydropower and neighboring wind and photovoltaic (PV) facilities, in order to reduce the output fluctuation and maximize the combined power generation of multi-energy system. In this model, a local complementary hydropower (LCH) model, hydropower with less regulating ability was dispatched to alleviate short-term fluctuations, whereas in the global complementary hydropower, (GCH), hydropower with greater regulating capacity was operated to alleviate long-term fluctuations. The results of a case study of a hydro-wind-PV cluster project in southwestern China showed that most short-term fluctuations can be smoothed by LCH with a minimum impact on the capacity factor. In contrast, alleviation of long-term fluctuations can be achieved with GCH, but this causes a large capacity factor reduction. The findings from the case study demonstrated that clustering of hydropower cascades with wind and solar generation facilities was a promising avenue for the de-carbonization of electricity systems.

Onokwai et al. (2020) presented optimum performance characteristics of Jebba hydropower plant, Nigeria, using response surface methodology (RSM) to enhance the estimated power generated from the plant. Results obtained showed that electricity generated, turbine speed, and efficiency of the hydropower plant were influenced by discharge, pressure drop, stator temperature, and water pressure. Also, the power output was optimum when the discharge and pressure drop were kept high. The RSM simulation of the experimental data showed that the optimal values for discharge, pressure drop, stator temperature, and water pressure were 8783.37 m^3, 28.59 m, 58.0 °C, and 31.00 N/m^2, respectively.

Hatamkhani et al. (2020) developed a simulation–optimization model for optimal design of hydropower systems with a systematic view of the basin. Then, PSO algorithm, linked to the simulation model and the developed optimization-simulation model, was used to solve the problem of optimal design of Garsha, Kuran Buzan, Sazbon and Tange mashoore power plant projects in Karkhe river basin. The objective function was maximization of net benefits of hydropower projects considering all resources, uses, and demands in the basin. Normal water level, minimum operation level, installed capacity of dams, and power plants were chosen as decision variables of the problem. Finally, the results of optimal design of hydropower system were discussed for both economic objective functions.

Chong et al. (2021) investigated the Jaya algorithm as an optimization method for reservoir operation. Also, when deriving the optimal operational rule a hedging strategy is introduced to attenuate the impact of reduced water supply. This strategy can effectively counterbalance the lack of water supply with reservoir storage requirements. The Jaya algorithm solution reported a higher amount of hydropower generated compared to other algorithms. Several reservoir performance indices, such as total hydropower generation, reliability, and resilience, were used to judge the proposed algorithm and other algorithms efficiency.

Ahmadianfar et al. (2021) developed an adaptive differential evolution with particle swarm optimization (A-DEPSO) algorithm to derive optimal operating rules for multi-reservoir systems with the purpose of hydropower production. The application of the proposed method was verified by solving a complex four-reservoir hydropower generation system, located in the southwest of Iran. The results achieved indicate that the A-DEPSO reduced the value of the objective function by an average of 57% compared with other well-known optimizers in the literature. Furthermore, the total power generated from the four-reservoir system using the proposed method was increased by an average of 11% compared to the other methods.

From the literature, it can be found that focus was on the operation of reservoir system as well as cascade system. These systems mainly discussed the operation of reservoir as well as interconnected system and developed the mathematical model for the optimum operation of reservoir for the power generation. These mathematical models generally considered the parameters such as reservoir level, time of release water, and flow of stream. The small hydropower plants are mostly run of river, and thus, further study is required for optimum operation of such plants.

Geothermal energy conversion

Geothermal energy has significant potential to reduce fossil fuel consumptions and environmental impacts. To improve energy conversion efficiency of geothermal energy systems, numerous systems designs have been proposed, and their optimization sought (Lee et al., 2019). At this point, it is worth reviewing current developed geothermal energy systems because understanding configurations and principles of basic and state-of-the-art technologies is important for developing advanced energy systems. A comprehensive review of the geothermal energy systems is carried out from the perspective of systems analysis, design, and optimization.

Tugcu and Arslan (2017) presented parametric analysis of an absorption refrigeration system designed using Simav geothermal resources. Also, a novel two-stage artificial neural network (ANN) model was built in which the outcome of the first network was comprised of the input of the second network and using the weights of the resultant networks and bias values, optimum system was determined. The optimum system determined with ANN was redesigned analytically, and the results were determined to be statistically consistent with ANN results.

Boyaghchi and Safari (2017) presented a quadruple energy production system integrated with geothermal energy involving a cascade organic Rankine cycle, liquefied natural gas vaporization process, proton exchange membrane electrolyzer to produce four types of energies, namely electricity power, heating load for vaporizing liquefied natural gas, cooling effect, and hydrogen. The proposed system parameters were then optimized using NSGA-II to achieve the maximum improvement potential for desired system. Optimization results showed that total avoidable exergy destruction rate and total avoidable exergy destruction cost rate get 3.27 and 4.9 times, respectively, and total avoidable investment cost rate was improved within 17.4% relative to the base point.

Cao et al. (2018) proposed a combined cooling and power system driven by geothermal energy for ice-making and hydrogen production. The proposed system combines geothermal flash cycle, Kalina cycle, ammonia-water absorption refrigeration cycle, and electrolyzer. Also, investigated the exergy destruction of different components and analyzed the effect of key parameters on system performance. Furthermore, an optimization was carried out with Jaya algorithm and genetic algorithm (GA) to achieve maximum exergy efficiencies under three different geothermal water temperatures. The results showed that the optimum exergy efficiencies calculated by the Jaya algorithm are about 23.59, 25.06 and 26.25% for geothermal water temperature of 150, 160 and 170 °C, respectively.

Miglani et al. (2018) presented a methodology for the optimization of a building energy system including a detailed thermal model of a borehole heat exchanger-based GSHP and solar thermal collectors. This approach is a bi-level, multi-objective optimization approach that minimizes the total costs and CO_2 emissions.

Ren et al. (2019) presented a hybrid combined cooling heating and power (CCHP) system integrated with natural gas, solar energy, geothermal energy, and electricity grid. Also, a multi-objective optimization model was proposed to optimize the configurations of the hybrid CCHP system in different operation strategies. The multi-objective optimization from the energy, economic, and environmental aspects was carried out using the NSGA-II algorithm. The results demonstrated that the configuration of the hybrid system that operates using the following electric load strategy achieves better performances than other modes.

Tian et al. (2019) proposed a novel superstructure of the geothermal binary power systems considering multiple heat source temperatures, working fluid selection, and alternative heat rejection systems. Based on the superstructure, a life cycle optimization model was formulated as a mixed-integer nonlinear fractional program (MINFP) to determine the optimal design, and a tailored global optimization algorithm was employed to optimize the model.

Yu et al. (2020) presented a novel combined system integrated with a Kalina and a transcritical CO_2 (T-CO_2) cycle and investigated for making effective use of geothermal resources. In the proposed system, the geothermal energy is first utilized by Kalina sub-cycle and then deeply recovered by T-CO_2 sub-cycle. A steady-state mathematical model was developed to further study the novel combined system. The result showed that under the given conditions, the system net power output was higher than both the single Kalina cycle and single T-CO_2 cycle. Furthermore, a multi-objective optimization considering both thermodynamic and economic aspects was carried out using the NSGA-II algorithm, and the optimal design condition was found by means of the TOPSIS method based on entropy weight.

Xie et al. (2020) presented a parametric collaborative optimization method for ground source heat pump-radiant ceiling (GSHP-RC) system to find the optimum parameters by maximizing system performance and reducing operating costs while ensuring indoor thermal comfort. This method integrates response surface method (RSM) and fast non-dominated sorting genetic algorithm (NSGA-II) to search the nonlinearity relationship between the controllable factors and the response factors and execute multi-objective optimization progressively.

Soltani et al. (2021) developed a new method for studying whole geothermal heating/cooling system on the basis of optimization of a ground heat pump (GHP) system and energy analysis of the optimized system using various circulating fluids for geothermal heat exchanger (GHE). Also, established the optimum configuration of the GHE using the genetic algorithm and investigated the effect of various circulating fluids on the system's energy consumption.

From the above literature, it can be observed that the geothermal energy systems can be expected to grow rapidly soon to meet the demand of growing worldwide primary energy demands coupled with need for fossil fuel replacement. For the better use of geothermal resources, improving conversion efficiency is the key issue of the advancement of geothermal energy systems. Understanding the configurations and conversion principles of the existing technologies would be helpful for efficient geothermal energy systems development.

Ocean thermal energy conversion

An advantage of ocean thermal energy conversion (OTEC) is that it can be used as a baseline power source due to constant ocean water temperatures and currents, a capability that both solar and wind power generation lack. Also, OTEC might be a good fit for island or coastal communities that have high electricity prices, are proximate to good OTEC resources, or are isolated from large-scale infrastructure.

An OTEC plant operates based on a Rankine cycle often with a refrigerant as the working fluid. Unlike traditional Rankine cycles, an OTEC plant operates at a very low temperature difference, usually about 20 °C. Therefore, the required water mass flow rates, as well as the heat exchanger areas of the evaporator and condenser, are very large. These requirements reduce the overall efficiency of OTEC systems.

To increase the efficiency of an OTEC plant, multiple cascaded stages can be added instead of only utilizing a single stage. The benefit of multiple stages is that they can use the available temperature difference more effectively, thus capturing more thermal energy per unit of pumped water. However, the diminishing returns of adding successive stages can eventually be outweighed by the added pumping power required to move the water though the added heat exchangers and piping. Therefore, an optimal amount of stages for a 20MWe plant depends on the design parameters of the system and is 2–5 depending on the thermal performance and pressure drop in the heat exchangers. Many engineering obstacles must be overcome in order to construct a large-scale OTEC plant, from manufacturing components to installation logistics. Upshaw et al. (2011) provided a methodology for analyzing major plant design parameters and a means for determining the optimal design of OTEC facilities.

Ahmadi et al. (2015) presented a comprehensive thermodynamic analysis and multi-objective optimization of hybrid solar OTEC system to produce hydrogen using electrolysis. Also, a fast and elitist non-dominated sorting genetic algorithm (NSGA-II) was applied to determine the best design parameters for the system. The objectives of the multi-objectives were minimization of total cost rate of the system and maximization of the cycle exergy efficiency. In addition, a closed form equation for the relationship between exergy efficiency and total cost rate was presented. Their results showed that system performance is notably affected by the mass flow rate of warm ocean surface water, solar radiation intensity, condenser temperature, and evaporator pinch point temperature difference.

Kim et al. (2016) developed fundamental theory and a software package for a dual-used open cycle OTEC plant with selective generation of electric power and desalinated water. Phenomenological equations were suggested to estimate steam generation and condensation rates, which were used to optimize the overall performance of the dual-used open cycle OTEC operations. Also, effects of system scales from the conventional 207 kW to designed 1000 kW plants were scrutinized in terms of steam and cold water flow rates and fractions, temperatures of input warm and cold seawater, and preset vacuum pressure of the evaporator.

Wang et al. (2018) presented a multi-objective optimization model for an ORC of an OTEC system. The multi-objective particle swarm optimization (MOPSO) algorithm has been used to solve the proposed model considering the levelized cost of energy (LCOE) and exergy efficiency as the objectives of optimization. Furthermore, the LINMAP technique has been used to identify Pareto-optimal solution. From the computational results, it was observed that, with the increase of cool seawater depth, LCOEs of six working fluids will all first decrease and then keep stable or slightly increase. The ranking of six working fluids in terms of LCOE and exergy efficiency was as follows: R717 and R601 have the best performance, followed by R152a, while R134a and R600a have relatively poor performance, and R227ea was the least desirable.

Wu et al. (2019) presented a constructal design of an evaporator in OTEC system by taking the minimum total dimensionless pumping power (DPP) as optimization objective and the effective volume and heat transfer rate as the constraints. The effects of the structure and flow parameters of the evaporator on the minimum DPP and optimal heat transfer plate width are analyzed. The optimal constructal results of the evaporator with eight different working fluids are compared. Their results showed that the DPP after constructal optimization reduces by 37.1% compared with that of the initial design point. The minimum DPP dramatically augments with the increases of heat transfer plate effective length, corrugation angle, and mass flow rate of the warm seawater and dramatically diminishes with the increases of corrugation wavelength and effective volume of the evaporator.

Wu et al. (2020) presented a constructal optimization of a plate condenser with fixed heat transfer rate and effective volume in an OTEC system based on constructal theory. Considering the entropy generation rate in heat transfer process and total pumping power due to friction loss as the objectives for the optimization was carried out using the multi-objective genetic algorithm. The results demonstrated that the effective volume of the plate condenser has a positive impact on the considered objectives.

Alawadhi et al. (2020) presented optimized design of the turbine geometry for an OTEC system by maximizing the efficiency of the turbine at low inlet temperature. Firstly, meanline design was computed that was later optimized using the response surface methodology. Furthermore, an improvement of 4.7% in power and 4.2% in efficiency is achieved through the optimization process.

Li et al. (2020) investigated the optimized thermodynamic performance of closed off-shore OTEC system based on Rankine cycle. To approach the ideal Lorenz cycle and obtain better thermal matching, a Rankine cycle using CO2-based binary zeotropic mixtures was considered. Pure working fluids, including NH_3 and CO_2, were also comparatively investigated with mixtures. Overall optimization to maximize the system thermal efficiency and specific work to warm seawater flow rate are carried out. The results indicate that CO_2-based binary zeotropic mixtures could improve thermodynamic coupling of cycle and external seawater. The performance of Rankine cycle at mixture composition has evaporating temperature glide of 7–8 °C is recommended.

Wu et al. (2020) presented a constructal thermodynamic optimization model for an OTEC system with a dual-pressure organic Rankine cycle (DPORC) by combining constructal theory with finite-time thermodynamics. The net power output of the OTECS is chosen as the optimization objective, and the results showed that the net power outputs after the primary, twice, triple, and sextuple constructal thermodynamic optimizations were improved by 2.80, 4.66, 9.95, and 14.95%, respectively, compared with the initial net power output.

Assareh et al. (2021) presented an energy system consisting of a combination of sub-systems, including a flat panel solar collector, OTEC system, wind turbine, organic Rankine cycle (ORC), and a thermoelectric. The proposed system was designed and evaluated based on the average yearly electricity consumption required by an Iranian household. The R227ea refrigerant was applied as an organic fluid in the ORC and water for the OTEC system. Engineering equation solver (EES) software was utilized to model the system and obtain thermodynamic results. After sensitivity analysis, the most significant and influential parameters were proved to be wind speed, ORC pump inlet temperature, solar irradiance, and collector area. Eventually, the system was optimized via NSGA-II.

From the above literature review, it can be observed that there is enormous potential for OTEC technology. Many researchers are working on developing new and efficient solutions, to take advantage of the potential of the ocean as an unlimited energy source. The optimization of heat exchangers, condensers, vaporizers, and turbines, among other devices, is an active line of research that must be developed from various points of view, including thermodynamic theory, new materials, and engineering designs. The following chapter presents the working of Jaya algorithm, Rao algorithms, and their modified versions.

References

Agarwal, A. K., Agarwal, R. A., Gupta, T., & Gurjar, B. R. (2017). Introduction to biofuels. Biofuels, Springer. https://doi.org/10.1007/978-981-10-3791-7_1

Ahmadi, M. H., Ahmadi, M. A., & Feidt, M. (2016a). Performance optimization of a solar-driven multi-step irreversible Brayton cycle based on a multi-objective genetic algorithm. *Oil and Gas Science and Technology—Revue d'IFP Energies Nouvelles, 71*(1), 16. https://doi.org/10.2516/ogst/2014028

Ahmadi, M. H., Ahmadi, M. A., Mellit, A., Pourfayaz, F., & Feidt, M. (2016). Thermodynamic analysis and multi objective optimization of performance of solar dish Stirling engine by the centrality of entransy and entropy generation. *International Journal of Electrical Power and Energy Systems, 78*, 88–95. https://doi.org/10.1016/j.ijepes.2015.11.042

Ahmadi, M. H., Ahmadi, M. A., Pourfayaz, F., Bidi, M., Hosseinzade, H., & Feidt, M. (2016c). Optimization of powered Stirling heat engine with finite speed thermodynamics. *Energy Conversion and Management, 108*, 96–105. https://doi.org/10.1016/j.enconman.2015.11.005

Ahmadi, M. H., Ahmadi, M. A., Pourfayaz, F., Hosseinzade, H., Acikkalp, E., Tlili, I., & Feidt, M. (2016). Designing a powered combined Otto and Stirling cycle power plant through multi-objective optimization approach. *Renewable and Sustainable Energy Reviews, 62*, 585–595. https://doi.org/10.1016/j.rser.2016.05.034

Ahmadi, M. H., Dehghani, S., Mohammadi, A. H., Feidt, M., & Barranco-Jimenez, M. A. (2013a). Optimal design of a solar driven heat engine based on thermal and thermo-economic criteria. *Energy Conversion and Management, 75*, 635–642. https://doi.org/10.1016/j.enconman.2013.07.078

Ahmadi, M. H., Mehrpooya, M., Abbasi, S., Pourfayaz, F., & Bruno, J. C. (2017). Thermo-economic analysis and multi-objective optimization of a transcritical CO_2 power cycle driven by solar energy and LNG cold recovery. *Thermal Science and Engineering Progress, 4*, 185–196. https://doi.org/10.1016/j.tsep.2017.10.004

Ahmadi, M. H., Mohammadi, A. H., Dehghani, S., & Barranco-Jiménez, M. A. (2013b). Multi-objective thermodynamic-based optimization of output power of solar dish-Stirling engine by implementing an evolutionary algorithm. *Energy Conversion and Management, 75*, 438–445. https://doi.org/10.1016/j.enconman.2013.06.030

Ahmadi, M. H., Sayyaadi, H., Mohammadi, A. H., & Barranco-Jimenez, M. A. (2013c). Thermo-economic multi-objective optimization of solar dish-Stirling engine by implementing evolutionary algorithm. *Energy Conversion and Management, 73*, 370–380.https://doi.org/10.1016/j.enconman.2013.05.031

Ahmadi, M. H., Sayyaadi, H., Dehghani, S., & Hosseinzade, H. (2013). Designing a solar powered Stirling heat engine based on multiple criteria: Maximized thermal efficiency and power. *Energy Conversion and Management, 75*, 282–291. https://doi.org/10.1016/j.enconman.2013.06.025

Ahmadi, P., Dincer, I., & Rosen, M. A. (2015). Multi-objective optimization of an ocean thermal energy conversion system for hydrogen production. *International Journal of Hydrogen Energy, 40*(24), 7601–7608. https://doi.org/10.1016/j.ijhydene.2014.10.056

Ahmadianfar, I., Kheyrandish, A., Jamei, M., & Gharabaghi, B. (2021). Optimizing operating rules for multi-reservoir hydropower generation systems: An adaptive hybrid differential evolution algorithm. *Renewable Energy, 167*, 774–790. https://doi.org/10.1016/j.renene.2020.11.152

Alawadhi, K., Alhouli, Y., Ashour, A., & Alfalah, A. (2020). Design and optimization of a radial turbine to be used in a rankine cycle operating with an otec system. *Journal of Marine Science and Engineering, 8*(11), 1–22. https://doi.org/10.3390/jmse8110855

Ali, M. Z., Awad, N. H., Suganthan, P. N., & Reynolds, R. G. (2017). An adaptive multipopulation differential evolution with dynamic population reduction. *IEEE Transactions on Cybernetics, 47*(9), 2768–2779. https://doi.org/10.1109/TCYB.2016.2617301

Arora, R., Kaushik, S. C., & Kumar, R. (2016). Multi-objective thermodynamic optimization of solar parabolic dish stirling heat engine with regenerative losses using NSGA-II and decision making. *Applied Solar Energy, 52*(4), 295–304. https://doi.org/10.3103/S0003701X16040046

Arora, R., Kaushik, S. C., & Kumar, R. (2017). Multi-objective thermodynamic optimisation of solar parabolic dish Stirling heat engine using NSGA-II and decision making. *International Journal of Renewable Energy Technology, 8*(1), 64. https://doi.org/10.1504/IJRET.2017.080873

Arora, R., Kaushik, S. C., Kumar, R., & Arora, R. (2016). Multi-objective thermo-economic optimization of solar parabolic dish Stirling heat engine with regenerative losses using NSGA-II and decision making. *International Journal of Electrical Power and Energy Systems, 74*, 25–35. https://doi.org/10.1016/j.ijepes.2015.07.010

Assareh, E., Assareh, M., Alirahmi, S. M., Jalilinasrabady, S., Dejdar, A., & Izadi, M. (2021). An extensive thermo-economic evaluation and optimization of an integrated system empowered by solar-wind-ocean energy converter for electricity generation—case study: Bandar Abas, Iran. *Thermal Science and Engineering Progress, 25*, 100965. https://doi.org/10.1016/j.tsep.2021.100965

Banerjee, A., Guria, C., & Maiti, S. K. (2016). Fertilizer assisted optimal cultivation of microalgae using response surface method and genetic algorithm for biofuel feedstock. *Energy, 115*, 1272–1290. https://doi.org/10.1016/j.energy.2016.09.066

Behzadi, A., Gholamian, E., Ahmadi, P., Habibollahzade, A., & Ashjaee, M. (2018). Energy, exergy and exergoeconomic (3E) analyses and multi-objective optimization of a solar and geothermal based integrated energy system. *Applied Thermal Engineering, 143*(August), 1011–1022. https://doi.org/10.1016/j.applthermaleng.2018.08.034

Behzadi, A., Habibollahzade, A., Ahmadi, P., Gholamian, E., & Houshfar, E. (2019). Multi-objective design optimization of a solar based system for electricity, cooling, and hydrogen production. *Energy, 169*, 696–709. https://doi.org/10.1016/j.energy.2018.12.047

Bellos, E., Tzivanidis, C., & Antonopoulos, K. A. (2017). Parametric analysis and optimization of a solar assisted gas turbine. *Energy Conversion and Management, 139*, 151–165. https://doi.org/10.1016/j.enconman.2017.02.042

Bharadwaz, Y. D., Rao, B. G., Rao, V. D., & Anusha, C. (2016). Improvement of biodiesel methanol blends performance in a variable compression ratio engine using response surface methodology. *Alexandria Engineering Journal, 55*(2), 1201–1209. https://doi.org/10.1016/j.aej.2016.04.006

Biswas, P. P., Suganthan, P. N., & Amaratunga, G. A. J. (2018). Decomposition based multi-objective evolutionary algorithm for windfarm layout optimization. *Renewable Energy, 115*, 326–337. https://doi.org/10.1016/j.renene.2017.08.041

Boyaghchi, F. A., & Safari, H. (2017). Parametric study and multi-criteria optimization of total exergetic and cost rates improvement potentials of a new geothermal based quadruple energy system. *Energy Conversion and Management, 137*, 130–141. https://doi.org/10.1016/j.enconman.2017.01.047

Cao, L., Lou, J., Wang, J., & Dai, Y. (2018). Exergy analysis and optimization of a combined cooling and power system driven by geothermal energy for ice-making and hydrogen production. *Energy Conversion and Management, 174*(May), 886–896. https://doi.org/10.1016/j.enconman.2018.08.067

Chong, K. L., Lai, S. H., Ahmed, A. N., Wan Jaafar, W. Z., & El-Shafie, A. (2021). Optimization of hydropower reservoir operation based on hedging policy using Jaya algorithm. *Applied Soft Computing, 106*, 107325. https://doi.org/10.1016/j.asoc.2021.107325

Chowdhury, S., Zhang, J., Messac, A., & Castillo, L. (2012). Unrestricted wind farm layout optimization (UWFLO): Investigating key factors influencing the maximum power generation. *Renewable Energy, 38*(1), 16–30. https://doi.org/10.1016/j.renene.2011.06.033

Chowdhury, S., Zhang, J., Messac, A., & Castillo, L. (2013). Optimizing the arrangement and the selection of turbines for wind farms subject to varying wind conditions. *Renewable Energy, 52*, 273–282. https://doi.org/10.1016/j.renene.2012.10.017

Dai, D., Yuan, F., Long, R., Liu, Z., & Liu, W. (2018). Performance analysis and multi-objective optimization of a Stirling engine based on MOPSOCD. *International Journal of Thermal Sciences, 124*, 399–406. https://doi.org/10.1016/j.ijthermalsci.2017.10.030

Damanik, N., Ong, H. C., Tong, C. W., Mahlia, T. M. I., & Silitonga, A. S. (2018). A review on the engine performance and exhaust emission characteristics of diesel engines fueled with biodiesel blends. *Environmental Science and Pollution Research, 25*(16), 15307–15325. https://doi.org/10.1007/s11356-018-2098-8

Dey, S., Reang, N. M., Deb, M., & Das, P. K. (2020a). Study on performance-emission trade-off and multi-objective optimization of diesel-ethanol-palm biodiesel in a single cylinder CI engine: ATaguchi-fuzzy approach. *Energy Sources, Part A: Recovery, Utilization and Environmental Effects*. https://doi.org/10.1080/15567036.2020.1767234

Dey, S., Reang, N. M., Majumder, A., Deb, M., & Das, P. K. (2020b). A hybrid ANN-Fuzzy approach for optimization of engine operating parameters of a CI engine fueled with diesel-palm biodiesel-ethanol blend. *Energy, 202*, 117813. https://doi.org/10.1016/j.energy.2020.117813

Dharma, S., Masjuki, H. H., Ong, H. C., Sebayang, A. H., Silitonga, A. S., Kusumo, F., & Mahlia, T. M. I. (2016). Optimization of biodiesel production process for mixed Jatropha curcas–Ceiba pentandra biodiesel using response surface methodology. *Energy Conversion and Management, 115*, 178–190. https://doi.org/10.1016/j.enconman.2016.02.034

Dhingra, S., Bhushan, G., & Dubey, K. K. (2014). Multi-objective optimization of combustion, performance and emission parameters in a jatropha biodiesel engine using non-dominated sorting genetic algorithm-II. *Frontiers of Mechanical Engineering, 9*(1), 81–94. https://doi.org/10.1007/s11465-014-0287-9

DuPont, B., & Cagan, J. (2016). A hybrid extended pattern search/genetic algorithm for multi-stage wind farm optimization. *Optimization and Engineering, 17*(1), 77–103. https://doi.org/10.1007/s11081-016-9308-3

DuPont, B., Cagan, J., & Moriarty, P. (2016). An advanced modeling system for optimization of wind farm layout and wind turbine sizing using a multi-level extended pattern search algorithm. *Energy, 106*, 802–814. https://doi.org/10.1016/j.energy.2015.12.033

González, J. S., Gonzalez Rodriguez, A. G., Mora, J. C., Santos, J. R., & Payan, M. B. (2010). Optimization of wind farm turbines layout using an evolutive algorithm. *Renewable Energy, 35*(8), 1671–1681. https://doi.org/10.1016/j.renene.2010.01.010

Grady, S. A., Hussaini, M. Y., & Abdullah, M. M. (2005). Placement of wind turbines using genetic algorithms. *Renewable Energy, 30*(2), 259–270. https://doi.org/10.1016/j.renene.2004.05.007

Habibollahzade, A., Gholamian, E., Houshfar, E., & Behzadi, A. (2018). Multi-objective optimization of biomass-based solid oxide fuel cell integrated with Stirling engine and electrolyzer. *Energy Conversion and Management, 171*, 1116–1133. https://doi.org/10.1016/j.enconman.2018.06.061

Hasni, K., Ilham, Z., Dharma, S., & Varman, M. (2017). Optimization of biodiesel production from Brucea javanica seeds oil as novel non-edible feedstock using response surface methodology. *Energy Conversion and Management, 149*, 392–400. https://doi.org/10.1016/j.enconman.2017.07.037

Hatamkhani, A., Moridi, A., & Yazdi, J. (2020). A simulation—optimization models for multi-reservoir hydropower systems design at watershed scale. *Renewable Energy, 149*, 253–263. https://doi.org/10.1016/j.renene.2019.12.055

Hirkude, J., & Belokar, V. (2020). Investigations on performance of CI engine with waste palm oil biodiesel-diesel blends using response surface methodology. In S. Singh & V. Ramadesigan (Eds.), *Advances in Energy Research, Vol. 2* (pp. 505–514). Springer.

Ituarte-Villarreal, C. M., & Espiritu, J. F. (2011). Optimization of wind turbine placement using a viral based optimization algorithm. *Procedia Computer Science, 6*, 469–474. https://doi.org/10.1016/j.procs.2011.08.087

Jaliliantabar, F., Ghobadian, B., Najafi, G., Mamat, R., & Carlucci, A. P. (2019). Multi-objective NSGA-II optimization of a compression ignition engine parameters using biodiesel fuel and exhaust gas recirculation. *Energy, 187*, 115970. https://doi.org/10.1016/j.energy.2019.115970

Jing, R., Zhu, X., Zhu, Z., Wang, W., Meng, C., Shah, N., & Zhao, Y. (2018). A multi-objective optimization and multi-criteria evaluation integrated framework for distributed energy system optimal planning. *Energy Conversion and Management, 166*, 445–462. https://doi.org/10.1016/j.enconman.2018.04.054

Khanmohammadi, S., Musharavati, F., Kizilkan, O., & Duc Nguyen, D. (2020). Proposal of a new parabolic solar collector assisted power-refrigeration system integrated with thermoelectric generator using 3E analyses: Energy, exergy, and exergo-economic. *Energy Conversion and Management, 220*, 113055. https://doi.org/10.1016/j.enconman.2020.113055

Khoobbakht, G., Najafi, G., Karimi, M., & Akram, A. (2016). Optimization of operating factors and blended levels of diesel, biodiesel and ethanol fuels to minimize exhaust emissions of diesel engine using response surface methodology. *Applied Thermal Engineering, 99*, 1006–1017. https://doi.org/10.1016/j.applthermaleng.2015.12.143

Kim, A. S., Kim, H. J., Lee, H. S., & Cha, S. (2016). Dual-use open cycle ocean thermal energy conversion (OC-OTEC) using multiple condensers for adjustable power generation and seawater desalination. *Renewable Energy, 85*, 344–358. https://doi.org/10.1016/j.renene.2015.06.014

Kim, T., Choi, B. I., Han, Y. S., & Do, K. H. (2018). A comparative investigation of solar-assisted heat pumps with solar thermal collectors for a hot water supply system. *Energy Conversion and Management, 172*(April), 472–484. https://doi.org/10.1016/j.enconman.2018.07.035

Kumar, A., Pathak, A. K., & Guria, C. (2015). NPK-10:26:26 complex fertilizer assisted optimal cultivation of Dunaliella tertiolecta using response surface methodology and genetic algorithm. *Bioresource Technology, 194*, 117–129. https://doi.org/10.1016/j.biortech.2015.06.082

Kumar, R. S., & Prasad, A. K. V. (2019). Environment friendly butyl ester biodiesel production from mahua oil: Optimization and characterization. *SN Applied Sciences, 1*(8), 872. https://doi.org/10.1007/s42452-019-0913-6

Kumar, V., & Yadav, S. M. (2018). Optimization of reservoir operation with a new approach in evolutionary computation using TLBO algorithm and jaya algorithm. *Water Resources Management, 32*(13), 4375–4391. https://doi.org/10.1007/s11269-018-2067-5

Kumari, A., Pathak, A. K., & Guria, C. (2015). Effect of light emitting diodes on the cultivation of Spirulina platensis using NPK-10:26:26 complex fertilizer. *Phycological Research, 63*(4), 274–283. https://doi.org/10.1111/pre.12099

Kumari, A., Sharma, V., Pathak, A. K., & Guria, C. (2014). Cultivation of Spirulina platensis using NPK-10:26:26 complex fertilizer and simulated flue gas in sintered disk chromatographic glass bubble column. *Journal of Environmental Chemical Engineering, 2*(3), 1859–1869. https://doi.org/10.1016/j.jece.2014.08.002

Lee, I., Tester, J. W., & You, F. (2019). Systems analysis, design, and optimization of geothermal energy systems for power production and polygeneration: State-of-the-art and future challenges. *Renewable and Sustainable Energy Reviews, 109*(April), 551–577. https://doi.org/10.1016/j.rser.2019.04.058

Li, C., Pan, L., & Wang, Y. (2020). Thermodynamic optimization of Rankine cycle using CO2-based binary zeotropic mixture for ocean thermal energy conversion. *Applied Thermal Engineering, 178*, 115617. https://doi.org/10.1016/j.applthermaleng.2020.115617

Li, F. F., & Qiu, J. (2016). Multi-objective optimization for integrated hydro-photovoltaic power system. *Applied Energy, 167*, 377–384. https://doi.org/10.1016/j.apenergy.2015.09.018

Li, W., Özcan, E., & John, R. (2017). Multi-objective evolutionary algorithms and hyper-heuristics for wind farm layout optimisation. *Renewable Energy, 105*, 473–482. https://doi.org/10.1016/j.renene.2016.12.022

Li, Y., Liao, S., & Liu, G. (2015). Thermo-economic multi-objective optimization for a solar-dish Brayton system using NSGA-II and decision making. *International Journal of Electrical Power and Energy Systems, 64*, 167–175. https://doi.org/10.1016/j.ijepes.2014.07.027

Liao, X., Zhou, J., Ouyang, S., Zhang, R., & Zhang, Y. (2014). Multi-objective artificial bee colony algorithm for long-term scheduling of hydropower system: A case study of China. *Water Utility Journal, 7*, 13–23.

Miglani, S., Orehounig, K., & Carmeliet, J. (2018). Integrating a thermal model of ground source heat pumps and solar regeneration within building energy system optimization. *Applied Energy, 218*(February), 78–94. https://doi.org/10.1016/j.apenergy.2018.02.173

Moorthy, C. B., & Deshmukh, M. K. (2013). A new approach to optimise placement of wind turbines using particle swarm optimisation. *International Journal of Sustainable Energy, 34*(6), 396–405. https://doi.org/10.1080/14786451.2013.860140

Moorthy, C. B., Deshmukh, M. K., & Mukherejee, D. (2014). New approach for placing wind turbines in a wind farm using genetic algorithm. *Wind Engineering, 38*(6), 633–642. https://doi.org/10.1260/0309-524X.38.6.633

Mosetti, G., Poloni, C., & Diviacco, B. (1994). Optimization of wind turbine positioning in large windfarms by means of a genetic algorithm. *Journal of Wind Engineering and Industrial Aerodynamics, 51*(1), 105–116. https://doi.org/10.1016/0167-6105(94)90080-9

Nayak, M., Dhanarajan, G., Dineshkumar, R., & Sen, R. (2018). Artificial intelligence driven process optimization for cleaner production of biomass with co-valorization of wastewater and flue gas in an algal biorefinery. *Journal of Cleaner Production, 201*, 1092–1100. https://doi.org/10.1016/j.jclepro.2018.08.048

Niu, W. jing, Feng, Z. kai, Cheng, C. tian, & Wu, X. yu. (2018). A parallel multi-objective particle swarm optimization for cascade hydropower reservoir operation in southwest China. *Applied Soft Computing Journal, 70*, 562–575. https://doi.org/10.1016/j.asoc.2018.06.011

Ogunjuyigbe, A. S. O., Ayodele, T. R., & Bamgboje, O. D. (2017). Optimal placement of wind turbines within a wind farm considering multi-directional wind speed using two-stage genetic algorithm. *Frontiers in Energy*. https://doi.org/10.1007/s11708-018-0514-x

Ong, H. C., Masjuki, H. H., Mahlia, T. M. I., Silitonga, A. S., Chong, W. T., & Leong, K. Y. (2014). Optimization of biodiesel production and engine performance from high free fatty acid Calophyllum inophyllum oil in CI diesel engine. *Energy Conversion and Management, 81*, 30–40. https://doi.org/10.1016/j.enconman.2014.01.065

Onokwai, A. O., Owamah, H. I., Ibiwoye, M. O., Ayuba, G. C., & Olayemi, O. A. (2020). Application of response surface methodology (RSM) for the optimization of energy generation from Jebba hydro-power plant Nigeria. *ISH Journal of Hydraulic Engineering, 00*(00), 1–9. https://doi.org/10.1080/09715010.2020.1806120

Patel, J., Savsani, V., & Patel, R. (2015). Maximizing energy output of a wind farm using teaching–learning-based optimization. *Volume 2: Photovoltaics; Renewable-Non-Renewable Hybrid Power System; Smart Grid, Micro-Grid Concepts; Energy Storage; Solar Chemistry; Solar Heating and Cooling; Sustainable Cities and Communities, Transportation; Symposium on Integrated/Sustainable Buil*. https://doi.org/10.1115/ES2015-49164

Patel, J., Savsani, V., Patel, V., & Patel, R. (2017). Layout optimization of a wind farm to maximize the power output using enhanced teaching learning based optimization technique. *Journal of Cleaner Production, 158*, 81–94. https://doi.org/10.1016/j.jclepro.2017.04.132

Pillai, A. C., Chick, J., Johanning, L., & Khorasanchi, M. (2018). Offshore wind farm layout optimization using particle swarm optimization. *Journal of Ocean Engineering and Marine Energy, 4*(1), 73–88. https://doi.org/10.1007/s40722-018-0108-z

Pookpunt, S., & Ongsakul, W. (2013). Optimal placement of wind turbines within wind farm using binary particle swarm optimization with time-varying acceleration coefficients. *Renewable Energy, 55*, 266–276. https://doi.org/10.1016/j.renene.2012.12.005

Rahman, M. A., Aziz, M. A., Al-khulaidi, R. A., Sakib, N., & Islam, M. (2017). Biodiesel production from microalgae Spirulina maxima by two step process: Optimization of process variable. *Journal of Radiation Research and Applied Sciences, 10*(2), 140–147. https://doi.org/10.1016/j.jrras.2017.02.004

Rashidi, H., & Khorshidi, J. (2018). Exergoeconomic analysis and optimization of a solar based multigeneration system using multiobjective differential evolution algorithm. *Journal of Cleaner Production, 170*, 978–990. https://doi.org/10.1016/j.jclepro.2017.09.201

Reddy, S. R. (2020). Wind Farm Layout Optimization (WindFLO): An advanced framework for fast wind farm analysis and optimization. *Applied Energy, 269*, 115090. https://doi.org/10.1016/j.apenergy.2020.115090

Ren, F., Wang, J., Zhu, S., & Chen, Y. (2019). Multi-objective optimization of combined cooling, heating and power system integrated with solar and geothermal energies. *Energy Conversion and Management, 197*, 111866. https://doi.org/10.1016/j.enconman.2019.111866

Sadatsakkak, S. A., Ahmadi, M. H., & Ahmadi, M. A. (2015). Thermodynamic and thermo-economic analysis and optimization of an irreversible regenerative closed Brayton cycle. *Energy Conversion and Management, 94*, 124–129. https://doi.org/10.1016/j.enconman.2015.01.040

Sanaye, S., & Taheri, M. (2018). Modeling and multi-objective optimization of a modified hybrid liquid desiccant heat pump (LD-HP) system for hot and humid regions. *Applied Thermal Engineering, 129*, 212–229. https://doi.org/10.1016/j.applthermaleng.2017.09.116

Santya, G., Maheswaran, T., & Yee, K. F. (2019). Optimization of biodiesel production from high free fatty acid river catfish oil (Pangasius hypothalamus) and waste cooking oil catalyzed by waste chicken egg shells derived catalyst. *SN Applied Sciences, 1*(2), 152. https://doi.org/10.1007/s42452-018-0155-z

Sessarego, M., Dixon, K. R., Rival, D. E., & Wood, D. H. (2014). A hybrid multi-objective evolutionary algorithm for wind-turbine blade optimization. *Engineering Optimization, 47*(8), 1043–1062. https://doi.org/10.1080/0305215x.2014.941532

Shakoor, R., Hassan, M. Y., Raheem, A., & Rasheed, N. (2015). The modelling of wind farm layout optimization for the reduction of wake losses. *Indian Journal of Science and Technology, 8*(17). https://doi.org/10.17485/ijst/2015/v8i17/69817

Shang, Y., Lu, S., Gong, J., Liu, R., Li, X., & Fan, Q. (2017). Improved genetic algorithm for economic load dispatch in hydropower plants and comprehensive performance comparison with dynamic programming method. *Journal of Hydrology, 554*, 306–316. https://doi.org/10.1016/j.jhydrol.2017.09.029

Sharma, R. N., Chand, N., Sharma, V., & Yadav, D. (2015). Decision support system for operation, scheduling and optimization of hydro power plant in Jammu and Kashmir region. *Renewable and Sustainable Energy Reviews, 43*, 1099–1113. https://doi.org/10.1016/j.rser.2014.11.005

Shirneshan, A., Samani, B. H., & Ghobadian, B. (2016). Optimization of biodiesel percentage in fuel mixture and engine operating conditions for diesel engine performance and emission characteristics by artificial bees colony algorithm. *Fuel, 184*, 518–526. https://doi.org/10.1016/j.fuel.2016.06.117

Soltani, M., Farzanehkhameneh, P., Moradi Kashkooli, F., Al-Haq, A., & Nathwani, J. (2021). Optimization and energy assessment of geothermal heat exchangers for different circulating fluids. *Energy Conversion and Management, 228*(November 2020), 113733. https://doi.org/10.1016/j.enconman.2020.113733

Song, Z., Liu, T., & Lin, Q. (2020). Multi-objective optimization of a solar hybrid CCHP system based on different operation modes. *Energy, 206*, 118125. https://doi.org/10.1016/j.energy.2020.118125

Tian, X., Meyer, T., Lee, H., & You, F. (2020). Sustainable design of geothermal energy systems for electric power generation using life cycle optimization. *AIChE Journal, 66*(4). https://doi.org/10.1002/aic.16898

Tugcu, A., & Arslan, O. (2017). Optimization of geothermal energy aided absorption refrigeration system—GAARS: A novel ANN-based approach. *Geothermics, 65*, 210–221. https://doi.org/10.1016/j.geothermics.2016.10.004

Turner, S. D. O., Romero, D. A., Zhang, P. Y., Amon, C. H., & Chan, T. C. Y. (2014). A new mathematical programming approach to optimize wind farm layouts. *Renewable Energy, 63*, 674–680. https://doi.org/10.1016/j.renene.2013.10.023

Upshaw, C. R., & Webber, M. E. (2011). Integrated thermal-fluids system modeling of an ocean thermal energy conversion power plant for analysis and optimization. In *ASME 2011 5th International Conference on Energy Sustainability, Parts A, B, and C*, (pp. 1255–1264). https://doi.org/10.1115/ES2011-54595

Wang, J., Huang, W., Ma, G., & Chen, S. (2015). An improved partheno genetic algorithm for multi-objective economic dispatch in cascaded hydropower systems. *International Journal of Electrical Power and Energy Systems, 67*, 591–597. https://doi.org/10.1016/j.ijepes.2014.12.037

Wang, L., Zuo, M. J., Xu, J., Zhou, Y., & Tan, A. C. (2019). Optimizing wind farm layout by addressing energy-variance trade-off: A single-objective optimization approach. *Energy, 189*, 116149. https://doi.org/10.1016/j.energy.2019.116149

Wang, M., Jing, R., Zhang, H., Meng, C., Li, N., & Zhao, Y. (2018). An innovative organic rankine cycle (ORC) based ocean thermal energy conversion (OTEC) system with performance simulation and multi-objective optimization. *Applied Thermal Engineering, 145*, 743–754. https://doi.org/10.1016/j.applthermaleng.2018.09.075

Wang, X., Virguez, E., Xiao, W., Mei, Y., Patiño-Echeverri, D., & Wang, H. (2019). Clustering and dispatching hydro, wind, and photovoltaic power resources with multiobjective optimization of power generation fluctuations: A case study in southwestern China. *Energy, 189*, 116250. https://doi.org/10.1016/j.energy.2019.116250

Wu, X., Hu, W., Huang, Q., Chen, C., Jacobson, M. Z., & Chen, Z. (2020). Optimizing the layout of onshore wind farms to minimize noise. *Applied Energy, 267*, 114896. https://doi.org/10.1016/j.apenergy.2020.114896

Wu, Z., Feng, H., Chen, L., & Ge, Y. (2020). Performance optimization of a condenser in ocean thermal energy conversion (OTEC) system based on constructal theory and a multi-objective genetic algorithm. *Entropy, 22*(6). https://doi.org/10.3390/E22060641

Wu, Z., Feng, H., Chen, L., Tang, W., Shi, J., & Ge, Y. (2020). Constructal thermodynamic optimization for ocean thermal energy conversion system with dual-pressure organic Rankine cycle. *Energy Conversion and Management, 210*, 112727. https://doi.org/10.1016/j.enconman.2020.112727

Wu, Z., Feng, H., Chen, L., Xie, Z., & Cai, C. (2019). Pumping power minimization of an evaporator in ocean thermal energy conversion system based on constructal theory. *Energy, 181*, 974–984. https://doi.org/10.1016/j.energy.2019.05.216

Xie, Y., Hu, P., Zhu, N., Lei, F., Xing, L., & Xu, L. (2020). Collaborative optimization of ground source heat pump-radiant ceiling air conditioning system based on response surface method and NSGA-II. *Renewable Energy, 147*, 249–264. https://doi.org/10.1016/j.renene.2019.08.109

Yang, K., Kwak, G., Cho, K., & Huh, J. (2019). Wind farm layout optimization for wake effect uniformity. *Energy, 183*, 983–995. https://doi.org/10.1016/j.energy.2019.07.019

Yatish, K. V., Lalithamba, H. S., Suresh, R., & Hebbar, H. R. H. (2018). Optimization of bauhinia variegata biodiesel production and its performance, combustion and emission study on diesel engine. *Renewable Energy, 122*, 561–575. https://doi.org/10.1016/j.renene.2018.01.124

Yu, Z., Su, R., & Feng, C. (2020). Thermodynamic analysis and multi-objective optimization of a novel power generation system driven by geothermal energy. *Energy, 199*. https://doi.org/10.1016/j.energy.2020.117381

Yusuff, A. S., Lala, M. A., Popoola, L. T., & Adesina, O. A. (2019). Optimization of oil extraction from Leucaena leucocephala seed as an alternative low-grade feedstock for biodiesel production. *SN Applied Sciences, 1*(4), 357. https://doi.org/10.1007/s42452-019-0364-0

Chapter 4
Working of Jaya and Rao Optimization Algorithms and Their Variants

Abstract This chapter presents the details of the working of the Jaya algorithm, Rao algorithms and their modified versions, namely multi-team perturbation-guiding Jaya algorithm, adaptive multi-team perturbation-guiding Jaya algorithm, elitist Rao algorithms and self-adaptive population Rao algorithm. Also, it presents the implementation of these algorithms to multi-objective optimization problems. In addition, multi-objective optimization performance indicators such as hypervolume, spacing, coverage and inverted generational distance are described. Furthermore, the results of application of the Jaya and Rao algorithms along with their modified versions to 53 unconstrained single-objective optimization benchmark problems and five multi-objective optimization benchmark problems are presented.

4.1 Working of the Jaya Algorithm and Its Modified Versions

The working of the Jaya algorithm and its modified versions in solving single- and multi-objective optimization problems is described in this section.

4.1.1 *Jaya Algorithm*

The Jaya algorithm (Rao, 2016) is a robust and easily applicable optimization approach that always tries to become victorious by reaching the best solution. Jaya algorithm has no algorithm-specific control parameters. It requires tuning of common control parameters only and has only one stage for updating the solutions. Thus, it is easier to employ in engineering applications. The Jaya algorithm uses both the best and the worst solutions to update the candidate solutions.

The flowchart of the Jaya algorithm is presented in Fig. 4.1. Let the objective be to minimize or maximize the function $O(u)$ with m number of design variables (i.e., $v = 1, 2, \ldots, m$). Begin the Jaya algorithm with the initialization of a random population of size n (i.e., $p = 1, 2, \ldots, n$). Next, identify the best and worst candidate

V. R. Ravipudi et al., *Design Optimization of Renewable Energy Systems Using Advanced Optimization Algorithms*, Green Energy and Technology,
https://doi.org/10.1007/978-3-030-95589-2_4

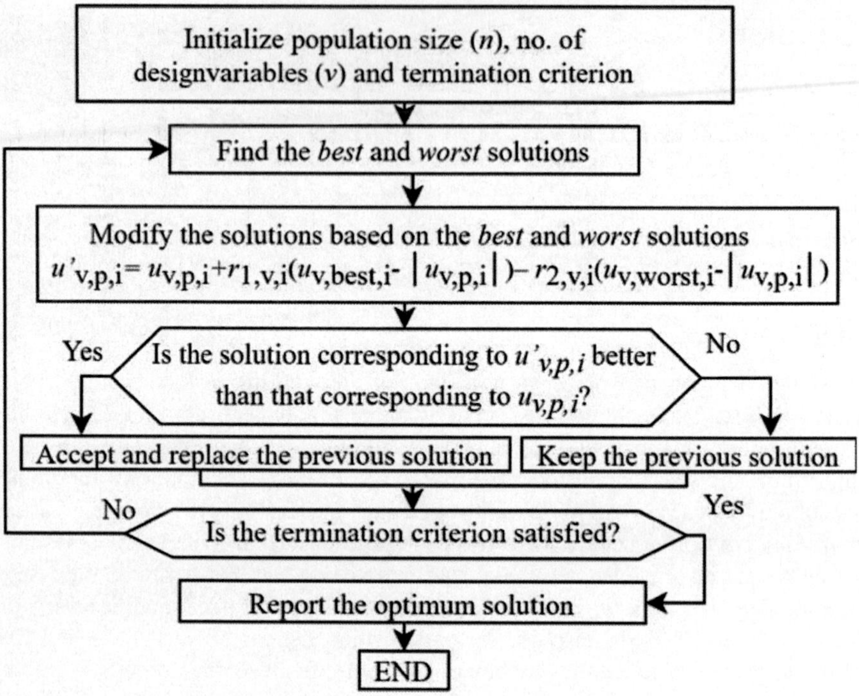

Fig. 4.1 Flowchart of the Jaya algorithm in single-objective optimization

solutions from the population based on the objective function value. Then, update all the candidate solutions using the following equation (Rao, 2016):

$$u'_{v,p,i} = u_{v,p,i} + r_{1,v,i}\left(u_{v,\text{best},i} - \left|u_{v,p,i}\right|\right) - r_{2,v,i}\left(u_{v,\text{worst},i} - \left|u_{v,p,i}\right|\right) \tag{4.1}$$

During any iteration *i*, $u_{v,p,i}$ is the value of the *v*th variable for the *p*th candidate, $u_{v,\text{best},i}$ is the value of *v*th variable for the best candidate solution, and $u_{v,\text{worst},i}$ is the value of the *v*th variable for the worst candidate solution. $u'_{v,p,i}$ is the updated value of $u_{v,p,i}$, and $r_{1,v,i}$ and $r_{2,v,i}$ are the two random numbers for the *v*th variable during the *i*th iteration in the range [0, 1]. Accept updated candidate solution if it is giving better objective value else keep the old candidate solution. At the end of the iteration, the accepted candidate solutions are moved to the subsequent iteration. This procedure is repeated until the termination criterion is satisfied. The best and worst solutions are updated during every iteration.

4.1.2 Multi-team Perturbation-Guiding Jaya (MTPG-Jaya) Algorithm

In the MTPG-Jaya algorithm (Rao & Keesari, 2018), the following modifications to the Jaya algorithm are implemented:

1. During the search process, the proposed algorithm modifies perturbation or movement equation based on solution quality produced by the movement equation. The movement equations considered here are the modified versions of the basic Jaya algorithm. The capability of these modified versions in finding the optimal solutions has already been proven by previous researchers (Ocłoń et al., 2018; Rao & Rai, 2017; Yu et al., 2017). Thus, these movement equations have been considered to take advantage of these modified versions of the Jaya algorithm.
2. The proposed algorithm uses multiple teams over the entire population to explore the search space. Unlike the other multi-population algorithms, the population is not divided into multiple teams. These teams are not multiple sets of the population searching in individual regions of the search space. Each team uses the same set of the population and has a different movement equation. As each team has a different movement equation, and in the search process, each team moves the current set of solutions to a different region in the search space. It means that although all teams are moving from the same candidate solutions, they may get different new candidate solutions, but may overlap in searching regions.
3. When exploiting search space, the population may generate almost identical or identical solutions. If the fitness value remains the same after a certain number of iterations, then the solution is said to be stagnated. Then, non-improved solutions with certain probability are selected to allow further exploration. Thus, the candidates resulting in lower-quality solutions can be inserted into the search plan, replacing higher-quality solutions. Convergence avoidance mechanisms may help the population to escape from local optima. The uses of stagnation treatments enable moving to regions far from the local neighborhood. While this may slow down the attainment of the optimum solution and reduce its accuracy, it can expand the algorithm's exploration capability to locate promising regions in the search space.

4.1.2.1 Perturbation Equations

The proposed algorithm uses six perturbation equations, including the basic Jaya algorithm movement equation, to find the new solution in the search space. The first perturbation equation is the same as that of the Jaya algorithm equation presented in Eq. 4.1. The second perturbation equation is similar to that of the Jaya algorithm. However, the random numbers r_1 and r_2 have been replaced by the chaotic random number. The chaotic sequence used in this study is the well-known logistic map defined by Eq. 4.2. Also, the effectiveness of this chaotic sequence generator is presented by Yu et al. (2017). Let c_m be a chaotic random number in the iteration i; then, the second perturbation equation is given in Eq. 4.3.

$$c_{m+1} = 4c_m(1 - c_m) \tag{4.2}$$

$$u'_{v,p,i} = u_{v,p,i} + c_{m,v,i}\left(u_{v,\text{best},i} - \left|u_{v,p,i}\right|\right) - c_{m,v,i}\left(u_{v,\text{worst},i} - \left|u_{v,p,i}\right|\right) \tag{4.3}$$

The third perturbation equation is taken from Rao and Rai (2017). This movement equation considers the concept of opposition-based learning. A quasi-opposite value of a variable of a candidate solution is a value randomly chosen between the center of the search space and the mirror point of the variable. Let LB_v and UB_v be the lower and upper bounds of variable v, respectively. Then, the quasi-opposite value of the candidate solution variable $u_{v,p,i}$ is given by Eqs. 4.4–4.6.

$$u^q_{v,p,i} = \text{rand}(a, b) \tag{4.4}$$

$$a = (\text{LB}_v + \text{UB}_v)/2 \tag{4.5}$$

$$b = \text{LB}_v + \text{UB}_v - u_{v,p,i} \tag{4.6}$$

The fourth movement equation is taken from Ocłon et al. (2018). In the original Jaya, the movement equation considers only the best candidate solution, whereas in this movement equation, instead of the best candidate, any one of the top three solutions based on the fitness value (best-1, best-2, and best-3) is considered. Let rb be a random integer among $\{1, 2, 3\}$, and then, the movement equation is as follows:

$$u'_{v,p,i} = u_{v,p,i} + r_{1,v,i}\left(u_{v,\text{best}(rb),i} - \left|u_{v,p,i}\right|\right) - r_{2,v,i}\left(u_{v,\text{worst},i} - \left|u_{v,p,i}\right|\right) \tag{4.7}$$

The fifth movement equation is taken from Yu et al. (2017). Let c_m be a chaotic random number generated by Eq. 4.2 and rand be a random number between [0, 1], and then, the new candidate solution is generated by the Eq. 4.8. The sixth movement equation is shown in the Eq. 4.9, which is inspired by the fifth movement equation. In this, the chaotic random number is replaced by a random number between [0, 1].

$$u'_{v,p,i} = u_{v,\text{best},i} + \text{rand}(2c_m - 1) \tag{4.8}$$

$$u'_{v,p,i} = u_{v,\text{best},i} + \text{rand}(2\text{rand} - 1) \tag{4.9}$$

4.1.2.2 Boundary Violation and Corrections

Boundary violations of the variables in the candidate solutions are of two types. First is a variable below the lower boundary. The second is a variable above the upper boundary. In the proposed approach, violations are handled in three ways in both cases. For each team, along with the corrections, violations are also calculated. These violations are later used in finding the quality of each team. At any iteration *i*, let $u_{v,p,i}$ be the value of the *v*th variable for the *p*th candidate. Violation of individual team is calculated using the following equation for all $u_{v,p,i,\text{Team}} < \text{LB}_v$ or $u_{v,p,i,\text{Team}} > \text{UB}_v$.

$$\text{Violation}_{\text{Team}} = \sum_{p=1}^{N} \sum_{v=1}^{D} \left(\left| u_{p,v,i_{\text{Team}}} - \text{UB}_v \right| + \left| \text{LB}_v - u_{p,v,i_{\text{Team}}} \right| \right) \tag{4.10}$$

If $u_{v,p,i} < \text{LB}_v$, then it is corrected in any one of the following methods randomly.

1. Method-1: $u_{v,p,i} = \text{LB}_v$
2. Method-2: $u_{v,p,i} = \text{LB}_v + 0.1r_1(\text{UB}_v - \text{LB}_v)$, where r_1 is a random number between 0 and 1.
3. Method-3: $u_{v,p,i} = \text{LB}_v + \left|\text{remainder}\left(u_{v,p,i}/(\text{UB}_v - \text{LB}_v)\right)\right|$

Similarly, if $u_{v,p,i} > \text{UB}_v$, then it is corrected in any one of the following methods randomly.

4. Method-1: $u_{v,p,i} = \text{UB}_v$
5. Method-2: $u_{v,p,i} = \text{UB}_v + 0.1r_1(\text{UB}_v - \text{LB}_v)$, where r_1 is a random number between 0 and 1.
6. Method-3: $u_{v,p,i} = \text{UB}_v + \left|\text{remainder}\left(u_{v,p,i}/(\text{UB}_v - \text{LB}_v)\right)\right|$.

4.1.2.3 Steps in Multi-team Perturbation-Guiding Jaya Algorithm

Figure 4.2 presents the flowchart of the MTPG-Jaya algorithm. The necessary steps of the proposed approach are as follows:

Step 1: Initialize control parameters such as population size (*N*), number of teams (*N_T*), number of design variables (*D*), lower boundary (LB), upper boundary (UB), the probability of accepting a worse solution if stagnated (*P_aws*), the maximum number of *movement_iterations* (i.e., *MaxMI*), and termination criteria. The termination criteria can be limited function evaluations (*Max_Fevs*) or the desired level of accuracy of the objective function value (*Termination_Value*).

Step 2: Initialize the candidate solutions of size *N*, and calculate the initial function values for the generated candidate solutions. Identify the global best solution based on the function values.

Step 3: For each team, assign a movement equation randomly to guide the current candidate solutions to new candidate solutions. All movement equations available to each team have an equal probability of being selected. Start the counter for the *movement_iterations* (*MI*).

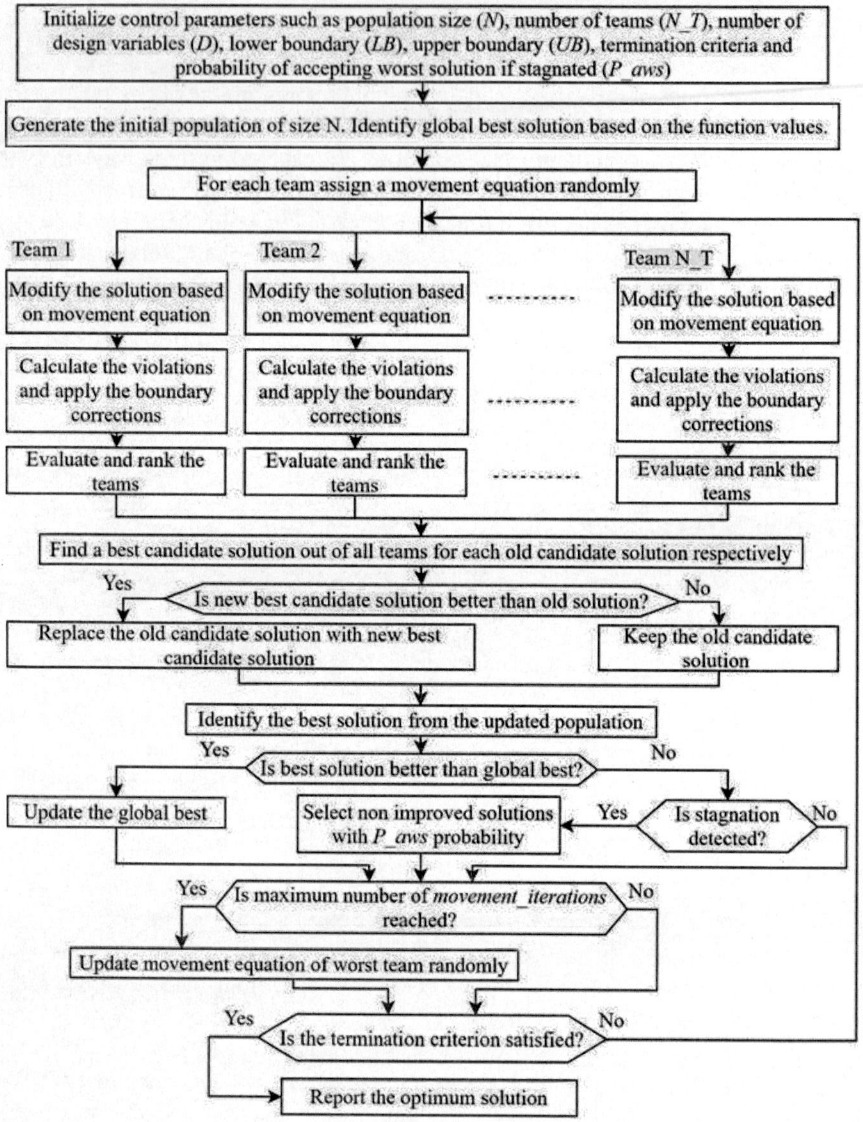

Fig. 4.2 Flowchart of the MTPG-Jaya algorithm

Step 4: Find the new solutions for all the candidate solutions in each team based on the movement equation of the respective team. If the new solutions violate the boundary conditions, then apply the boundary corrections, as explained in the previous section.

Step 5: Evaluate the function values for each solution in each team. Rank the teams according to the fitness of function values. The team whose function value for a candidate solution is better than the remaining teams' function value of respective candidate solution is ranked number one, and the team whose function value for a candidate solution is worse than remaining teams' function value of respective candidate solution is ranked number N_T.

Step 6: Update the old solutions with new solutions in the following way. For each old candidate solution, each team generates a new candidate solution. Find the best candidate solution out of all teams for each old candidate solution, respectively. If this best candidate is better than the old candidate, substitute the old candidate with the best candidate. In this manner, update all old candidate solutions.

Step 7: Update the global best solution if the best solution for updated candidate solutions is better than the previous global best solution.

Step 8: If the global best solution is stagnated, select non-improved solutions with P_aws probability.

Step 9: Update the perturbation or movement equation by the quality of the team. Based on the average teams ranking and boundary condition violations, the quality of a team is measured. Thus, for $i = 1 \ldots N\_T$, the team quality (T_Q) is calculated by $T\_Q_i = T\_AR_i + T\_V_i$. Where T_AR is an average team ranking, and T_V is a team boundary violation. Sometimes just by chance, good solutions may be generated by correction procedures. So, violations are also considered in finding team quality. Consider the team with minimum T_Q as the best team and the team with maximum T_Q as the worst team. Team quality is calculated during every iteration and accumulated until *movement_iterations* (*MI*) reaches the *MaxMI*. Here, *MI* is the iterations after selecting or changing the movement equation for any team. As soon as the *MI* reaches the *MaxMI*, the worst team movement equation is randomly replaced by a new movement equation.

Step 10: Verify the termination criterion. If the search process satisfies the termination criterion, then stop the program by reporting a global optimal solution, else repeat the procedure from **Step 4**. The general framework of the proposed approach is described in Algorithm 1.

Algorithm 1: framework of the multi-team perturbation guiding Jaya algorithm

BEGIN

Initialize the population size '*N*', Number of design variables '*D*', Teams '*N_T*' probability of accepting worst solution '*P_aws*', maximum movement iterations '*M_Itr*' and termination criterion '*FE_max*'. Maximum iterations *Max_I* = *FE_max* / (*N*×*N_T*);

Generate the initial population, Identify global best solution '*G_Best*' and set current iteration i=0

For each team assign a movement equation randomly and set *Move_Itr*=1

While *i* < *Max_I*

    i = *i* +1;

    For *j* = 1→*N_T*

        For *k* = 1→*N*

            $u'_{j,k}$→ Update the variables of $u_{j,k}$ solutions using the equation assigned for the jth team.

            $u'_{j,k}$→ Check the boundary condition violations and apply corrections.

        End For

    End For

    For *j* = 1→*N*

        Rank the teams based on objective function fitness value.

        $u_{j,best}$ →Find the best candidate solution for the $u_{j,old}$candidate solution out of all teams.

        If $O(u_{j,best})$better than $O(u_{j,old})$

            $u_{j,old} = u_{j,best}$

        Else

            $u_{j,old} = u_{j,old}$

        End If

```
        End For

        u_{i,best} & O(u_{i,best}) → Find the best solution from updated solutions.

        If O(u_{j,best}) better than G_Best

                G_Best = O(u_{j,best})

        Else If (stagnation reached)

                u_{j,old} = u_{j,new} Select non-improved solutions with P_aws
                probability.

        End If

        If (Move_Itr=M_Itr)

                For each team assign a movement equation randomly and
                        set Move_Itr=1

        Else

                Move_Itr=Move_Itr+1

        End

End While

END
```

4.1.3 *Adaptive Multi-team Perturbation-Guiding Jaya (AMTPG-Jaya) Algorithm*

In the MTPG-Jaya algorithm, the population size and number of teams need to be tuned according to the optimization problem. Also, the MTPG-Jaya algorithm uses stagnation treatment, which includes the probability of accepting the worst solutions (P_a) and the criterion for stagnation. This stagnation treatment method enables the population to move away from the local optima but may lead to more computational efforts and slow down the attainment of the optimum solution. Initial studies of the MTPG-Jaya had revealed that the value of P_a has a significant effect on the stagnation treatment, which in turn affects the algorithm's performance. However, as the complexity of the problem increases, tuning of the population size, P_a, and the number of teams concurrently can become a tiresome work. These parameters require appropriate tuning and may make the algorithm time-consuming. Hence, by implementing a few modifications to the MTPG-Jaya algorithm, the AMTPG-Jaya algorithm is proposed.

The AMTPG-Jaya algorithm (Rao et al., 2019; 2020; Rao & Keesari, 2019) is proposed by eliminating the adjustment of the number of teams and the probability of accepting the worst solutions in the MTPG-Jaya algorithm. In the AMTPG-Jaya algorithm, an autonomously updating scheme for the number of teams is implemented. Additionally, the stagnation treatments used in the MTPG-Jaya algorithm are removed and adopted a new scheme for stagnation treatment. During the search process, in line with the improvement of the best-so-far candidate solution, the AMTPG-Jaya adapts the number of teams. The AMTPG-Jaya algorithm increases the number of teams if the solution gets stagnated and removes the worst performing team if there is an improvement in the best-so-far solution fitness value.

The perturbation schemes used in the AMTPG-Jaya algorithm are the same as those used in the MTPG-Jaya algorithm. The stepwise procedure is also the same. However, Step-8 of the MTPG-Jaya algorithm is removed, and an additional step to update the number of teams is added. The AMTPG-Jaya updates the number of teams when movement iterations (*MI*) become equal to the maximum number of movement iterations (*MaxMI*). After updating the number of teams, it allows the teams to explore the search space until *MI* reaches *MaxMI*. If the global best solution recorded at *MaxMI* is better than that recorded at the beginning of movement iterations, then the number of teams is reduced by one, else the number of teams is increased by one. After updating the number of teams, it allows the teams to explore search space until *MI* reaches *MaxMI*. The flowchart of the AMTPG-Jaya algorithm is presented in Fig. 4.3.

Generally, the number of function evaluations is taken as the algorithm-computational expenditure. Let TI be the total number of iterations; then, the computational expense of the AMTPG-Jaya algorithm can be calculated by $N \times N\_T \times \mathrm{TI}$. However, the number of teams of the AMTPG-Jaya algorithm will be varying as the search process is in progress. Thus, in the computational experiments, computational expense in iterations can be calculated by multiplying the number of population with the number of teams in the respective iteration, and the summation of computational expense of all iterations will give the total computational expense.

In the proposed approach, the terms $r_{1,v,i}\left(u_{v,\mathrm{best},i} - \left|u_{v,p,i}\right|\right)$ and $c_{m1,v,i}\left(u_{v,\mathrm{best},i} - \left|u_{v,p,i}\right|\right)$ in Eqs. 4.1 and 4.3 represent the movement of the solution towards the best solution, respectively. Similarly, the terms $-r_{2,v,i}\left(u_{v,\mathrm{worst},i} - \left|u_{v,p,i}\right|\right)$ and $-c_{m2,v,i}\left(u_{v,\mathrm{worst},i} - \left|u_{v,p,i}\right|\right)$ in Eqs. 4.1 and 4.3 represent the movement of the solution to avoid the worst solution, respectively. The amount of the movements towards the best solution and away from the worst solution is controlled by the random numbers present in the respective terms. In Eq. 4.1, the random numbers are selected from the uniform random distribution between 0 and 1, whereas in Eq. 4.3, the random numbers are selected from a chaotic sequence generated using Eq. 4.2. This chaotic sequence has randomicity and ergodicity features, which will introduce more diversity in the population. The quasi-oppositional movement (Eq. 4.4) further diversifies the population and improves the convergence rate by simultaneously exploring the quasi-oppositional region of the current population. During the search process, sometimes the best solution in the current iteration may be the local optimum solution, and subsequent

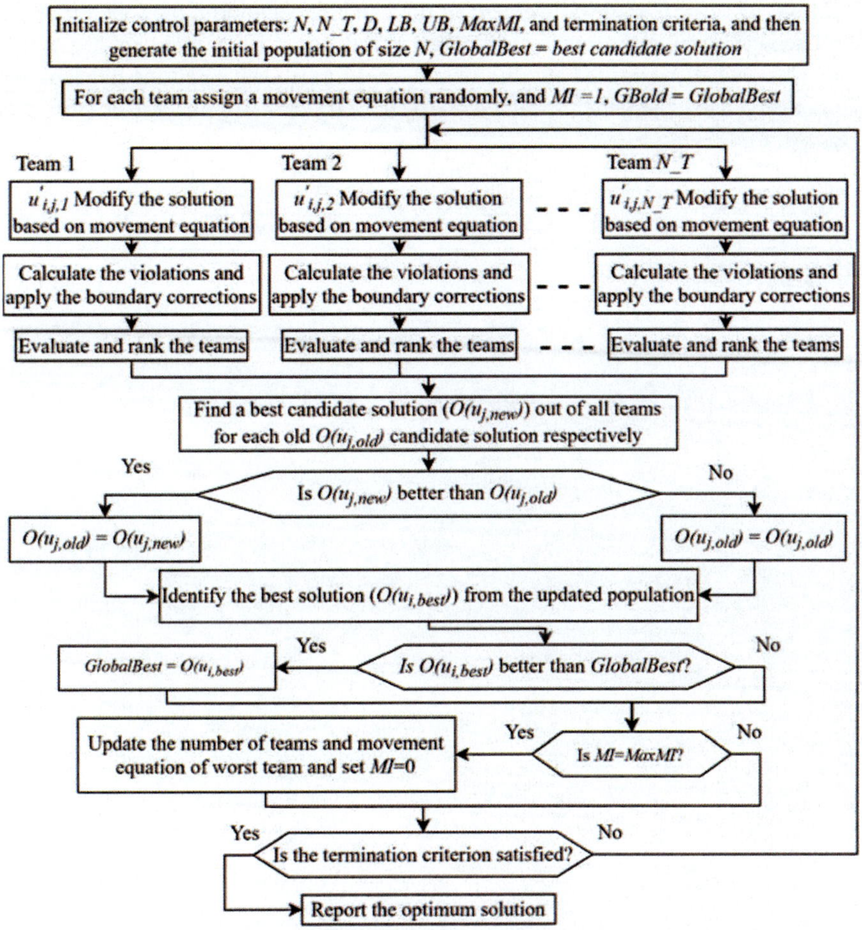

Fig. 4.3 Flowchart of the AMTPG-Jaya algorithm in single-objective optimization

best solutions may be located near to global optima. Thus, to further enhance the capability to examine the search space, multiple best solutions concept is employed in Eq. 4.7. Furthermore, to improve the exploitation capability, local search movement is introduced in Eqs. 4.8 and 4.9. These movements will improve the quality of solutions by producing solutions around it.

The proposed algorithm uses a single set of the population during the search process. Multiple movement equations simultaneously guide this single set of the population. The single set of population with a selected movement equation is considered as a team in this book. As each team has its own movement equation, each team will move the population towards different regions of the search space. Furthermore, based on the improvement in the best candidate, the number of teams will be updated. If there is an improvement in the candidate solution, then the proposed AMTPG-Jaya approach reduces the number of teams, thereby reducing the computational expense.

Similarly, if there is no improvement in the candidate solution, then the proposed approach increases the teams, which in turn introduces more diversity in the population. This way, the proposed algorithm updates the candidate solutions and achieves the convergence. The general framework of the proposed approach is described in Algorithm 2.

Algorithm 2: the framework of the AMTPG-Jaya algorithm

```
BEGIN

Initialize: N, D, LB, UB, N_T, MaxMI, and set the termination crite-
    rion as MaxFE.

Generate the initial population

GlobalBest = best candidate solution,

Function evaluations FE = 0; Iterations i=0;

Allocate perturbation equation arbitrarily for each team and set MI
    =1, GBold = GlobalBest

While FE < MaxFE

        i=i+1

        For k= 1→N_T

                For j= 1→N

                        u'j,k→ Update variables of uj,k candidate using
                                equation assigned to kth team.

                        u'j,k→ Check the boundary condition violations
                        and apply corrections.

                End For

                FE=FE+N

        End For
```

```
        For j= 1→N

                Rank the teams based on objective function value.

                u_j,best → Find the best candidate out of all teams.

                If O(u_j,best) better than O(u_j,old)

                        u_j,old = u_j,best

                Else

                        u_j,old = u_j,old

                End If
        End For

        u_i,best & O(u_i,best) → Find the best candidate solution from the up
        dated candidate solutions.

        If O(u_j,best) better than GlobalBest

                GlobalBest = O(u_j,best)

        End If

        GBnew = GlobalBest

        If MI=MaxMI

                For each team assign a movement equation randomly and
                        MI=1

                If GBnew is better than GBold

                        N_T=N_T-1

                Else

                        N_T=N_T+1

                End If

                GBold = GBnew

        Else

                MI= MI+1

        End If

End While

END
```

4.1.4 Multi-objective Jaya and Multi-objective AMTPG-Jaya Algorithms

Identifying a global best solution that gives optimum values for all the objectives of the multi-objective optimization problem is difficult when the objectives are conflicting in nature. The values of decision variables that produce the most optimum value for one objective may produce non-optimal values for the other objectives. In such situations, the decision-maker can convert the multi-objective optimization problem into a single-objective optimization problem using the priority information among the objectives and can find the global best solution respective to the priorities. If the priorities of the objectives are unknown, then Pareto-optimal solutions can be found using a *posteriori* approach (Deb et al., 2002) for multi-objective optimization problems. The Pareto-optimal solutions are the solutions in which at least one objective function value of each solution is better than the other solutions in the Pareto-front.

The flowcharts of the multi-objective Jaya and AMTPG-Jaya algorithms are presented in Figs. 4.4 and 4.5, respectively. In multi-objective optimization using the Jaya algorithm and its variant AMTPG-Jaya algorithm, a *posteriori* articulation of preferences is followed to handle multiple objectives simultaneously. However, the flow of the algorithms remains the same as those in the single-objective optimization scenario. In the multi-objective optimization scenario, solutions are ranked using dominance principles and crowding distance measurements (Deb et al., 2002) during the iterative process of the algorithms. The candidate solution with the best rank value (*i.e.,* 1) and maximum crowding distance (ξ) is considered as the best candidate. On the other hand, the solution having the worst rank value and lowest ξ value is considered as the worst candidate. Such a selection scheme is adopted so

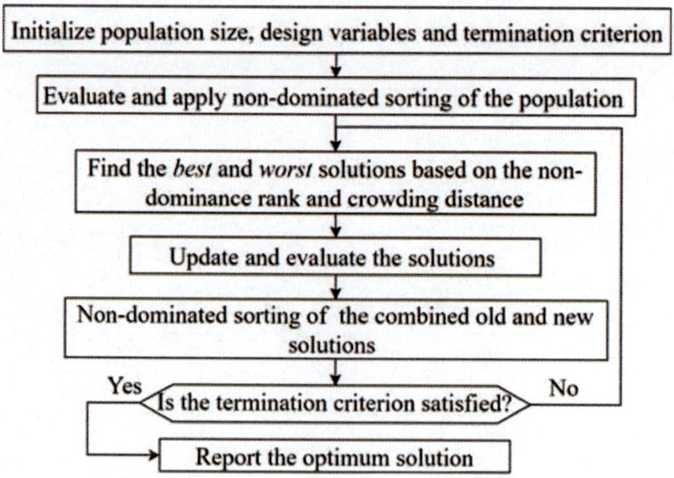

Fig. 4.4 Flowchart of the Jaya algorithm in multi-objective optimization

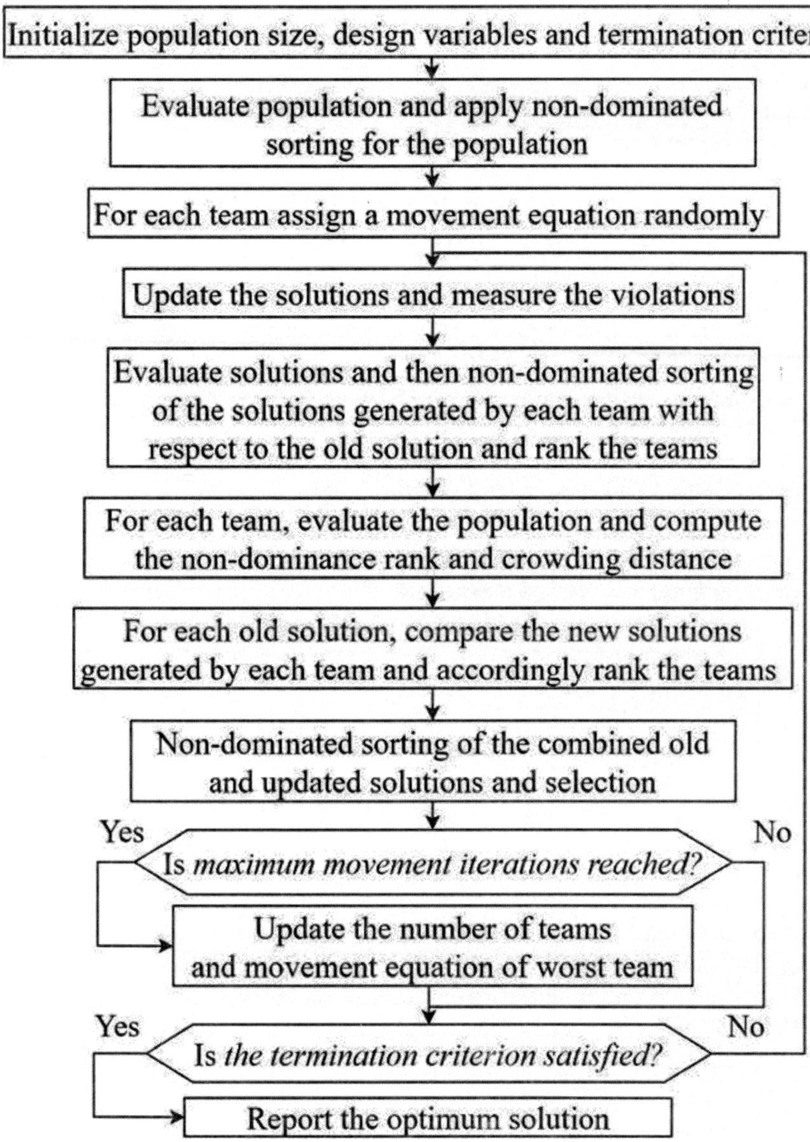

Fig. 4.5 Flowchart of the AMTPG-Jaya algorithm in multi-objective optimization

that the solution in the less populous region of the objective space may guide the search process.

4.1.4.1 Ranking Based on Non-dominance Principles

Let N be a set of candidate solutions to be ranked for an optimization problem consisting of M number of objectives. For a minimization objective, a candidate solution x_k is considered as dominating another candidate solution x_l if and only if $G_j(x_k) \leq G_j(x_l)\ \forall\ 1 \leq j \leq M$ and $G_j(x_k) < G_j(x_l)$ for at least one j, where $j \in \{1, 2, \ldots, M\}$. A candidate solution x_k in population N is considered as a non-dominated candidate if there is no candidate solution x_l in population N, which dominates x_k. In this manner, every candidate solution in N is compared with other candidate solutions in N, and the non-dominated candidates are ranked one. The remaining (excluding the ranked candidates) candidates are also ranked in the same manner and ranked two. This procedure is continued until all the candidate solutions are ranked. All the candidate solutions with the same rank will be considered as the front (F).

4.1.4.2 Crowding Distance (ξ) Calculation

The crowding distance (ξ_k) is an approximate concentration of the candidate solutions in the neighborhood of a particular candidate solution k. For a selected front F, the number of solutions in front $l = |F|$ is determined, and $\xi_k = 0\ \forall\ k \in l$ is assigned. Then, for each objective function $j = 1, 2, \ldots, M$, the set is sorted in the worst order of G_m. The largest crowding distance is assigned to boundary solutions in the sorted list ($\xi_1 = \xi_l = \infty$), and for all the other candidates in the sorted list $k = 2$ to $(l - 1)$, the crowding distance is assigned as follows:

$$\xi_k = \xi_k + (G_j^{k+1} - G_j^{k-1})/(G_j^{\max} - G_j^{\min}) \tag{4.11}$$

where G_j is the objective function value of jth objective, $G_j^{\max}$ and $G_j^{\min}$ are the population-maximum and population-minimum values of the jth objective function.

In the multi-objective optimization cases, more than one optimal candidate solution exists. Thus, to find efficient Pareto-frontier using the proposed algorithm, a candidate from the remote area of the search space is given more priority than the candidate in the packed area of the search area. In the proposed algorithm, the candidate having a better rank is given higher priority, and among the two competing candidates having an equal rank, the solution with a higher ξ value is preferred. This will avoid converging towards a single optimum candidate solution and ensure diversity among the candidate solutions.

Furthermore, in multi-objective optimization through the AMTPG-Jaya algorithm, the scheme used for updating the number of teams is changed for multi-objective optimization. Because after a few iterations, both new global best and old global best solutions become non-dominated solutions. Due to this, calculating the improvement in the global best solutions becomes difficult. Hence, in the multi-objective version of the AMTPG-Jaya algorithm, the number of teams is updated based on the available number of function evaluations. Initially, the search starts with a maximum number of teams. As the search progresses, the number of teams is randomly reduced or increased by one for every time *MI* = *MaxMI*. After completing 60% of function evaluations, the number of teams is reduced by one whenever *MI* = *MaxMI* and maintained a minimum number of teams (two). After updating the number of teams, it allows the teams to explore and exploit search space until *MI* reaches *MaxMI*. The next section presents the working of the Rao algorithms and their modified versions.

4.2 Working of the Rao Algorithms and Their Modified Versions

The working of the Rao algorithms (Rao, 2020) and their modified versions in solving single- and multi-objective optimization problems is described in this section.

4.2.1 Rao Algorithms

The Rao algorithms (Rao-1, Rao-2, and Rao-3) are population-based algorithms and are simple and easy to implement for optimization applications. These algorithms have no algorithm-specific parameters and have no metaphorical explanation. During the iterative process, these algorithms use iteration best solution, iteration worst solution, and random interactions among the population to explore and exploit the search region. The flow of these three algorithms is similar, but the movement equation used is different for each algorithm. The steps of the Rao algorithms are given below:

Step 1: Define the quantity of population (*P*); define the quantity design variables (*N*) and their boundaries: Lower (LB), Upper (UB); termination criterion: it can be the number of function evaluations or iterations.

Step 2: Randomly initialize the population of size *P*, and evaluate the objective function *Z* for all the population.

Step 3: Select the *best* and *worst* solutions from the current population based on their *Z* value. If the *Z* is a minimization function, then the solution with the smallest *Z* value is the *best*, and the solution with the largest *Z* value is the *worst* solution and vice versa if the *Z* is a maximization function.

Step 4: Locate new solutions for all the population ($m = 1, 2, \ldots, P$): During the *i*th iteration, let $s_{n,m,i}$ be the value of the *n*th variable of *m*th solution, $s_{n,b,i}$ be the value of the *n*th variable of the *best* solution, $s_{n,w,i}$ be the value of the *n*th variable of the *worst* solution, and $s'_{n,m,i}$ be the newly located value of $s_{n,m,i}$. Then,

As per the Rao-1 algorithm, the new solutions are found using the following equation:

$$s'_{n,m,i} = s_{n,m,i} + r_{1,n,i}\left(s_{n,b,i} - s_{n,w,i}\right) \tag{4.12}$$

As per the Rao-2 algorithm, the new solutions are located using the following equation:

$$s'_{n,m,i} = s_{n,m,i} + r_{1,n,i}\left(s_{n,b,i} - s_{n,w,i}\right) + r_{2,n,i}\left(\left|s_{n,m,i} \text{ or } s_{n,l,i}\right| - \left|s_{n,l,i} \text{ or } s_{n,m,i}\right|\right) \tag{4.13}$$

As per the Rao-3 algorithm, the new solutions are located using the following equation:

$$s'_{n,m,i} = s_{n,m,i} + r_{1,n,i}\left(s_{n,b,i} - \left|s_{n,w,i}\right|\right) + r_{2,n,i}\left(\left|s_{n,m,i} \text{ or } s_{n,l,i}\right| - \left(s_{n,l,i} \text{ or } s_{n,m,i}\right)\right) \tag{4.14}$$

where $r_{1,n,i}$ and $r_{2,n,i}$ are random numbers in the range [0, 1] for the *n*th variable during the *i*th iteration. In Eqs. 4.13 and 4.14, the third term on the right-hand side represents the interaction between the current solution (*m*th) and a random solution (*l*th) selected from the current population. These two terms are dependent on the *Z* values of the current (*m*th) and randomly selected (*l*th) solutions. If the current solution *Z* value is superior to the randomly selected solution *Z* value, then the third term in Eq. 4.13 becomes $r_{2,n,i}\left(\left|s_{n,m,i}\right| - \left|s_{n,l,i}\right|\right)$, and the third term in Eq. 4.14 becomes $r_{2,n,i}\left(\left|s_{n,m,i}\right| - \left(s_{n,l,i}\right)\right)$. Similarly, if the randomly selected solution *Z* value is superior to the current solution *Z* value, then the third term in Eq. 4.13 becomes $r_{2,n,i}\left(\left|s_{n,l,i}\right| - \left|s_{n,m,i}\right|\right)$, and the third term in Eq. 4.14 becomes $r_{2,n,i}\left(\left|s_{n,l,i}\right| - \left(s_{n,m,i}\right)\right)$.

Step 5: Evaluate *Z* values for the new population, and apply a greedy selection process. If the *Z* value corresponding to the new solution ($s'_{n,m,i}$) is superior to that of the old solution ($s_{n,m,i}$), then replace the old solution with the new solution, if not discard the new solution.

Step 6: Verify the stopping criterion. If the termination criterion is satisfied, report the optimum solution from the final population, else go to **Step 3**. Figure 4.6 presents the flowchart of Rao algorithms.

4.2.2 *Multi-objective Rao Algorithms*

In this book, a *posteriori* version of Rao algorithms is proposed for solving multi-objective optimization problems (Rao & Keesari, 2020a). In this version of the Rao algorithms, the new solutions are located in the same manner as in Rao algorithms. However, the superiority among the solutions is identified based on non-dominance rank and crowding distance evaluation approach (Deb et al., 2002; Rao et al., 2017).

In this version of Rao algorithms, the set of *N* solutions are ranked using dominance principles, and proximity of the solutions with each other is calculated using crowding distance measurement. The solution with the best rank (*rank* = *1*) and largest crowding distance is regarded as the *best* solution. On the other hand, the

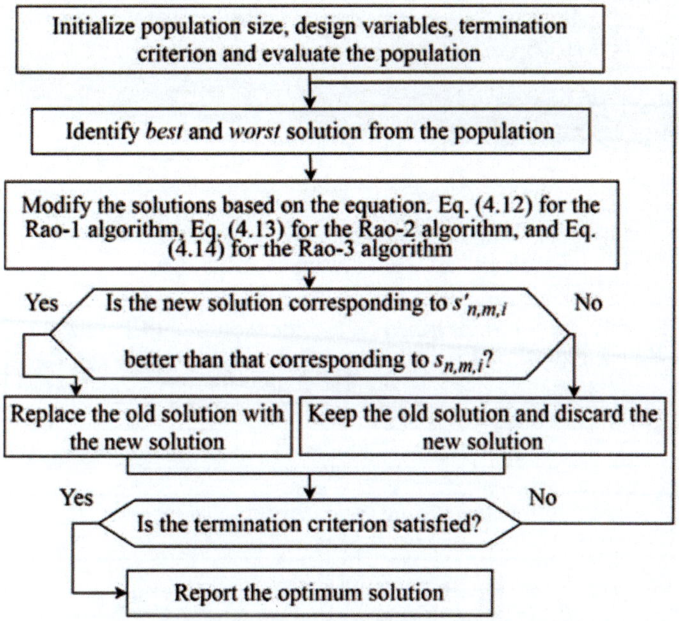

Fig. 4.6 Flowchart of the Rao algorithms in single-objective optimization

solution with the worst rank and least crowding distance is regarded as the *worst* solution. After identifying the *best* and *worst* solutions, a new set of N solutions are located using movement equations of the respective Rao algorithm. Now, the set of new solutions is combined with the set of earlier solutions forming a set of $2N$ solutions. Then, the combined solutions are again ranked using dominance principles, and the crowding distance is calculated for every solution. Based on the new ranking and crowding distance value, a set of N solutions will be selected for the next iteration. Figure 4.7 presents the flowchart of Rao algorithms for multi-objective optimization through a *posteriori* approach.

4.2.3 Elitist Rao Algorithms

Three elitist Rao algorithms based on the elitism concept for solving optimization problems are proposed (Rao et al., 2020b). The elitist Rao (ERao-1, ERao-2, and ERao-3) algorithms have no algorithm-specific parameters. The key features added in the elitist Rao algorithms are elitism concept and duplicate solutions removal. In the proposed algorithms to avoid premature convergence, the worst solutions in the current population are replaced by elite solutions in the current population. However, after several iterations, it may lead to the replacement of the entire population with elite solutions, which may result in trapping into a local optimum solution. Hence,

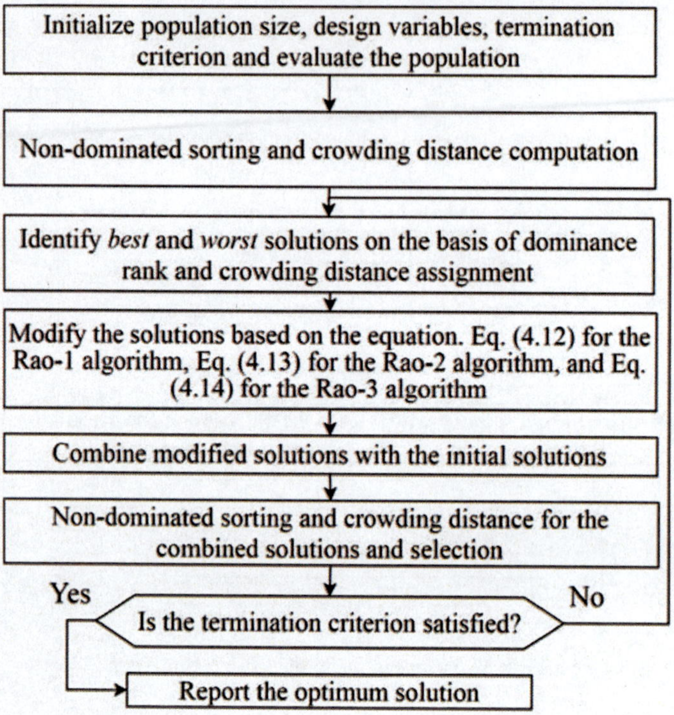

Fig. 4.7 Flowchart of the Rao algorithms in multi-objective optimization

to avoid local trapping, duplicate solutions are replaced by randomly generated solutions. Like the Rao algorithms, these algorithms use iteration best solution, iteration worst solution, and random interactions among the population to exploit and explore the solution space.

Similar to the Rao algorithms, the flow of these three algorithms is similar, but the perturbation equation used is different for each algorithm. Let Z be a minimization (or maximization) objective function. During any iteration i of the search process, the ERao-1 algorithm updates the solution (s) in the current population using the Eq. 4.12. Similarly, the ERao-2 and ERao-3 algorithms update the solutions in the current population using the Eqs. 4.13, 4.14, respectively. The flowchart of the elitist Rao algorithms is presented in Fig. 4.8. The essential steps of elitist Rao algorithms for optimization are as follows:

Step 1: Define the population size (P), elite size (ES), design variables (N), and their boundaries (Lower: LB; Upper: UB) and termination criterion.

Step 2: Randomly initialize the population of size P, and calculate the fitness value (Z) for all the population.

Step 3: Select the *best* and *worst* solutions from the current population based on their fitness value.

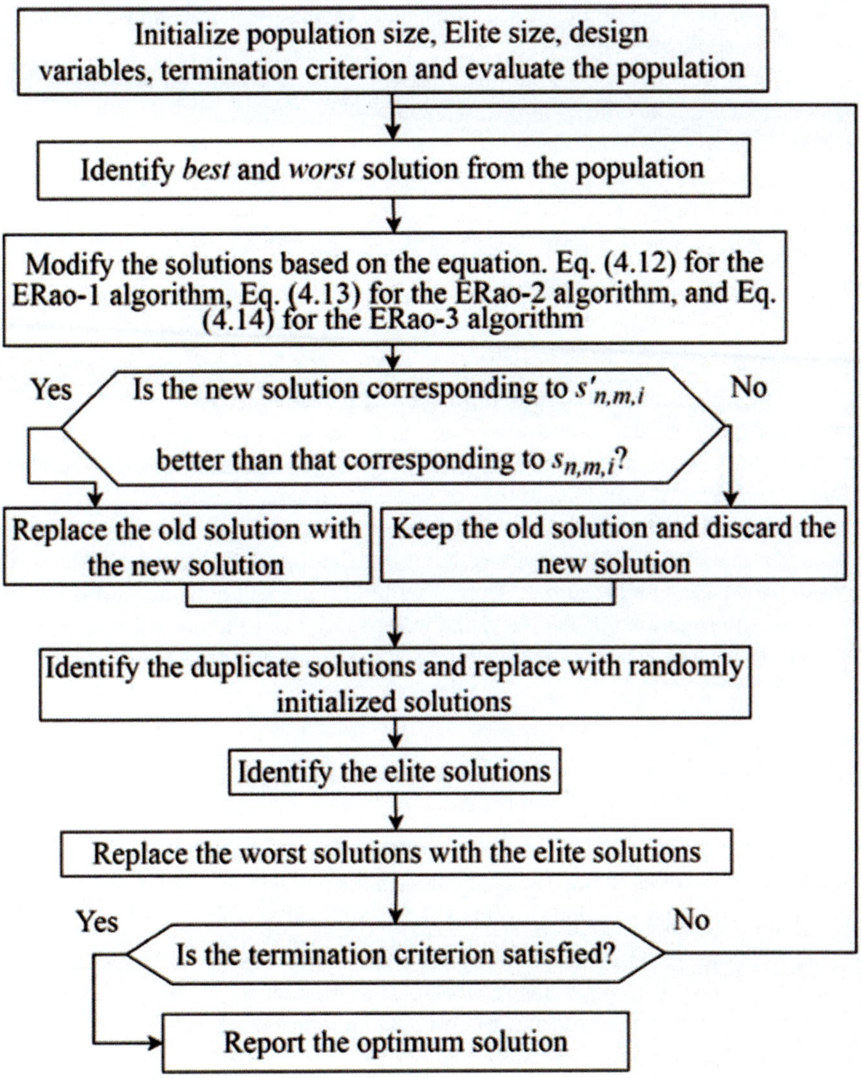

Fig. 4.8 Flowchart of the elitist Rao algorithms in single-objective optimization

Step 4: Update the solutions in the current population using the perturbation equation. Equation 4.12 is used for the ERao-1 algorithm, Eq. 4.13 is used for the ERao-2 algorithm, and Eq. 4.14 is used for the ERao-3 algorithm.

Step 5: Evaluate the fitness values of the updated solutions, and apply a greedy selection process. If the fitness value corresponding to the updated solution (s') is superior to that of the corresponding old solution (S), then replace the old solution with the updated solution, if not discard the new solution.

Step 6: Identify the duplicate solutions, and randomly generate new solutions in place of the duplicate solutions.

Step 7: Select the worst solutions of size *ES* based on the fitness values, and replace them with the elite solutions of size *ES*.

Step 8: Verify the termination criterion. If the termination criterion is satisfied, report the optimum solution from the final population, else go to **Step 3**.

4.2.4 Multi-objective Elitist Rao Algorithms

Similar to the Rao algorithms in multi-objective optimization, a *posteriori* version of elitist Rao algorithms is proposed for solving multi-objective optimization problems. Figure 4.9 presents the flowchart of elitist Rao algorithms for multi-objective optimization through a *posteriori* approach. In this version of the elitist Rao algorithms, the *best* and *worst* solutions are located using the same procedure used in the multi-objective Rao algorithms. Furthermore, after applying the non-dominated sorting to the current solutions, the solutions at the top of the sorted solutions are considered as the elite solutions. Correspondingly, the solutions at the bottom of the sorted solutions are considered as the worst solutions. After identifying the elite populations as per the elite size, the worst solutions are replaced by elite solutions.

4.2.5 Self-Adaptive Population Rao (SAP-Rao) Algorithm

The SAP-Rao algorithm (Rao & Keesari, 2021) has no algorithm-specific control parameters, and it does not require adjustment of the population size. The key features of the proposed algorithm are as follows:

7. In addition to the Rao algorithms perturbation equations, one more perturbation equation is proposed in the SAP-Rao algorithm. Inspired by the Rao algorithms perturbation equations, a new equation is proposed, and it is given by the following equation:

$$s'_{n,m,i} = s_{n,m,i} + r_{1,n,i}\left(s_{n,b,i} - s_{n,w,i}\right) + 0.5\left\{r_{2,n,i}\left(s_{n,w,i} - s_{n,m,i}\right) + r_{3,n,i}\left(s_{n,b,i} - s_{n,m,i}\right)\right\} - r_{4,n,i}\left(s_{n,b,i} - s_{n,m,i}\right) \tag{4.15}$$

8. where, $r_{1,n,i}$, $r_{2,n,i}$, $r_{3,n,i}$, and $r_{4,n,i}$ are random numbers in the range [0, 1] for the *n*th variable during the *i*th iteration.
9. During the iterative search process, the population is randomly divided into four sub-population groups. For each sub-population, a unique perturbation equation from Eqs. 4.12–4.15 is allocated. As these equations have different performance characteristics, each sub-population is moved towards a different region of search space. Furthermore, in all iterations, different solutions will

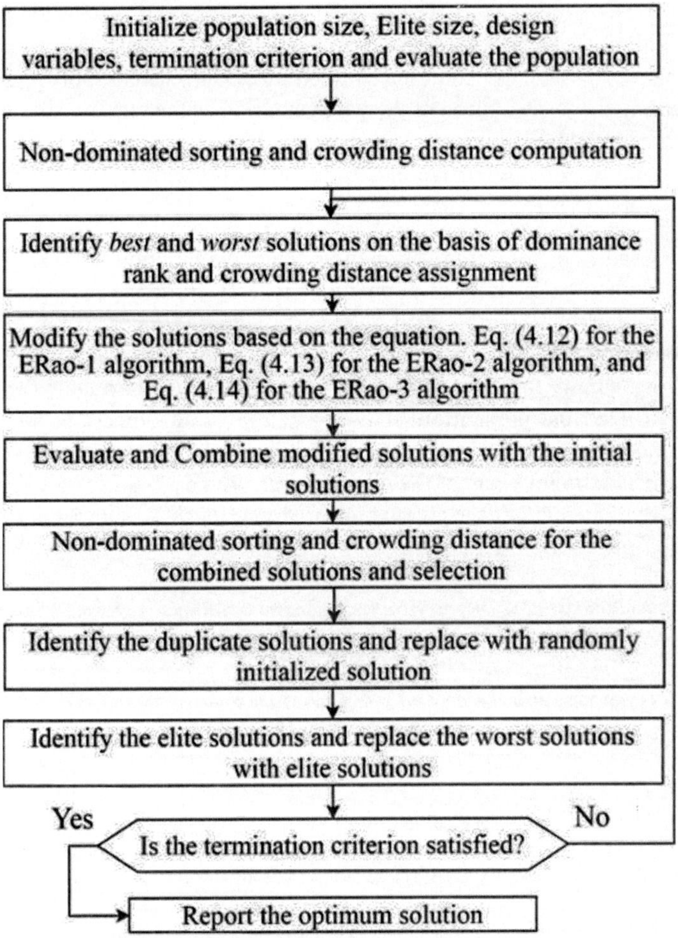

Fig. 4.9 Flowchart of the elitist Rao algorithms in multi-objective optimization

enter into different sub-population sets, and perturbation equations will also change. This will ensure diversity in the exploration and exploitation of the search space.

10. During the search process, the proposed algorithm adapts the population size based on the improvement in the fitness value. Let Z be a minimization objective function, Z_{best} be the minimum value of the objective function in the previous iteration, and Z'_{best} be the minimum value of the objective function in the current iteration. If $\left|\text{modulo}\left(Z_{\text{best}}, Z'_{\text{best}}\right)\right| < 0.1$, then the current population is reduced by 10%, else the current population is increased by 10%. However, as these equations use the *best*, *worst*, and *random* solutions to move the population in search space, a minimum of population size 20 is maintained in all iterations. This is because when the population gets divided into four sub-population groups, each sub-population can at least have a population of

size 5. Then, the *best*, *worst*, and *random* solutions can be selected from these five solutions. If the population size falls below 20, at least one sub-population will have less than five solutions. In such situations, there is more probability of selecting the best solution or the worst solution as the random solution, which will reduce the algorithm's effectiveness. Hence, it is suggested to maintain a minimum population size of 20. Figure 4.10 presents the flowchart of the SAP-Rao algorithm.

The essential steps for optimization of an objective function Z are as follows:

Step 1: Define the quantity of population ($P = P_{old}$); define the quantity design variables (N) and their boundaries: Lower (LB), Upper (UB); termination criterion: it can be the number of function evaluations or iterations.

Step 2: Randomly initialize the population of size P, and evaluate the objective function Z for all the population. If Z is a minimization function, then take the minimum value of the objective function as Z_{best}. If Z is a maximization function, then take the maximum value of the objective function as Z_{best}.

Step 3: Randomly divide the population into four groups. Assign a unique equation for each group from Eqs. 4.12–4.15.

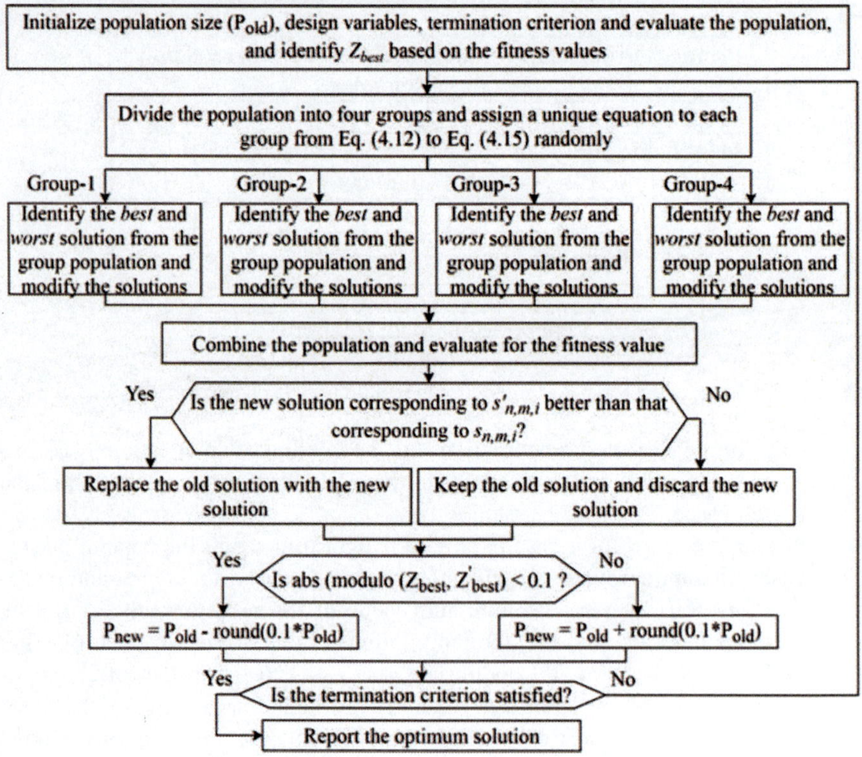

Fig. 4.10 Flowchart of the SAP-Rao algorithm in single-objective optimization

Step 4: For each group, select the *best* and *worst* solutions from the group population based on their Z value.

Step 5: Locate new solutions for each group population ($m = 1, 2\ldots (P/4)$) using the equation assigned to it.

Step 6: Combine the population of all groups, evaluate Z values for the new population, and apply a greedy selection process. If the Z value corresponding to the new solution $s'_{n,m,i}$ is superior to that of the old solution, then replace the old solution with the new solution, if not discard the new solution.

Step 7: Similar to the Z_{best} value presented in Step-II, identify the Z'_{best} value from the new population. Then, compare the Z'_{best} with Z_{best}. If $\left|\text{modulo}\left(Z_{\text{best}}, Z'_{\text{best}}\right)\right| < 0.1$, then reduce the current population by 10% ($P_{\text{new}} = P_{\text{old}} - \text{round}(0.1 \times P_{\text{old}})$), else increase the current population by 10% ($P_{\text{new}} = P_{\text{old}} + \text{round}(0.1 \times P_{\text{old}})$).

Step 8: If $P_{\text{new}} < P_{\text{old}}$, then select P_{new} amount of best solutions for the next iteration from the current population P_{old}. If $P_{\text{new}} > P_{\text{old}}$, then the extra solutions ($P_{\text{extra}} = P_{\text{new}} - P_{\text{old}}$) needed are selected from the current population itself. The P_{extra} amounts of elite solutions are duplicated in the current population, and the total population will be considered for the next iteration. Else the current population will be considered in the next iteration.

Step 9: Verify the stopping criterion. If the stopping criterion is satisfied, report the best solution from the final population, else go to **Step 3**. The general framework of the SAP-Rao algorithm is given in Algorithm 3.

Algorithm 3: the framework of the SAP-Rao algorithm

BEGIN

Initialize P=P_{old}, N, LB, UB, and FE_max- termination criterion;

Generate the initial candidate solutions and find fitness values (Z), find Z_{best}, FE = 0, i=0;

While FE<FE_Max

  i=i +1; FE = FE +P;& Divide the population into 4 groups ly:P_{s1}, P_{s2}, P_{s3}, and P_{s4}

  Assign a unique equation to each subpopulation.

  For SP= P_{s1}→P_{s4}

    For m=1→P_s

      For n= 1→N

        Identify $s_{n,b,i}$ and $s_{n,w,i}$ solutions from the subpopulation group SP

        $s'_{n,m,i}$→ Update the variables of $s_{n,m,i}$ using the assigned equation to group SP

```
            End For
        End For
    End For
    Merge the updated solutions of all groups (P_s1, P_s2, P_s3, and P_s4)
       into P
    Z' - Evaluate the fitness values for the updated solutions
    For m= 1→P
        If Z'(m) better than Z(m)
                Z(m)=Z'(m)
        End If
    End For
    Find the best fitness value from the updated solutions - Z'_best
    If |modulo(Z_best, Z'_best)| < 0.1
        P_new = P_old − round(0.1 × P_old)
    Else
        P_new = P_old + round(0.1 × P_old)
    End If
    If P_new < P_old
        Select P_new number of best solutions from the current population
           to the next iteration
    Else If P_new > P_old
        P_extra = P_new − P_old
        P_extrapop= Select P_extra number of best solutions from the current
           population
        P_newpop = P_currentpop + P_extrapop
    Else
        Solutions in the current population will be considered for the
           next iteration
    End If
End While
END
```

4.2.6 Multi-objective SAP-Rao Algorithms

Figure 4.11 presents the flowchart of the SAP-Rao algorithm for solving multi-objective optimization problems

After identifying the *best* solution and *worst* solution for each group, a new set of *P*/4 solutions are located for each group. Then, these new solutions are combined and evaluated for objective function values. Then, the non-dominance ranks are found, and the crowding distances are calculated for each solution. Now, the population size is updated based on the number of solutions ranked one. Let S_{R1} be the number of solutions ranked one in the combined population before updating the solutions and S'_{R1} be the number of solutions ranked one in the combined population after updating the solutions. If $S_{R1} > S'_{R1}$, then increase the current population by 10% ($P_{\text{new}} = P_{\text{old}} + \text{round}(0.1 \times P_{\text{old}})$), else reduce the current population by 10% ($P_{\text{new}} = P_{\text{old}} - \text{round}(0.1 \times P_{\text{old}})$). Now, the set of new solutions is combined with the set of earlier

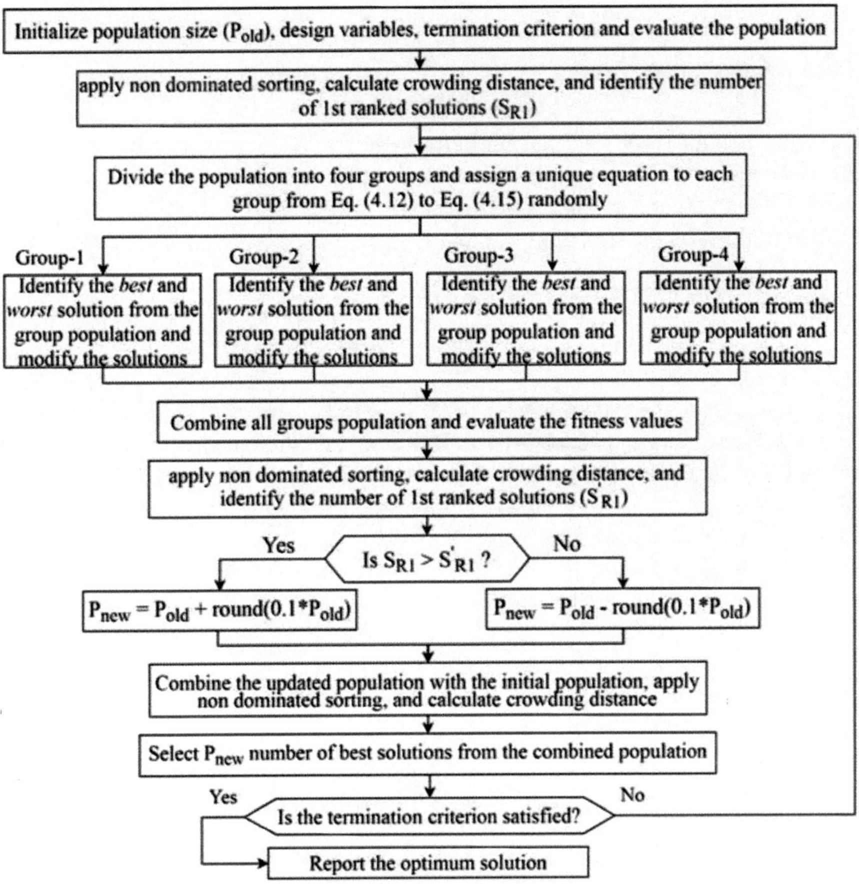

Fig. 4.11 Flowchart of the SAP-Rao algorithm in multi-objective optimization

solutions forming solutions set of size $2P$. Then, these combined solutions are again ranked using dominance principles, and crowding distance values are calculated for all solutions. Based on the new ranks and crowding distance values, a set of P_{new} solutions will be selected for the next iteration. The next subsection presents a brief description of the multi-objective optimization performance indicators considered in this book.

4.3 Performance Indicators

In multi-objective optimization, the algorithms obtain a diverse set of non-dominated solutions called as a Pareto-front. Hence, to assess the performances of the algorithms based on the Pareto-fronts achieved, four performance indicators are adopted in this book, and they are as follows.

4.3.1 Coverage

Zitzler et al. (2000) proposed that this performance indicator to compare and calculate the percentage of solutions in a Pareto-front (B) is dominated by the solutions in another Pareto-front (A). That is the coverage of the Pareto-front A over the Pareto-front B is given by the following equation:

$$\mathrm{Cov}(A, B) = \frac{|\{b \in B | \exists a \in A : a <= b\}|}{|B|} \tag{4.16}$$

where $a <= b$ means a dominates b or is equal to b.

If all the solutions of Pareto-front B are either dominated or equal to all the solutions of the Pareto-front A then the value of $\mathrm{Cov}(A, B)$ is equal to 1. Similarly, if none of the solutions in Pareto-front B are covered by the Pareto-front A then the value of $\mathrm{Cov}(A, B)$ is equal to zero.

4.3.2 Spacing

This performance indicator proposed by Schott (1995) quantifies the distribution of solutions along the Pareto-front. The spacing of a Pareto-front consisting of n non-dominated solutions is given by the following equation:

$$S = \sqrt{\frac{1}{|n-1|} \sum\nolimits_{i=1}^{n} \left(\overline{d} - d_i\right)^2} \tag{4.17}$$

$$d_i = \min_{i, i \neq j} \sum_{m=1}^{k} \left| f_m^i - f_m^j \right|, \quad i, j = 1, 2, \ldots, n \tag{4.18}$$

$$\overline{d} = \sum_{i=1}^{n} (d_i / n) \tag{4.19}$$

where k is the number of objectives, and f_m is the objective function value of the mth objective. The spacing indicator determines the uniformity of the distribution of the solutions along the Pareto-front. For a Pareto-front, zero spacing value indicates that the solutions in the Pareto-front are equidistantly spread.

4.3.3 Hypervolume

The hypervolume indicator (Rao, 2019) for a Pareto-front provides the search space volume, which is dominated by it with respect to a specified reference point. Therefore, in multi-objective optimization when two or more algorithm-performances are compared, a higher value of hypervolume indicates superior performance. For a Pareto-front consisting of N solutions, the hypervolume indicator is given by the following equation:

$$\text{Hypervolume} = \text{volume}\left(\bigcup_{i=1}^{N} v_i \right) \tag{4.20}$$

where v_i is the hypervolume constructed with reference point *Ref*, and the solution i as the diagonal corners of the hypercube.

4.3.4 Inverted Generational Distance (IGD)

The IGD indicator measures the convergence and diversity performance of the algorithms in objective space. The IGD values are given by the following equation Zhou et al. (2009):

$$\text{IGD}\left(P^*, P\right) = \frac{1}{|P^*|} \sum_{f \in P^*} \min_{f'} \left| f - f' \right| \tag{4.21}$$

where P^* is the number of solutions in the true Pareto-front, P is the number of Pareto solutions obtained by the algorithm, and $\left| f - f' \right|$ is the Euclidean distance between the two solutions from the Pareto solutions sets of P^* and P. The algorithm with the least IGD value can be considered as the best.

4.4 Computational Results Analysis on Single-objective Optimization of Unconstrained Benchmark Problems

The Jaya algorithm, Rao algorithms, and their modified versions are coded in MATLAB-R2016b, and computational tests are performed by using a CPU with 3.40 GHz Intel (R) Core i5-7500 processor and 8 GB RAM. The performance of the Jaya and Rao algorithms and their modified versions are compared with various optimization algorithms: genetic algorithm (GA) and its variants, particle swarm optimization (PSO) and its variants, grasshopper optimization algorithm (GOA), artificial bee colony (ABC), dragonfly algorithm (DA), differential evolution (DE), ant lion optimizer (ALO), and teaching–learning-based optimization (TLBO) algorithm. The performance of the proposed modified algorithms is evaluated on 53 unconstrained single-objective optimization benchmark problems (including unimodal and multimodal functions), and five multi-objective optimization benchmark problems considered from literature.

4.4.1 Computational Results Analysis on 30 Unconstrained Standard Benchmark Problems

The performances of the proposed modified algorithms are tested using 30 unconstrained standard benchmark problems. A detailed description, including the number of independent variables and their ranges, is presented in Appendix A1 (Rao, 2016). Firstly, the performance of the proposed modified algorithms in solving the 25 unconstrained problems is compared with the performances of the basic Jaya and Rao algorithms. Then, the performance of proposed modified algorithms is compared with those of other algorithms from the literature for all 30 unconstrained problems. In computational experiments on these test problems, all the algorithms compared from the literature are tested by taking 500,000 function evaluations as the termination criterion. The statistical results are presented for 30 consecutive runs. Hence, for a fair comparison of the performances of the proposed modified algorithms with other algorithms, the proposed algorithms are tested for 500,000 function evaluations, and the statistical results are presented for 30 consecutive runs. In computational experiments, different population sizes ranging from 10 to 100 are used to solve these problems by keeping function evaluations fixed. Also, different elite sizes are used for elitist Rao algorithms based on population size. The elite size is varied from 10 to 40% of the population size.

In this optimization scenario, the performances of the proposed modified algorithms are compared with those of the PSO, DE, GA, ABC, TLBO, Jaya, and Rao algorithms in solving 30 unconstrained standard benchmark problems. Friedman's statistical test is conducted for all the algorithms' results to judge the performances of the proposed algorithms with those of the other algorithms. The algorithms are ranked based on their statistical results achieved in solving the unconstrained benchmark problems. In ranking the algorithms for each problem, the priority is given to the best value achieved. If the best value achieved is the same for two or more algorithms, then the mean values achieved are considered for the ranking. Similarly, the next priority is given to the worst value achieved and then standard deviation value. If all these values are the same for the algorithms, then mean function evaluations value is considered for ranking the algorithms. However, the mean function evaluations taken by the PSO, DE, GA, ABC, TLBO, and Jaya algorithms to reach the global optimum solution are not available. However, the mean function evaluations taken by the Rao algorithms are available (Rao, 2020). Hence, the performances of the proposed modified algorithms are compared with Rao algorithms separately. In comparison with the Rao algorithms' results, if the statistical values (best, worst, mean, and standard deviation) are the same for two or more algorithms, then the algorithms are ranked based on the mean function evaluations. In comparison with the results of PSO, DE, GA, ABC, TLBO, and Jaya algorithms, if the statistical values (best, worst, mean, and standard deviation) are same for two or more algorithms, then the algorithms are given average ranks as per the Friedman's test.

The algorithms are ranked based on their statistical results achieved in solving the unconstrained benchmark problems. In ranking the algorithms for each problem, the priority is given to the best value achieved. If the best value achieved is the same for two or more algorithms, then the mean values achieved are considered for the ranking. Similarly, the next priority is given to the worst value achieved and then standard deviation value. If all these values are the same for the algorithms, then mean function evaluations value is considered for ranking the algorithms.

Table 4.1 presents a comparison of the statistical results achieved by the proposed modified algorithms with Rao algorithms. In all the tables, the term 'B' refers to the best solution, the term 'W' refers to the worst solution, the term 'M' refers to the mean value of the final objective function values in 30 runs, the term 'SD' refers to the standard deviation of the final objective function values in 30 runs, and the term 'MFE' refers to the mean of the function evaluations taken by the algorithm to reach the final reported solution in each run. Also, the MFE taken by the Jaya algorithm was not available in Rao (2016). Hence, the results of the Jaya algorithm are not considered in Table 4.1. However, the Jaya algorithm's computational results are compared with the proposed modified algorithms in Table 4.3. Here, the performance of the SAP-Rao algorithm is compared with the performances of the Rao algorithms. The performance of the ERao-1 algorithm is compared with that of the Rao-1 algorithm. Similarly, the ERao-2 and ERao-3 algorithms' performances are compared with those of the Rao-2 and Rao-3 algorithms, respectively. Similarly, the performance of the AMTPG-Jaya algorithm is compared with that of the MTPG-Jaya algorithm.

Table 4.1 Comparison of the results of the basic and improved versions in the optimization of the unconstrained standard benchmark problems

S. No	Function (Optimum)		Rao-1	Rao-2	Rao-3	ERao-1	ERao-2	ERao-3	SAP-Rao	MTPG-Jaya	AMTPG-Jaya
F1	Sphere (0)	B	0	0	0	0	0	0	0	0	0
		W	0	0	0	0	0	0	0	0	0
		M	0	0	0	0	0	0	0	0	0
		SD	0	0	0	0	0	0	0	0	0
		MFE	499,976	499,791	277,522	381,309	499,935	***199,177***	306,759	416,661	390,316
F2	Sum Squares (0)	B	0	0	0	0	0	0	0	0	0
		W	0	0	0	0	0	0	0	0	0
		M	0	0	0	0	0	0	0	0	0
		SD	0	0	0	0	0	0	0	0	0
		MFE	499,975	499,851	276,556	378,232	499,968	***234,084***	304,779	416,401	380,789
F3	Beale (0)	B	0	0	0	0	0	0	0	0	0
		W	0	0	0	0	0	0	0	0	0
		M	0	0	0	0	0	0	0	0	0
		SD	0	0	0	0	0	0	0	0	0
		MFE	9805	7612	7325	9591	5124	5152	***3653***	7632	5588
F4	Easom (−1)	B	−1	−1	−1	−1	−1	−1	−1	−1	−1
		W	0	**−1**	**−1**	**−1**	**−1**	**−1**	**−1**	**−1**	**−1**
		M	−0.5667	**−1**	**−1**	**−1**	**−1**	**−1**	**−1**	**−1**	**−1**
		SD	0.504	**0**	**0**	**0**	**0**	**0**	**0**	**0**	**0**
		MFE	3010	11,187	14,025	1261	1209	***1152***	1998	4989	2870
F5	Matyas (0)	B	0	0	0	0	0	0	0	0	0

(continued)

Table 4.1 (continued)

S. No	Function (Optimum)		Rao-1	Rao-2	Rao-3	ERao-1	ERao-2	ERao-3	SAP-Rao	MTPG-Jaya	AMTPG-Jaya
		W	0	0	0	0	0	0	0	0	0
		M	0	0	0	0	0	0	0	0	0
		SD	0	0	0	0	0	0	0	0	0
		MFE	77,023	110,544	143,088	22,511	24,948	20,349	35,274	13,147	***8556***
F6	Colville (0)	B	0	0	0	0	0	0	0	0	0
		W	**0**	5.4E−23	1E−25	**0**	2E−30	1E−29	**0**	**0**	**0**
		M	**0**	1.8E−24	7E−27	**0**	2E−31	8E−31	**0**	**0**	**0**
		SD	**0**	9.8E−24	2E−26	**0**	4E−31	2.E−30	**0**	**0**	**0**
		MFE	385,066	477,753	488,127	369,779	103,949	106,320	145,962	239,087	***93,134***
F7	Trid 6 (−50)	B	−50	−50	−50	−50	−50	−50	−50	−50	−50
		W	−50	−50	−50	−50	−50	−50	−50	−50	−50
		M	−50	−50	−50	−50	−50	−50	−50	−50	−50
		SD	0	0	0	0	0	0	0	0	0
		MFE	17,485	37,209	34,796	12,866	17,628	16,690	***4916***	26,531	16,670
F8	Trid 10 (−210)	B	−210	−210	−210	−210	−210	−210	−210	−210	−210
		W	**−210**	1171	**−210**	**−210**	**−210**	**−210**	**−210**	**−210**	**−210**
		M	**−210**	−30.858	**−210**	**−210**	**−210**	**−210**	**−210**	**−210**	**−210**
		SD	0	413	0	0	0	0	0	0	0
		MFE	48,231	144,156	142,253	***8336***	37,091	30,062	31,684	55,708	39,770
F9	Zakharov (0)	B	0	0	0	0	0	0	0	0	0
		W	0	0	0	0	0	0	0	0	0

(continued)

Table 4.1 (continued)

S. No	Function (Optimum)		Rao-1	Rao-2	Rao-3	ERao-1	ERao-2	ERao-3	SAP-Rao	MTPG-Jaya	AMTPG-Jaya
		M	0	0	0	0	0	0	0	0	0
		SD	0	0	0	0	0	0	0	0	0
		MFE	345,615	499,767	258,451	271,356	472,115	316,339	***217,552***	380,925	378,262
F10	Schwefel 1.2 (0)	B	0	0	0	0	0	0	0	0	0
		W	0	0	0	0	0	0	0	0	0
		M	0	0	0	0	0	0	0	0	0
		SD	0	0	0	0	0	0	0	0	0
		MFE	301,513	499,849	144,367	198,157	343,190	102,493	159,572	99,119	***75,501***
F11	Rosenbrock (0)	B	9.0E−26	1.9E−16	1.40E−14	3.0E−29	**0**	**0**	6.7E−29	**0**	1.3E−28
		W	3.9866	2.2E+01	2.22E+01	1.1E−27	2.0E−29	1.5E−26	9.1E−26	**0**	1.4E−25
		M	0.6644	7.4E−01	7.40E−01	2.5E−28	1.5E−30	7.7E−28	4.4E−27	**0**	7.6E−27
		SD	1.51	4.05	4.05E+00	2.4E−28	4.3E−30	2.7E−27	1.7E−26	**0**	2.5E−26
		MFE	489,811	478,410	478,420	***57,680***	124,044	495,743	97,659	94,401	84,985
F12	Dixon-Price (0)	B	0.667	2.8E−30	0.667	1.5E−29	**1.8E−30**	9.8E−29	1.4E−29	0.667	0.667
		W	**0.667**	**0.667**	0.667	**0.667**	**0.667**	**0.667**	**0.667**	**0.667**	**0.667**
		M	0.667	**0.289**	0.667	0.644	0.311	0.644	0.644	0.667	0.667
		SD	**0**	0.336	7.4E−05	1.2E−01	3.4E−01	1.2E−01	1.2E−01	**0**	**0**
		MFE	75,427	113,638	159,231	40,471	129,255	***35,742***	73,259	82,335	95,163
F13	Shekel's Foxholes (0.998004)	B	0.998	0.998	0.998	0.998	0.998	0.998	0.998	0.998	0.998
		W	0.998	0.998	0.998	0.998	0.998	0.998	0.998	0.998	0.998
		M	0.998	0.998	0.998	0.998	0.998	0.998	0.998	0.998	0.998

(continued)

Table 4.1 (continued)

S. No	Function (Optimum)		Rao-1	Rao-2	Rao-3	ERao-1	ERao-2	ERao-3	SAP-Rao	MTPG-Jaya	AMTPG-Jaya
		SD	0	0	0	0	0	0	0	0	0
		MFE	18,839	95,983	243,748	21,530	24,357	154,277	***12,933***	25,664	20,071
F14	Branin (0.397887)	B	0.397887	0.397887	0.39788	0.39788	0.39788	0.39788	0.39788	0.39788	0.39788
		W	0.397931	0.397933	0.39789	0.39788	0.39788	0.39788	0.39788	0.39788	0.39788
		M	0.397892	0.397891	0.39788	0.39788	0.39788	0.39788	0.39788	0.39788	0.39788
		SD	1.05E−05	1.03E−05	1.44E−07	**0**	**0**	**0**	**0**	**0**	**0**
		MFE	102,785	41,263	80,683	67,660	13,930	16,213	***1354***	28,740	29,489
F15	Bohachevsky 1 (0)	B	0	0	0	0	0	0	0	0	0
		W	0	0	0	0	0	0	0	0	0
		M	0	0	0	0	0	0	0	0	0
		SD	0	0	0	0	0	0	0	0	0
		MFE	3129	4751	3435	1036	2719	***878***	1669	1831	1222
F16	Booth (0)	B	0	0	0	0	0	0	0	0	0
		W	0	0	0	0	0	0	0	0	0
		M	0	0	0	0	0	0	0	0	0
		SD	0	0	0	0	0	0	0	0	0
		MFE	5583	4485	4312	1860	1847	***1824***	2893	4481	3688
F17	Michalewicz-2 (−1.8013)	B	−1.8013	−1.8013	−1.8013	−1.8013	−1.8013	−1.8013	−1.8013	−1.8013	−1.8013
		W	−1.8013	−1.8013	−1.8013	−1.8013	−1.8013	−1.8013	−1.8013	−1.8013	−1.8013
		M	−1.8013	−1.8013	−1.8013	−1.8013	−1.8013	−1.8013	−1.8013	−1.8013	−1.8013
		SD	0	0	0	0	0	0	0	0	0

(continued)

Table 4.1 (continued)

S. No	Function (Optimum)		Rao-1	Rao-2	Rao-3	ERao-1	ERao-2	ERao-3	SAP-Rao	MTPG-Jaya	AMTPG-Jaya
		MFE	3863	2694	2751	***419***	871	437	1965	2921	1650
F18	Michalewicz-5 (−4.6877)	B	−4.6877	−4.6877	−4.6877	−4.6877	−4.6877	−4.6877	−4.6877	−4.6877	−4.6877
		W	−4.5377	−3.1168	−3.4959	**−4.6536**	−4.4959	−4.4959	−3.5992	−4.6459	−4.6459
		M	−4.6743	−4.4299	−4.4922	**−4.6865**	−4.6370	−4.6466	−4.5376	−4.6835	−4.6799
		SD	0.0309	0.3600	0.2790	**0.0062**	0.0514	0.0412	0.2045	0.0127	0.0161
		MFE	39,710	67,252	58,401	247,154	164,310	138,840	***34,031***	84,480	203,453
F19	Bohachevsky-2 (0)	B	0	0	0	0	0	0	0	0	0
		W	0	0	0	0	0	0	0	0	0
		M	0	0	0	0	0	0	0	0	0
		SD	0	0	0	0	0	0	0	0	0
		MFE	2963	4272	3191	1081	1589	***931***	1733	1141	1093
F20	Bohachevsky-3 (0)	B	0	0	0	0	0	0	0	0	0
		W	0	0	0	0	0	0	0	0	0
		M	0	0	0	0	0	0	0	0	0
		SD	0	0	0	0	0	0	0	0	0
		MFE	4725	12,337	6821	1331	1673	***1289***	2207	3164	1459
F21	GoldStein-Price (3)	B	3	3	3	3	3	3	3	3	3
		W	**3**	84	**3**	**3**	**3**	**3**	**3**	**3**	**3**
		M	**3**	5.7	**3**	**3**	**3**	**3**	**3**	**3**	**3**
		SD	**0**	14.8	**0**	**0**	**0**	**0**	**0**	**0**	**0**
		MFE	180,121	176,933	353,893	2185	***1881***	301,287	22,158	16,768	6391

(continued)

Table 4.1 (continued)

S. No	Function (Optimum)		Rao-1	Rao-2	Rao-3	ERao-1	ERao-2	ERao-3	SAP-Rao	MTPG-Jaya	AMTPG-Jaya
F22	Perm (0)	B	0	0	0	0	0	0	0	0	0
		W	3.7E−09	**0**	**0**	8.9E−31	**0**	**0**	**0**	**0**	**0**
		M	1.5E−10	**0**	**0**	1.1E−31	**0**	**0**	**0**	**0**	**0**
		SD	6.8E−10	**0**	**0**	2.9E−31	**0**	**0**	**0**	**0**	**0**
		MFE	82,792	***3139***	4453	3192	18,004	8325	3749	22,072	17,524
F23	Hartmann-3 (0)	B	−3.863	−3.863	−3.863	−3.863	−3.863	−3.863	−3.863	−3.863	−3.863
		W	−3.863	−3.863	−3.863	−3.863	−3.863	−3.863	−3.863	−3.863	−3.863
		M	−3.863	−3.863	−3.863	−3.863	−3.863	−3.863	−3.863	−3.863	−3.863
		SD	0	0	0	0	0	0	0	0	0
		MFE	4459	3022	3271	4256	1938	1900	85,136	2859	***1557***
F24	Ackley (0)	B	1.5E−14	8.0E−15	4.4E−15	8.0E−15	8.9E−16	4.4E−15	4.4E−15	**0**	8.9E−16
		W	2.2E+00	1.5E−14	1.5E−14	6.2E00	2.0E+01	4.4E−15	4.4E−15	**0**	8.9E−16
		M	5.7E−01	1.0E−14	6.7E−15	1.3E+00	4.0E+00	4.4E−15	4.4E−15	**0**	8.9E−16
		SD	7.4E−01	3.1E−15	2.4E−15	1.3E+00	8.1E+00	**0**	**0**	**0**	**0**
		MFE	129,392	417,741	76,352	232,012	267,183	73,578	81,369	***2463***	3612
F25	Penalized-2 (0)	B	1.4E−32	1.4E−32	1.4E−32	7.3E−17	7.7E−17	8.0E−17	6.3E−17	**0**	7.6E−17
		W	1.1E−02	1.6E+00	1.4E−01	1.0E−16	1.0E−16	1.0E−01	1.0E−16	**0**	1.0E−16
		M	1.5E−03	5.8E−02	1.6E−02	9.4E−17	9.2E−17	9.4E−03	8.7E−17	**0**	9.2E−17
		SD	3.8E−03	2.9E−01	3.5E−02	6.2E−18	6.2E−18	2.9E−02	9.6E−18	**0**	6.4E−18
		MFE	173,661	115,593	55,637	219,860	120,748	61,684	127,018	***39,767***	67,741

Source Rao-1, Rao-2, Rao-3: Rao (2020)
Result in boldface indicates better values

From Table 4.1, it can be observed that the SAP-Rao algorithm is converging faster than the Rao algorithms. In 24 out of 25 test problems, the SAP-Rao algorithm has achieved global optimum solutions in fewer MFEs when compared to those of the Rao-1 algorithm. Similarly, the SAP-Rao algorithm consumed fewer MFEs for 23 out of 25 test problems compared to the Rao-2 algorithm, and in 19 out of 25 problems compared to the Rao-3 algorithm. Furthermore, the SAP-Rao algorithm has attained better or the same results in terms of the best, worst, mean, and standard deviation values in 23 problems compared to those of the Rao algorithms. For the test problems 11, 24, and 25, the SAP-Rao algorithm has achieved better performance compared to that of Rao algorithms. Furthermore, it can be observed that the AMTPG-Jaya algorithm is converging faster than the MTPG-Jaya algorithm. In 19 out of 25 test problems, the AMTPG-Jaya algorithm achieved global optimum solutions in fewer MFEs compared to those of the MTPG-Jaya algorithm in all the test problems.

The performance of the ERao-1 algorithm is superior or the same when compared to the performance of the Rao-1 algorithm in solving 24 out of 25 problems in terms of better, worst, mean, and standard deviation values. However, in all the test problems except for the benchmark problems 13, 18, 24, and 25, the ERao-1 algorithm has consumed relatively lesser function evaluations than the Rao-1 algorithm to achieve better or the same solutions. For the benchmark problems 4, 11, 12, 14, 18, 22, and 25, the ERao-1 algorithm has achieved superior performance compared to that of the Rao-1 in terms of best, worst, mean, and standard deviation values.

Similarly, the performance of the ERao-2 algorithm is better or the same in solving 24 out of 25 problems when compared to the performance of the Rao-2 algorithm in terms of better, worst, mean, and standard deviation values. However, the ERao-2 algorithm has consumed relatively lesser function evaluations than the Rao-2 algorithm to achieve better or the same solutions in all the test problems except for the benchmark problems 1, 2, 12, 18, 22, and 25. Also, for the benchmark problems 6, 8, 11, 12, 14, 18, 21, and 25, the ERao-2 algorithm has achieved superior performance in terms of best, worst, mean, and standard deviation values than that of the Rao-2 algorithm. The performance of the ERao-3 algorithm is superior or the same in solving all benchmark problems when compared to the performance of the Rao-3 algorithm in terms of better, worst, mean, and *SD* values. However, the elitist Rao-3 algorithm has consumed relatively lesser function evaluations than the Rao-3 algorithm to achieve better or the same solutions in all the test problems except for the benchmark problems 9, 11, 18, 22, and 25. For the benchmark problems 6, 11, 12, 14, 18, 24, and 25, the ERao-3 algorithm has achieved superior performance in terms of best, worst, mean, and standard deviation values.

From the above analysis, it can be observed that the proposed modified algorithms are faster in converging and more effective in achieving global optimum solutions. However, the performances of these algorithms in solving the unconstrained benchmark problems are evaluated using the Friedman statistical test. The algorithms are ranked based on their statistical results achieved in solving the unconstrained benchmark problems. In ranking the algorithms for each problem, the priority is given to the best value achieved. If the best value achieved is the same for two or more algorithms, then the mean values achieved are considered for the ranking. Similarly, the next priority is given to the worst value achieved and then standard deviation value. If all these values are the same for the algorithms, then mean function evaluations value is considered for ranking the algorithms. After assigning the ranks to algorithms based on their statistical results, each algorithm's average rank is calculated. As nine algorithms are compared in this test, the χ^2 distribution with eight degrees of freedom is considered. The average ranks, χ^2 value, and *p*-value achieved in the Friedman test are presented in Table 4.2. The ranks of these algorithms and their average rank values are plotted in Fig. 4.12.

Here, an observation can be made that the *p*-value of the Friedman test is much less than 0.01, which indicates that the results are highly significant. The least value of the average rank indicates superior performance. The ERao-3 algorithm has achieved the least average rank of 3.40. The ranking of these algorithms in solving unconstrained benchmark problems is ERao-3, AMTPG-Jaya, SAP-Rao, ERao-1, ERao-2, MTPG-Jaya, Rao-3, Rao-1, and Rao-2. Here, an observation can be made that the proposed modified algorithms are converging faster than the respective basic algorithms.

In addition, the performance of the proposed modified algorithms is compared with those of the PSO, DE, GA, ABC, and TLBO algorithms in solving 30 unconstrained standard benchmark problems. The computational results are presented in Table 4.3 in terms of mean and standard deviation values achieved in 30 runs. The results of the TLBO algorithm are taken from Rao and Patel (2013), and the results of the ABC, DE, PSO, and GA algorithms are taken from Karaboga and Akay (2009).

Here, an observation can be made that the performances of the modified versions are better or competitive compared to the basic Jaya and Rao algorithms as well as the GA, PSO, DE, ABC, and TLBO algorithms. In 24 out of 30 problems, the SAP-Rao algorithm has achieved better or the same performance. The AMTPG-Jaya algorithm has achieved better or the same performance in 22 out of 30 problems. Similarly, the MTPG-Jaya algorithm has achieved better or the same performance in 25 out of 30 problems. In 19 out of 30 problems, the elitist Rao algorithms (ERao-1, ERao-2, and ERao-3) have achieved better or the same performance.

However, to judge the performance of the proposed algorithms in comparison with the other algorithms, the Friedman statistical test is conducted. The mean values are given priority and then to the standard deviation values for ranking the algorithms in all the problems. Here, 12 algorithms are compared in this test with 11° of freedom for the χ^2 distribution. The average ranks, χ^2 value, and *p*-value achieved in the Friedman test are presented in Table 4.4. The ranks of these algorithms and their average rank values are plotted in Fig. 4.13.

The *p*-value of the Friedman test is 8.46E−06, which indicates that the results are highly significant. The MTPG-Jaya algorithm has achieved the least average rank of 5.20. The ranking of these algorithms in solving the unconstrained benchmark problems is MTPG-Jaya, SAP-Rao, AMTPG-Jaya, Jaya, ERao-1, ERao-2, ERao-3, ABC, TLBO, DE, PSO, and GA. Here, an observation can be made that the proposed algorithms have better performance than those of the ABC, PSO, DE, GA, and TLBO algorithms. The next subsection presents the analysis of the proposed modified algorithms' computational results in solving some additional unimodal, multimodal, and fixed dimension multimodal optimization problems.

Table 4.2 Friedman statistical test results for the results presented in Table 4.1

Algorithm	Rao-1	Rao-2	Rao-3	ERao-1	ERao-2	ERao-3	SAP-Rao	MTPG-Jaya	AMTPG-Jaya
Avg. Rank	6.88	7.16	6.36	3.96	4.56	3.40	3.88	4.84	3.84
χ^2	51.035								
p-value	The *p*-value is **3.09519e−8**. The result is significant at $p < 0.01$								

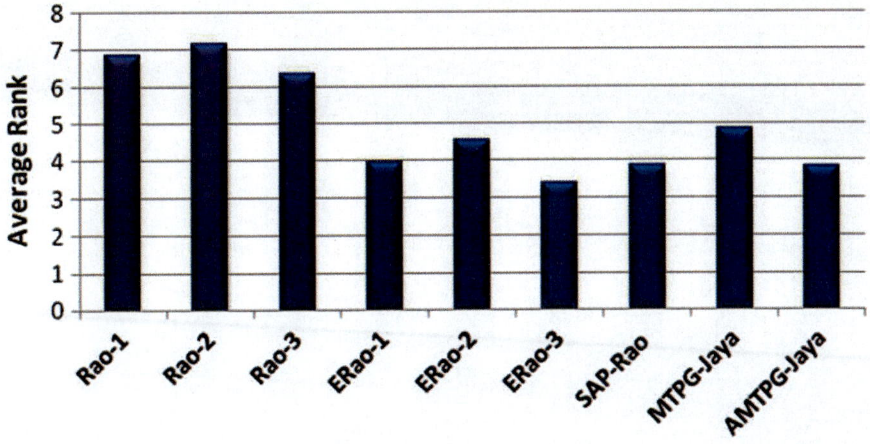

Fig. 4.12 Average ranks of Rao algorithms and modified versions in the Friedman statistical test for the unconstrained standard benchmark problems

4.4.2 Computational Results Analysis on Unconstrained Unimodal and Multimodal Standard Benchmark Problems

In addition to the unconstrained benchmark test problems, modified algorithms' performance is tested using 23 unimodal and multimodal optimization test problems. The algorithms' performance in the unimodal test problems reveals the exploitation capability of the algorithm, and the performance in the multimodal problems reveals the algorithm's exploration capability. Firstly, the performances of the proposed modified algorithms in solving 23 unimodal and multimodal test problems are compared with the performances of the Jaya and Rao algorithms. A detailed description, including the number of independent variables and their ranges of these test problems, is presented in Appendix A2 (Rao, 2020). Then, the performances of the proposed modified algorithms are compared with GSA, RGA, PSO, and its variants, Jaya algorithm, and its variant using ten unimodal and multimodal optimization test functions. A detailed description, including the number of independent variables and their ranges of these test functions, is presented in Appendix A3 (Rao & Saroj, 2017). In computational experiments, different population sizes ranging from 10 to 100 are used for solving these problems by keeping function evaluations as constant. Also, different elite sizes are used for elitist Rao algorithms based on population size. The elite size is varied from 10 to 40% based on the population size.

Table 4.3 Comparison of the results of improved versions with the results of other algorithms achieved in the optimization of the unconstrained standard benchmark problems

Function		GA	PSO	DE	ABC	TLBO	Jaya	ERao-1	ERao-2	ERao-3	MTPG-Jaya	AMTPG-Jaya	SAP-Rao
F1	M	1110	**0**	**0**	**0**	**0**	**0**	**0**	**0**	**0**	**0**	**0**	**0**
	SD	74.214474	**0**	**0**	**0**	**0**	**0**	**0**	**0**	**0**	**0**	**0**	**0**
F2	M	148	**0**	**0**	**0**	**0**	**0**	**0**	**0**	**0**	**0**	**0**	**0**
	SD	12.409289	**0**	**0**	**0**	**0**	**0**	**0**	**0**	**0**	**0**	**0**	**0**
F3	M	**0**	**0**	**0**	**0**	**0**	**0**	**0**	**0**	**0**	**0**	**0**	**0**
	SD	**0**	**0**	**0**	**0**	**0**	**0**	**0**	**0**	**0**	**0**	**0**	**0**
F4	M	**−1**	**−1**	**−1**	**−1**	**−1**	**−1**	**−1**	**−1**	**−1**	**−1**	**−1**	**−1**
	SD	**0**	**0**	**0**	**0**	**0**	**0**	**0**	**0**	**0**	**0**	**0**	**0**
F5	M	**0**	**0**	**0**	**0**	**0**	**0**	**0**	**0**	**0**	**0**	**0**	**0**
	SD	**0**	**0**	**0**	**0**	**0**	**0**	**0**	**0**	**0**	**0**	**0**	**0**
F6	M	0.014938	**0**	0.0409122	0.0929674	**0**	**0**	**0**	2.72732E−31	8.12321E−31	**0**	**0**	**0**
	SD	0.007364	**0**	0.081979	0.066277	**0**	**0**	**0**	4.82641E−31	2.20373E−30	**0**	**0**	**0**
F7	M	−49.9999	**−50**	**−50**	**−50**	**−50**	**−50**	**−50**	**−50**	**−50**	**−50**	**−50**	**−50**
	SD	2.25E−5	**0**	**0**	**0**	**0**	**0**	**0**	**0**	**0**	**0**	**0**	**0**
F8	M	0.193417	0	0	0	0	**−210**	**−210**	**−210**	**−210**	**−210**	**−210**	**−210**
	SD	0.035313	**0**	**0**	**0**	**0**	**0**	**0**	**0**	**0**	**0**	**0**	**0**
F9	M	0.013355	**0**	**0**	0.0002476	**0**	**0**	**0**	**0**	**0**	**0**	**0**	**0**
	SD	0.004532	**0**	**0**	0.000183	**0**	**0**	**0**	**0**	**0**	**0**	**0**	**0**
F10	M	7400	**0**	**0**	**0**	**0**	**0**	**0**	**0**	**0**	**0**	**0**	**0**
	SD	1140	**0**	**0**	**0**	**0**	**0**	**0**	**0**	**0**	**0**	**0**	**0**
F11	M	196,000	15.088617	18.203938	0.0887707	0.0000162	**0**	2.5149E−28	1.47911E−30	7.6729E−28	**0**	7.61941E−27	4.41602E−27

(continued)

Table 4.3 (continued)

Function		GA	PSO	DE	ABC	TLBO	Jaya	ERao-1	ERao-2	ERao-3	MTPG-Jaya	AMTPG-Jaya	SAP-Rao
	SD	38,500	24.170196	5.036187	0.07739	0.0000364	**0**	2.3951E−28	4.27474E−30	2.73293E−27	**0**	2.50556E−26	1.66875E−26
F12	M	1220	0.6666667	0.6666667	**0**	0.6666667	**0**	0.64444444	0.311111111	0.644444444	0.666667	0.666666667	0.644444444
	SD	266	1E–8	1E–9	**0**	0	**0**	0.12171612	0.338277509	0.121716124	**0**	**0**	0.121716124
F13	M	**0.998004**	**0.998004**	**0.998004**	**0.998004**	**0.998004**	**0.998004**	**0.998004**	**0.998004**	**0.998004**	**0.9980038**	**0.998004**	**0.998004**
	SD	**0**	**0**	**0**	**0**	**0**	**0**	**0**	**0**	**0**	**0**	**0**	**0**
F14	M	**0.397887**	**0.3978874**	**0.3978874**	**0.3978874**	**0.3978874**	**0.397887**	**0.397887**	**0.397887**	**0.397887**	**0.397887**	**0.397887**	**0.397887**
	SD	**0**	**0**	**0**	**0**	**0**	**0**	**0**	**0**	**0**	**0**	**0**	**0**
F15	M	**0**	**0**	**0**	**0**	**0**	**0**	**0**	**0**	**0**	**0**	**0**	**0**
	SD	**0**	**0**	**0**	**0**	**0**	**0**	**0**	**0**	**0**	**0**	**0**	**0**
F16	M	**0**	**0**	**0**	**0**	**0**	**0**	**0**	**0**	**0**	**0**	**0**	**0**
	SD	**0**	**0**	**0**	**0**	**0**	**0**	**0**	**0**	**0**	**0**	**0**	**0**
F17	M	**−1.8013**	−1.572869	**−1.801303**	**−1.801303**	**−1.801303**	**−1.801303**	**−1.801303**	**−1.801303**	**−1.801303**	**−1.801303**	**−1.801303**	**−1.801303**
	SD	**0**	0.11986	**0**	**0**	**0**	**0**	**0**	**0**	**0**	**0**	**0**	**0**
F18	M	−4.64483	−2.490872	−4.683482	**−4.687658**	−4.672657	−4.68013	−4.6865216	−4.63703182	−4.64660672	−4.683482	−4.67987586	−4.53757116
	SD	0.09785	0.256952	0.012529	**0**	0.0474	0.0158	0.00622509	0.051374466	0.041245533	0.012743	0.016094139	0.204546577
F19	M	0.06829	**0**	**0**	**0**	**0**	**0**	**0**	**0**	**0**	**0**	**0**	**0**
	SD	0.078216	**0**	**0**	**0**	**0**	**0**	**0**	**0**	**0**	**0**	**0**	**0**
F20	M	**0**	**0**	**0**	**0**	**0**	**0**	**0**	**0**	**0**	**0**	**0**	**0**
	SD	**0**	**0**	**0**	**0**	**0**	**0**	**0**	**0**	**0**	**0**	**0**	**0**
F21	M	5.870093	**3**	**3**	**3**	**3**	**3**	**3**	**3**	**3**	**3**	**3**	**3**
	SD	1.071727	**0**	**0**	**0**	**0**	**0**	**0**	**0**	**0**	**0**	**0**	**0**

(continued)

Table 4.3 (continued)

Function		GA	PSO	DE	ABC	TLBO	Jaya	ERao-1	ERao-2	ERao-3	MTPG-Jaya	AMTPG-Jaya	SAP-Rao
F22	M	0.302671	0.0360516	0.0240069	0.0411052	0.0006766	**0**	1.1176E−31	**0**	**0**	**0**	**0**	**0**
	SD	0.193254	0.048927	0.046032	0.023056	0.0007452	**0**	2.9037E−31	**0**	**0**	**0**	**0**	**0**
F23	M	**−3.86278**	−3.633352	**−3.86278**	**−3.86278**	**−3.86278**	**−3.86278**	**−3.86278**	**−3.86278**	**−3.86278**	**−3.86278**	**−3.86278**	**−3.86278**
	SD	**0**	0.116937	**0**	**0**	**0**	**0**	**0**	**0**	**0**	**0**	**0**	**0**
F24	M	14.67178	0.1646224	**0**	**0**	**0**	**0**	1.28742222	3.992006819	4.44089E−15	**0**	8.88178E−16	4.44089E−15
	SD	0.178141	0.493867	**0**	**0**	**0**	**0**	1.31175382	8.120502433	**0**	**0**	**0**	**0**
F25	M	125.0613	0.0076754	0.0021975	**0**	2.34E−08	**0**	9.4079E−17	9.23047E−17	0.009419631	**0**	9.18662E−17	8.73247E−17
	SD	12.0012	0.016288	0.004395	**0**	**0**	**0**	6.206E−18	6.15941E−18	0.02878998	**0**	6.37641E−18	9.56381E−18
F26	M	**−1.08094**	−0.67927	**−1.08094**	**−1.08094**	**−1.08094**	**−1.08094**	**−1.08094**	**−1.08094**	**−1.08094**	**−1.08094**	**−1.08094**	**−1.08094**
	SD	**0**	0.274621	**0**	**0**	**0**	**0**	**0**	**0**	**0**	**0**	**0**	**0**
F27	M	0.287548	0.213626	0	0.000208	0.000016	−1.2405	−1.4357000	−1.35794285	−1.39999731	−1.343692	−1.49359886	**−1.49571339**
	SD	0.052499	0.039003	0	0.000038	0.000003	0.312691	0.17171383	0.249983625	0.212284697	0.266023	0.002454328	0.325329651
F28	M	−0.63644	−0.002566	−1.0528	−0.446093	−0.64906	−0.62050	−0.6631870	−0.76402499	−0.73180160	−1.003922	**−1.49425443**	−1.4485531
	SD	0.374682	0.003523	0.302257	0.133958	0.172862	0.319588	0.2236297	0.32784579	0.236382006	0.352104	**0.163604635**	0.162849789
F29	M	0.004303	1457.8834	5.988783	0.17355	2.203813	0.000159	5.5974E−05	0.000823461	0.000356285	**0**	6.9003E−08	**0**
	SD	0.009469	1269.3624	7.334731	0.068175	4.386321	0.00052	8.7041E−05	0.001987679	0.000677345	**0**	3.27883E−07	**0**
F30	M	29.57348	1364.4556	781.55028	8.23344	35.971004	0.000543	9.6657E−05	0.000667392	0.000199627	**0**	**0**	**0**
	SD	16.02108	1325.3797	1048.8135	8.092742	71.284369	0.000987	0.00023549	0.001469405	0.000449935	**0**	**0**	**0**

Source TLBO: Rao and Patel (2013); ABC, DE, PSO, and GA: Karaboga and Akay (2009), Jaya: Rao (2016)
Bold values indicate best values

Table 4.4 Friedman statistical test results for the results presented in Table 4.3

Algorithm	GA	PSO	DE	ABC	TLBO	Jaya	ERao-1	ERao-2	ERao-3	SAP-Rao	MTPG-Jaya	AMTPG-Jaya
Avg. rank	9.6	8.35	7	6.82	6.87	5.48	5.8	6.15	6.2	5.23	5.2	5.27
χ^2	43.621											
p-value	The p-value is **8.46E−06**. The result is significant at $p < 0.01$											

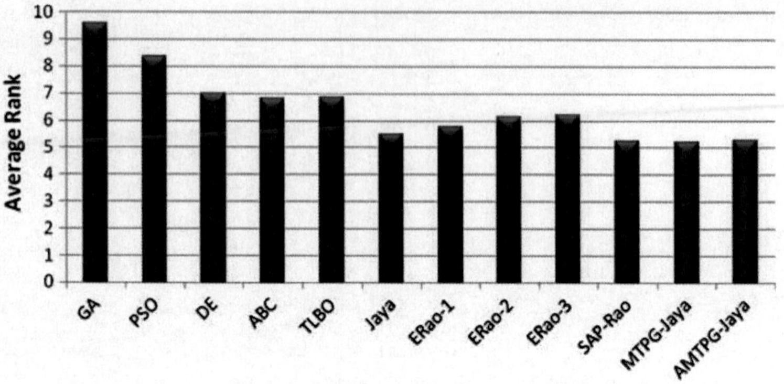

Fig. 4.13 Average ranks of various algorithms in the Friedman statistical test for the unconstrained standard benchmark problems

4.4.2.1 Experiment-1

In the first test, the Rao algorithms were tested on 23 unimodal and multimodal optimization test problems by taking 30,000 function evaluations as the termination criterion, and the statistical results were presented for 30 consecutive runs (Rao, 2020). Hence, for a fair comparison among the performances of the basic and the modified algorithms, the modified algorithms along with the basic Jaya algorithm are tested by taking 30,000 function evaluations as termination criterion, and statistical results are presented for 30 consecutive runs. In Table 4.5, the performances of the proposed modified algorithms are compared with those of the Jaya and Rao algorithms in terms of the mean (M) values, standard deviation (SD), and the mean of the function evaluations (MFE) taken by the algorithm to reach the global optimum solution. The problems P1 to P7 are unimodal test problems, P8 to P13 are multimodal test problems, and P14 to P23 are multimodal fixed dimension test problems. Here, the performance of the SAP-Rao algorithm is compared with the performances of the Rao algorithms. The performance of the ERao-1 algorithm is compared with that of the Rao-1 algorithm. Similarly, the performances of the ERao-2 and ERao-3 algorithms are compared with those of the Rao-2 and Rao-3 algorithms, respectively. Similarly, the performances of the MTPG-Jaya and AMTPG-Jaya algorithms are compared with the Jaya algorithm's performance.

Here, an observation can be made that the SAP-Rao algorithm has better exploration capability when compared to that of the Rao algorithms. The SAP-Rao algorithm has achieved better mean values in multimodal problems (P9 to P19 and P21 to P23) compared to those of the Rao algorithms. Similarly, the standard deviation values of the SAP-Rao algorithm in multimodal problems are better or competitive when compared to those of the Rao algorithms. Furthermore, the exploitation capability of the SAP-Rao algorithm is better or competitive to the Rao algorithms. In all unimodal test problems, the SAP-Rao algorithm has achieved better mean and standard deviation values when compared to those of the Rao-1 and Rao-2 algorithms.

Table 4.5 Comparison of the results of the basic and improved algorithms in the experiment-1 of the unimodal and multimodal benchmark problems

Problem Number	(Optimum Value)	Rao-1	Rao-2	Rao-3	ERao-1	ERao-2	ERao-3	SAP-Rao	Jaya	MTPG-Jaya	AMTPG-Jaya
P1 (0)	M	3.59E−22	3.57E−12	6.71E−42	7.1653E−21	8.31983E−13	1.21844E−42	2.497E−27	1.93353E−11	**0**	**0**
	SD	7.33E−22	7.95E−12	1.56E−41	2.72917E−20	2.01641E−12	4.26759E−42	8.947E−27	9.42658E−11	**0**	**0**
	MFE	29,998	29,953	29,991	30,000	29,972.66667	29,991	29,985	29,990	16,075	**8659**
P2 (0)	M	4.07E−12	6.78E−01	9.33E−21	3.36004E−12	0.334779	2.27072E−19	1.315E−18	0.001589666	1.0535E−266	**0**
	SD	1.4E−11	2.53E + 00	3.84E−20	1.23803E−11	1.825483173	1.1745E−18	2.678E−18	0.007652932	**0**	**0**
	MFE	29,994	29,882	29,983	29,995	29,923.4	29,974.93333	29,981	29,990	29,855	**16,378**
P3 (0)	M	8.34E−40	1.27E−16	1.68E−53	2.5186E−39	1.03267E−21	6.17941E−73	9.878E−55	2.10485E−16	**0**	**0**
	SD	2.9E−39	6.93E−16	9.12E−53	1.00807E−38	5.42299E−21	3.3432E−72	4.350E−51	1.15055E−15	**0**	**0**
	MFE	29,993	29,975	29,959	29,995	29,965	29,982.33333	29,991	29,991	9481	**5015**
P4 (0)	M	2.19522	16.56395	0.081469	2.196461584	14.96336474	0.070440448	0.032234	35.13809811	8.4471E−156	**3.5266E−296**
	SD	1.150517	5.632224	0.078402	1.011003951	6.393191671	0.102924171	0.020639	6.830014968	4.6266E−155	**0**
	MFE	29,882	28,845	29,899	29,933	29,623.2	29,873.33333	29,925	29,991	29,988	**17,122**
P5 (0)	M	31.6044	11.47408	29.2063	29.17288578	11.34218421	23.48199954	**0.886832**	25.0963	24.22616415	7.360013893
	SD	28.4067	16.68387	29.0933	26.22552675	14.93837212	28.410953	1.163335	**0.6335**	0.830964354	6.581159362
	MFE	29,609	28,925	**28,922**	29,477	29,716	29,822.83333	29,908	29,992	29,991	29,653
P6 (0)	M	2.63E−21	1.09E−07	2.919904	**2.12382E−21**	5.53027E−09	2.556397649	5.923E−12	42.89921891	1.74742E−06	1.2626E−09
	SD	7.87E−21	3.09E−07	0.39977	**2.94981E−21**	1.5451E−08	0.341602663	1.383E−11	39.59748517	5.49956E−06	6.7466E−09
	MFE	29,993	29,945	20,023	29,989	29,984	**18,843.5**	29,981	24,427	29,949	29,981
P7 (0)	M	0.058328	0.087804	0.01577	0.070803102	0.079307038	0.013706853	0.011698	0.287414385	0.286096194	**9.72796E−05**
	SD	0.027453	0.044495	0.008669	0.026594612	0.04049604	0.008316235	0.004410	0.145627618	0.163397722	**0.00016837**
	MFE	26,785	25,355	24,044	26,331.66667	25,270	24,839	26,448	28,552	28,580	**17,103**

(continued)

Table 4.5 (continued)

Problem Number	(Optimum Value)	Rao-1	Rao-2	Rao-3	ERao-1	ERao-2	ERao-3	SAP-Rao	Jaya	MTPG-Jaya	AMTPG-Jaya
P8 (−12,569)	M	−8685.17	−8757.58	−9664.70	−9168.326455	−9738.589736	−9745.532039	−9416.37	−5644.26	−4430.431501	**−10,877.30491**
	SD	1690.55	1896.34	1544.66	1647.69389	923.4525188	1629.643643	**589.57**	1181.28	928.8317877	1592.909786
	MFE	21,166	22,377	28,385	29,090	27,573.33333	28,282.8	29,292	24,072	**20,786**	27,577
P9 (0)	M	87.0136	148.9495	84.1229	91.65656892	98.91672012	34.53251765	41.8233	76.7882	**0**	**0**
	SD	32.3175	41.5267	38.1792	44.50224678	49.82896302	33.7673293	14.1592	24.0204	**0**	**0**
	MFE	26,015	24,754	27,934	29,021.5	28,196.66667	28,948	29,771	28,965	2963	**1857**
P10 (0)	M	0.619739	0.170688	7.97E−08	1.08289E−10	7.03105E−06	2.13755E−14	2.718E−14	4.238427857	**8.88178E−16**	**8.88178E−16**
	SD	0.695792	0.31832	8.69E−08	2.45761E−10	3.00679E−05	1.17898E−14	3.695E−14	6.20468532	**0**	**0**
	MFE	29,929	29,881	29,919	29,991	29,951	21,731.5	28,758	27,512	3003	**1761**
P11 (0)	M	0.011455	0.044885	0.028906	0.00886436	0.028550199	0.007132844	5.884E−06	0.341604681	**0**	**0**
	SD	0.014397	0.066572	0.042806	0.009729053	0.027959171	0.01247169	1.227E−05	0.974177478	**0**	**0**
	MFE	29,971	29,406	21,654	29,961	29,892.5	20,597.33333	29,885	29,743	2004	**1444**
P12 (0)	M	1.549523	6.222186	0.791997	1.29548785	0.632098973	0.32481914	0.005735	45.30554917	31.84073202	**0.003214338**
	SD	1.49792	7.075035	0.372832	2.02404146	0.757373886	0.082793378	0.019997	52.69414276	10.70135226	**0.013707242**
	MFE	29,957	28,537	26,432	29,979	29,977.2	**24,858.16667**	29,692	28,012	27,704	29,405
P13 (0)	M	0.024281	0.458132	0.009724	1.351972689	0.771735	0.008753439	**0.002564**	4.394105597	0.984793828	0.494615064
	SD	0.078964	0.638728	0.026098	3.692304456	1.244696081	0.026151805	**0.008503**	7.706212251	1.276655337	1.029536115
	MFE	29,927	29,996.33333	29,947	29,655.83333	29,998	29,984.16667	29,946	29,596	**27,959**	29,927
P14 (0.998)	M	**0.998004**	**0.998004**	0.998116	**0.998004**	**0.998004**	0.998100837	**0.998004**	0.99828498	1.064140548	1.097275325
	SD	8.25E−17	2.43E−08	0.000251	7.1417E−17	**0**	0.000162056	**0**	0.000955316	0.362245683	0.399531422
	MFE	12,013	24,069	14,583	**6474**	16,239	14,333.33333	8979	18,213	24,152	22,596

(continued)

Table 4.5 (continued)

Problem Number	(Optimum Value)	Rao-1	Rao-2	Rao-3	ERao-1	ERao-2	ERao-3	SAP-Rao	Jaya	MTPG-Jaya	AMTPG-Jaya
P15 (0.0003)	M	0.0014295	0.0006656	0.0004858	0.001237618	0.00053442	0.000438635	**0.000418**	0.0010676	0.000720215	0.00042325
	SD	0.0035890	0.0005148	0.0003264	0.003618665	0.000428917	0.000179217	**0.000157**	0.0036482	0.001477502	0.000248732
	MFE	21,826.66667	23,386	21,737	25,300	20,273.33333	20,101.66667	24,297	24,158	**19,112**	25,975
P16 (−1.0316)	M	−1.031627	−1.031626	−1.031628	**−1.031628453**	**−1.031628453**	−1.03162813	**−1.031628**	−1.031626471	−1.031628403	−1.031628205
	SD	0.00000436	0.00000739	8.39E−08	**4.87871E−16**	5.83118E−16	5.11231E−07	6.775E−16	5.21578E−06	5.18055E−13	1.04625E−06
	MFE	2577	4612	20,283	**1035.666667**	4483	18,786	3731	16,643	15,087	14,024
P17 (0.397887)	M	**0.397887**	**0.397887**	**0.397887**	**0.397887**	**0.397887**	**0.397887**	**0.397887**	**0.397887**	**0.397887**	**0.397887**
	SD	**0**	**0**	**0**	**0**	**0**	**0**	**0**	**0**	**0**	**0**
	MFE	995	695	692	791	773	766	**682**	3387	2972	2342
P18 (3)	M	3	3	3.000021	3	3	3.000017892	3	3.000043566	5.700060063	**3**
	SD	9E−16	6.06E−16	0.000033	**1.21E−16**	5.71336E−16	3.08598E−05	1.293E−15	4.22765E−05	14.78849771	1.80296E−15
	MFE	10,031	18,098	22,145.6	9774.5	9419	20,921.2	**9398**	20,723	18,815	16,730
P19 (−3.86)	M	**−3.86278**	**−3.86278**	**−3.86278**	**−3.86278**	**−3.86278**	**−3.86278**	**−3.86278**	**−3.86278**	**−3.86278**	**−3.86278**
	SD	**1.56E−15**	3.11E−15	3.06E−15	3.17786E−15	3.16177E−15	3.16177E−15	3.148E−15	3.14776E−15	3.16177E−15	3.16177E−15
	MFE	**575**	4093	6680	1173	4048	3207.5	2094	3296	3236	3191
P20 (−3.32)	M	−3.286657	**−3.29792**	−3.278659	−3.286487924	−3.282632647	−3.286487924	−3.296739	−3.242896976	−3.246870819	−3.258630447
	SD	0.05664	0.05719	0.058427	**0.055748152**	0.057154947	0.055748152	0.060487	0.057155167	0.058426783	0.060646486
	MFE	8003	**2799**	6916	4120	5043	3713.2	10,065	8204	8105	8692
P21 (−10.1532)	M	−7.566177	−8.405803	−8.168698	−8.263266404	**−9.485500142**	−9.016394549	−9.463294	−5.121234807	−5.832555732	−7.135195843
	SD	2.413688	2.391694	2.693478	2.81732706	**1.499163896**	2.102883561	1.676048	3.216961611	3.418373019	2.507042827
	MFE	11,371	11,016	13,321	10,927	26,130	20,521.66667	20,252	**3159**	4879	13,767

(continued)

Table 4.5 (continued)

Problem Number	(Optimum Value)	Rao-1	Rao-2	Rao-3	ERao-1	ERao-2	ERao-3	SAP-Rao	Jaya	MTPG-Jaya	AMTPG-Jaya
P22	M	−8.760775	−10.108301	−9.976039	−10.40294056	**−10.40294057**	−10.40108248	**−10.402941**	−6.275389632	−10.08284131	−10.36102021
(−10.4029)	SD	2.146664	1.004131	0.626313	8.55657E−09	**3.29861E−16**	0.009767544	1.189E−15	3.538623556	1.418549935	0.129930888
	MFE	13,592	17,633	22,713	29,208	20,097	29,420	17,914	**5794**	15,048	14,672.33333
P23	M	−9.570118	−10.470286	−10.486057	−10.53640982	**−10.53640982**	−10.53639413	**−10.536410**	−6.253818899	−10.53042573	**−10.536410**
(−10.5364)	SD	1.598056	0.212811	0.275792	2.84401E−12	2.21278E−15	8.59026E−05	**1.776E−15**	3.661325977	0.032776206	2.28535E−15
	MFE	16,652	26,983	18,602	25,374	15,089.33333	29,745	15,853	**7077.333333**	14,370.66667	16,153

Source Rao-1, Rao-2, Rao-3: Rao (2020)
Result in boldface indicates better values

For the test problems P4, P5, P7, P9, P10, P11, P12, P13, P15, P16, P21, P22, and P23, the SAP-Rao algorithm has achieved better mean and standard deviation values when compared to those of the Rao algorithms.

For the problems P14, P17, P18, and P19, the SAP-Rao algorithm has achieved better or the same performance in terms of the mean and standard deviation when compared to those of the Rao algorithms. For the test problems P1, P2, P3, P6, P8, and P20, the SAP-Rao algorithm has achieved better or competitive performance in terms of the mean and standard deviation when compared to those of the Rao algorithms. In addition, the AMTPG-Jaya algorithm has achieved better results in 21 out of 23 problems when compared to those of the MTPG-Jaya and Jaya algorithms. Similarly, the MTPG-Jaya algorithm has achieved better results in 20 out of 23 problems compared to those of the Jaya algorithm.

Furthermore, the elitist Rao algorithms have better exploration and exploitation capabilities than those of the Rao algorithms. In unimodal test problems P1 to P7, except for the problem P2, the ERao-3 algorithm has outperformed the Rao-3 algorithm in terms of the mean values and MFE. Similarly, in all unimodal test problems except for the problem P5, the ERao-2 algorithm has achieved better results in terms of the mean and MFE values when compared to those of the Rao-2 algorithm. The ERao-1 algorithm has achieved better or competitive results compared to those of the Rao-1 algorithm in unimodal test problems.

Similarly, in multimodal test problems P8 to P13, the ERao-3 algorithm has achieved better results than those achieved by the Rao-3 algorithm in terms of the mean and MFE values. Similarly, the ERao-2 algorithm has achieved better results in terms of mean values when compared to those of the Rao-2 algorithm for the multimodal problems P9 to P12. The ERao-1 algorithm has achieved better or competitive results when compared to those of the Rao-1 algorithm in solving multimodal test problems.

Furthermore, the ERao-3 algorithm has outperformed the Rao-3 algorithm in solving the multimodal fixed dimension problems P14 to P23. For all the multimodal fixed dimension problems, the ERao-3 algorithm has achieved superior results in terms of the mean and MFE values compared to those of the Rao-3 algorithm. Similarly, in all multimodal fixed dimension test problems except for the problem P20, the ERao-1 algorithm has achieved better results in terms of best and mean when compared to those of the Rao-1 algorithm. The ERao-2 algorithm has achieved better results in terms of best and mean when compared to those of the Rao-2 algorithm in solving all multimodal fixed dimension test problems except for the problems P15 and P20.

Here, an observation can be made that the modified versions are converging faster than the respective basic algorithms. However, the performances of these algorithms in solving these benchmark problems are evaluated using the Friedman statistical test. The algorithms are ranked based on their statistical results achieved in solving these problems. Here, while ranking the algorithms, priority is given to the mean value achieved, then the MFEs value achieved, and then the standard deviation value. Ten algorithms are compared in this test, with nine degrees of freedom for the χ^2 distribution. The average ranks, χ^2 value, and p-value achieved in the Friedman test are presented in Table 4.6. The ranks of these algorithms and their average rank values are plotted in Fig. 4.14.

Here, an observation can be made that the *p*-value of the Friedman test is much lesser than 0.01, which indicates that the results are highly significant. The least value of the average rank indicates superior performance. The SAP-Rao algorithm has achieved the least average rank of 2.57. The ranking of these algorithms in solving unimodal and multimodal benchmark problems is SAP-Rao, AMTPG-Jaya, ERao-3, ERao-1, ERao-2, Rao-3, MTPG-Jaya, Rao-1, Rao-2, and Jaya algorithms. Furthermore, from the mean function evaluations (*MFE*) consumed by algorithms, it can be observed that the proposed algorithms are converging faster than the respective basic algorithms.

4.4.2.2 Experiment-2

In the second experiment, the performances of the proposed modified algorithms are evaluated with respect to the performances of the real-coded genetic algorithm (RGA), gravitational search algorithm (GSA), PSO and its variants, Jaya, and self-adaptive multi-population Jaya (SAMP-Jaya) algorithms in solving ten unimodal and multimodal test functions. The description, along with the ranges of design variables of these functions, is presented in Appendix A3. For all the functions, 30 design variables are considered. The computational results presented by various algorithms are with either 50,000 function evaluations or 200,000 function evaluations. Hence, the proposed modified algorithms are tested separately by taking 50,000 and 200,000 function evaluations as the termination criterion. The algorithms are executed for 30 consecutive runs, and statistical results are presented in terms of mean (M) and standard deviation (SD).

The test functions TF1 to TF7 are unimodal test functions, and TF8 to TF10 are multimodal test functions. All these test functions are minimization problems. The performances are compared with PSO (Eberhart & Kennedy, 1995), cooperative PSO (CPSO) (van den Bergh and Engelbrecht, 2004), fully informed particle swarm (FIPS) (Mendes et al., 2004), RGA (Haupt and Haupt, 2003), comprehensive learning PSO (CLPSO) (Liang et al., 2006), Frankenstein's PSO (F-PSO) (Montes de Oca et al., 2009), GSA (Rashedi et al., 2009), adaptive inertia weight PSO (AIWPSO) (Nickabadi et al., 2011), extraordinariness particle swarm optimizer (EPSO) (Ngo et al., 2016), Jaya, and its variant SAMP-Jaya (Rao & Saroj, 2017) algorithms. Table 4.7 presents the comparison of the results achieved by the proposed modified algorithms with those of other algorithms in 50,000 function evaluations.

From Table 4.7, it can be observed that the AMTPG-Jaya algorithm has outperformed all the algorithms compared in solving 9 out of 10 (TF1–TF7, TF9, and TF10) functions. Similarly, the ERao-3, SAP-Rao, and MTPG-Jaya algorithms have achieved better or the same results in solving 8 out of 10 problems when compared to the performances of the other algorithms from the literature. The ERao-1 and ERao-2 algorithms have achieved better or the same performances in solving 6 out of 10 problems compared to other algorithms from the literature. Similar to experiment-1,

Table 4.6 Friedman statistical test results for the results presented in Table 4.5

Algorithm	Rao-1	Rao-2	Rao-3	ERao-1	ERao-2	ERao-3	SAP-Rao	Jaya	MTPG-Jaya	AMTPG-Jaya
Avg. Rank	6.26	6.96	5.96	5.17	5.26	4.13	2.57	9.13	6.04	3.48
χ^2	76.89249012									
p-value	The p-value is **<0.00001**. The result is significant at $p < 0.01$									

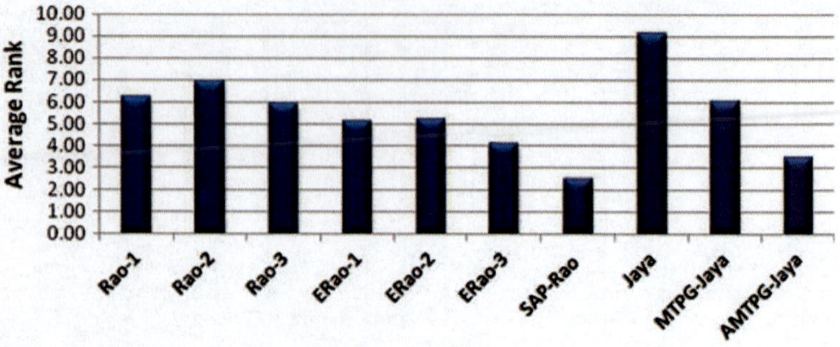

Fig. 4.14 Average ranks of Jaya and Rao algorithms along with their modified versions in the Friedman statistical test of the unimodal and multimodal benchmark problems of experiment-1

to judge the performance of the proposed modified algorithms with the other algorithms, the Friedman statistical test is conducted. The mean values are given priority and then to the standard deviation values for ranking the algorithms in all the problems. Here, 11 algorithms are compared in this test with ten degrees of freedom for the χ^2 distribution. The average ranks, χ^2 value, and *p*-value achieved in the Friedman test are presented in Table 4.8. The ranks of these algorithms and their average rank values are plotted in Fig. 4.15.

The *p*-value of the Friedman test is 8.04989e−7, which indicates that the results are highly significant. The AMTPG-Jaya algorithm has achieved the least average rank of 2.85. The ranking of these algorithms in solving the unimodal and multimodal test functions is AMTPG-Jaya, MTPG-Jaya, ERao-3, SAP-Rao, ERao-1, SAMP-Jaya, EPSO, ERao-2, Jaya, RGA, and GSA. Here, an observation can be made that the proposed algorithms have better or competitive performance compared to those of the GSA, RGA, EPSO, Jaya, and SAMP-Jaya algorithms.

Table 4.9 presents the comparison of the results of the proposed modified algorithms with those of other algorithms in 200,000 function evaluations. Here, an observation can be made that the AMTPG-Jaya, SAP-Rao, and MTPG-Jaya algorithms have achieved better or the same results in 8 out of 10 problems. Similarly, the elitist Rao algorithms have achieved better or the same results for the problems TF1 to TF7 and achieved competitive results for the problems TF8 to TF10, when compared to the results achieved by the other algorithms from the literature. In addition, to judge the performance of the proposed modified algorithms with the other algorithms, the Friedman statistical test is conducted. The mean values are given priority and then to the standard deviation values for ranking the algorithms in all the problems. Here, 15 algorithms are compared in this test with 14 degrees of freedom for the χ^2 distribution. The average ranks, χ^2 value, and *p*-value achieved in the Friedman test are presented in Table 4.10. The ranks of these algorithms and their average rank values are plotted in Fig. 4.16.

The *p*-value of the Friedman test is 4.049E−05, which indicates that the results are highly significant. The AMTPG-Jaya algorithm has achieved the least average rank of 4.3. The ranking of these algorithms in solving the unimodal and multimodal test

Table 4.7 Comparison of the results of proposed modified algorithms with the results of other algorithms achieved in 50,000 function evaluations for the optimization of the unimodal and multimodal test functions

Test function		RGA	GSA	EPSO	Jaya	SAMP-Jaya	ERao-1	ERao-2	ERao-3	MTPG-Jaya	AMTPG-Jaya	SAP-Rao
TF1	M	23.13	6.8000E−17	7.7760E−18	7.3215E−19	2.7998E−19	1.1515E−38	3.6207E−21	1.2821E−73	**0**	**0**	5.9000E−46
	SD	12.15	2.1800E−17	6.0100E−18	2.8206E−18	8.2707E−19	2.6622E−38	1.3790E−20	5.3552E−73	**0**	**0**	2.2825E−45
TF2	M	1.073	6.0600E−08	6.7870E−12	4.5329E−11	1.3439E−12	5.3155E−17	4.6506E−02	2.1707E−36	**0**	**0**	1.0639E−29
	SD	0.2666	1.1900E−08	3.0080E−12	1.0544E−10	2.1634E−11	1.5376E−16	1.1723E−01	1.0270E−35	**0**	**0**	2.6690E−29
TF3	M	561.7	942.7	0.2121	5.6234E−26	1.5464E−35	8.6098E−72	1.2778E−40	1.438E−138	**0**	**0**	3.6007E−90
	SD	125.6	246.6	0.5461	1.6346E−25	4.4840E−35	4.5044E−71	4.0768E−40	7.875E−138	**0**	**0**	1.8095E−89
TF4	M	11.78	4.207	9.9410E−03	5.060152	1.152804	0.304880134	8.394887097	1.19E−03	1.86E−296	**0**	8.3423E−04
	SD	1.576	1.122	9.8550E−03	7.352134	2.808808	0.154216849	4.899670188	2.85E−03	**0**	**0**	1.8778E−03
TF5	M	1180	47.95	1.7850E−02	1.7527E−06	1.8006E−09	**2.2083E−20**	1.1122E−12	1.9122E−19	1.0092E−07	1.1459E−12	1.2442E−12
	SD	548.1	3.956	2.1360E−02	1.7527E−05	1.8443E−08	**9.2372E−20**	4.4456E−12	4.2324E−19	3.9963E−07	2.8337E−12	4.3990E−12
TF6	M	24.01	0.931	**0**	**0**	**0**	**0**	**0**	**0**	**0**	**0**	**0**
	SD	10.17	2.51	**0**	**0**	**0**	**0**	**0**	**0**	**0**	**0**	**0**
TF7	M	0.0675	0.0782	6.4700E−04	7.4146E−19	3.0770E−28	1.1053E−23	5.4179E−13	1.0485E−54	5.233E−197	**0**	8.7612E−36
	SD	0.0287	0.041	4.5420E−04	4.0529E−18	1.1412E−27	2.7420E−23	1.5510E−12	3.0905E−54	**0**	**0**	3.9283E−35
TF8	M	−12,480	−3604	**−12,570.00**	−11,466.34	−11,763.96	−9124.64	−9224.39	−10,343.10	−6599.25	−6234.76	−10,136.59
	SD	53.26	564.1	**3.85E−12**	1288.14	838.95	1297.42	1557.99	1008.49	1244.76	1073.75	1042.49
TF9	M	5.902	29.4	2.2740E−14	116.247	68.517	89.413	166.264	86.420	**0**	**0**	28.139
	SD	1.171	4.727	2.8320E−14	58.292	32.544	21.211	38.819	45.287	**0**	**0**	8.828
TF10	M	2.14	4.8000E−09	1.2840E−09	9.3528E−11	7.0737E−11	5.9626E−14	1.0171E−10	1.2257E−14	**8.8818E−16**	**8.8818E−16**	6.9278E−15

(continued)

Table 4.7 (continued)

Test function		RGA	GSA	EPSO	Jaya	SAMP-Jaya	ERao-1	ERao-2	ERao-3	MTPG-Jaya	AMTPG-Jaya	SAP-Rao
	SD	0.4014	5.4200E−10	7.2800E−10	1.4691E−10	6.7168E−11	2.2867E−14	5.1388E−10	3.5404E−15	**0**	**0**	1.6559E−15

Source RGA, GSA, EPSO, Jaya, and SAMP-Jaya: Rao and Saroj (2017)
Result in boldface indicates better values

Table 4.8 Friedman statistical test results for the results presented in Table 4.7

Algorithm	RGA	GSA	EPSO	Jaya	SAMP-Jaya	ERao-1	ERao-2	ERao-3	MTPG-Jaya	AMTPG-Jaya	SAP-Rao
Avg. rank	9.2	9.6	6.6	7.4	5.9	5.5	7.4	4	3.25	2.85	4.3
χ^2	47.377										
p-value	The *p*-value is **8.04989e−7**. The result is significant at $p < 0.01$										

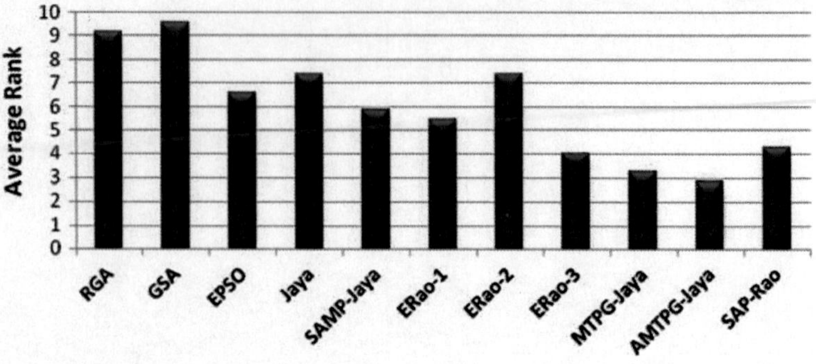

Fig. 4.15 Average ranks of various algorithms in the Friedman statistical test for the unimodal and multimodal test functions results with 50,000 function evaluations

functions is AMTPG-Jaya, MTPG-Jaya, SAP-Rao, ERao-3, SAMP-Jaya, ERao-1, ERao-2, Jaya, EPSO, AIWPSO, CLPSO, PSO, FIPSO, CPSO, and F-PSO. Here, an observation can be made that the proposed algorithms have better or competitive performance when compared to those of the EPSO, AIWPSO, CLPSO, PSO, FIPSO, CPSO, F-PSO, Jaya, and SAMP-Jaya algorithms.

For all the unimodal and multimodal test functions, the proposed algorithms have attained better or competitive results in terms of mean and standard deviation values. It indicates that the exploration and exploitation capability of the proposed modified algorithms are better or competitive than that of the other algorithms compared. In addition, the Friedman statistical test confirms that the performances of the proposed modified algorithms are better or competitive than the performances of basic algorithms as well as the other algorithms compared.

From the computational results of the proposed modified algorithms in single-objective optimization benchmark problems, it can be observed that the proposed modified algorithms are more robust in finding better solutions than the basic Jaya and Rao algorithms. Also, the modified versions are converging faster than the basic Jaya and Rao algorithms. In addition, the performances of the modified versions are better than or competitive to those of the other algorithms compared from the literature. Hence, the modified versions of the Jaya and Rao algorithms can be considered as the improved versions of the Jaya and Rao algorithms, respectively. Here, it can be noted that the AMTPG-Jaya algorithm is an improved version of the MTPG-Jaya algorithm. Furthermore, the performance of the AMTPG-Jaya algorithm precedes the performance of the MTPG-Jaya algorithm in solving almost all the problems. Hence, the MTPG-Jaya algorithm is excluded in solving the multi-objective optimization problems and case studies. The next subsection presents the analysis of the computational results of the proposed modified algorithms in solving multi-objective optimization benchmark problems.

Table 4.9 Comparison of the results of proposed modified algorithms with the results of other algorithms achieved in 200,000 function evaluations for the optimization of the unimodal and multimodal test functions

Test function		PSO	CPSO	CLPSO	FFIPS	F-PSO	AIWPSO	EPSO	Jaya	SAMP-Jaya	ERao-1	ERao-2	ERao-3	MTPG-Jaya	AMTPG-Jaya	SAP-Rao
TF1	M	5.2E−70	5.1E−13	4.9E−39	4.6E−27	2.4E−16	3.3E−133	1.7E−74	7.1E−90	4.9E−89	2.5E−167	2.5E−144	3.9E−279	**0**	**0**	2E−203
	SD	1.1E−74	7.8E−25	6.8E−39	2.0E−53	2.0E−31	5.2E−267	2.8E−75	3.4E−89	1.9E−88	**0**	7.7E−144	**0**	**0**	**0**	**0**
TF2	M	2.1E−25	1.3E−07	8.9E−24	2.3E−16	1.6E−11	1.4E−54	1.9E−47	6.9E−69	9.5E−96	6.2E−124	1.2E−109	3.2E−155	**0**	**0**	2E−124
	SD	1.4E−49	1.2E−14	7.9E−49	1.1E−32	1.0E−22	7.4E−119	2.2E−47	3.8E−68	4.4E−95	1.7E−123	3.2E−109	1.3E−154	**0**	**0**	1E−123
TF3	M	1.458	1889	192.2	9.463	173.2	1.9E−10	0.002014	1.2E−132	**0**	**0**	**0**	**0**	**0**	**0**	**0**
	SD	1.78	9,911,000	384.3	25.98	9158	1.2E−19	0.001934	2.9E−132	**0**	**0**	**0**	**0**	**0**	**0**	**0**
TF4	M	–	–	–	–	–	–	–	7.16E−05	6.91E−08	3.07E−12	8.48E−09	6.46E−15	1E−296	**0**	2E−17
	SD	–	–	–	–	–	–	–	2.00E−04	6.91E−08	3.38E−10	9.18E−08	2.24E−14	**0**	**0**	4E−17
TF5	M	25.4	0.8265	13.22	26.71	28.16	2.5	2.8E−05	1.8E−17	2.3E−25	1.9E−30	**2.5E−31**	1.2E−29	1E−07	1.1E−12	9E−27
	SD	590.3	2.345	214.8	200.3	231.3	16	3.7E−05	9.7E−17	6.2E−25	6.5E−30	**1.4E−30**	2.4E−29	4E−07	2.8E−12	4E−26
TF6	M	**0**	**0**	**0**	**0**	**0**	**0**	**0**	**0**	**0**	**0**	**0**	**0**	**0**	**0**	**0**
	SD	**0**	**0**	**0**	**0**	**0**	**0**	**0**	**0**	**0**	**0**	**0**	**0**	**0**	**0**	**0**
TF7	M	1.2E−02	1.1E−02	4.1E−03	3.3E−03	4.2E−03	5.5E−03	2.6E−04	1.0E−106	2.2E−111	9.8E−125	1.6E−115	9.7E−221	5E−197	**0**	1E−143
	SD	2.3E−05	2.8E−05	9.6E−07	8.4E−03	2.4E−06	1.5E−05	1.9E−04	5.7E−106	1.2E−110	3.7E−124	6.9E−113	**0**	**0**	**0**	8E−143
TF8	M	−11,000	−12,130	−12,550	−11,050	−11,220	**−12,570**	**−12,570**	−11,225	−12,276	−9467	−10,732	−11,199	−6599	−6235	−10,432
	SD	137,500	33,800	4257	944,200	222,700	**1.1E−25**	2.5E−12	1070	570	1189	1684	749	1245	1074	1131
TF9	M	34.76	3.6E−13	**0**	58.5	73.84	0.1658	**0**	76.982	59.246	92.410	116.493	53.707	**0**	**0**	24.01
	SD	106.4	1.5E−24	**0**	191.9	370.6	0.2105	**0**	26	24	43	43	23	**0**	**0**	9
TF10	M	1.5E−14	1.6E−07	9.2E−15	1.4E−14	2.2E−09	7.0E−15	1.2E−14	5.2E−15	4.4E−15	3.1E−14	2.4E−14	1.4E−14	**8E−16**	**8.9E−16**	4E−15

(continued)

Table 4.9 (continued)

Test function		PSO	CPSO	CLPSO	FFIPS	F-PSO	AIWPSO	EPSO	Jaya	SAMP-Jaya	ERao-1	ERao-2	ERao-3	MTPG-Jaya	AMTPG-Jaya	SAP-Rao
	SD	1.9E−29	7.9E−14	6.6E−30	2.3E−29	1.7E−18	4.2E−31	3.1E−15	2.7E−15	**0.0E + 00**	9.6E−15	1.3E−14	5.4E−15	**0**	**0**	**0**

Source PSO, CPSO, CLPSO, FFIPS, FPSO, AIWPSO, EPSO, Jaya, and SAMP-Jaya: Rao and Saroj (2017)
Result in boldface indicates better values

Table 4.10 Friedman statistical test results for the results presented in Table 4.9

Algorithm	PSO	CPSO	CLPSO	FFIPS	F-PSO	AIWPSO	EPSO	Jaya	SAMP-Jaya	ERao-1	ERao-2	ERao-3	MTPG-Jaya	AMTPG-Jaya	SAP-Rao
Avg. Rank	11.00	11.40	9.35	11.00	12.10	8.25	8.00	7.80	6.55	7.40	7.50	5.40	4.60	4.30	5.35
χ^2	45.020														
p-value	The *p*-Value is **0.0000404896**. The result is significant at $p < 0.01$														

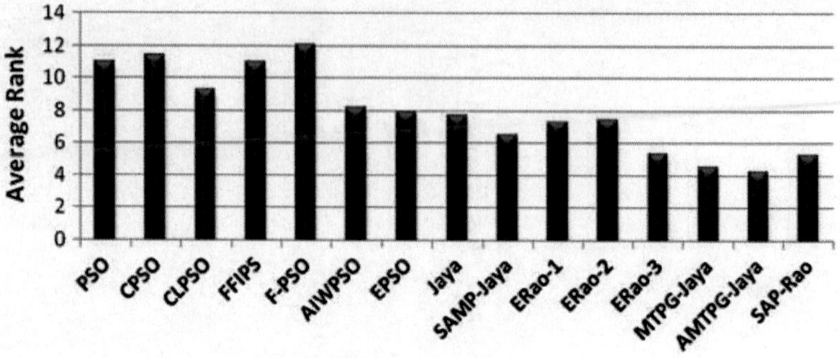

Fig. 4.16 Average ranks of various algorithms in the Friedman statistical test for the unimodal and multimodal test functions results with 200,000 function evaluations

4.5 Computational Results Analysis on Multi-objective Optimization Benchmark Problems

The performances of the proposed modified algorithms are tested using five challenging multi-objective optimization test problems known as ZDT (Zitzler, Deb, and Thiele's) test problems (Zitzler et al., 2000). These problems are ZDT-1, ZDT-2, ZDT-3, ZDT-1 with linear Pareto-front (ZDT-1L), and ZDT-2 with 3 objectives (ZDT2-3O). The ZDT-1 problem has a convex-shaped Pareto-front. The ZDT-2 problem has a concave-shaped Pareto-front, and the ZDT-3 problem has a discrete convex Pareto-front. A detailed description, including the number of independent variables and their ranges of the considered multi-objective optimization problems, is presented in Appendix B1 (Mirjalili, 2016).

In multi-objective optimization, a *posteriori* approach is implemented to handle all the objectives simultaneously. In computational experiments on these test problems, the proposed algorithms are tested for 10,000 function evaluations, and the statistical results are presented for ten consecutive runs. In computational experiments, population size is taken as 100 for solving these problems. In the execution of the SAP-Rao algorithm, a minimum of 100 population size is preserved to maintain consistency with other algorithms. Also, different elite sizes are used for elitist Rao algorithms based on population size. The elite size is varied from 10 to 40% of the population size. The computational results of the proposed algorithms are presented in terms of IGD and spacing values. The performances of the proposed modified algorithms in terms of IGD values are compared with those of the basic algorithms as well as the multi-objective dragonfly algorithm (MODA), multi-objective grasshopper optimization algorithm (MOGOA), multi-objective ant lion optimizer (MOALO), multi-objective particle swarm optimization (MOPSO), and non-dominated sorting genetic algorithm (NSGA-II). The spacing values achieved by the MODA, MOGOA, MOALO, MOPSO, and NSGA-II were not available in the literature. Hence, the performances of the proposed modified algorithms in terms of spacing values are compared with those of the basic algorithms.

Table 4.11 compares the statistical results achieved by the proposed modified algorithms with other algorithms to solve the ZDT-1 problem. The statistical results in all the tables are presented in terms of best, worst, mean values, median, and standard deviation (*SD*). Here, an observation can be made that the proposed modified algorithms and basic Jaya and Rao algorithms have outperformed the other algorithms from the literature in terms of IGD values. The SAP-Rao algorithm has achieved the least mean IGD and spacing values. The next best mean IGD values are achieved by ERao-3, ERao-1, AMTPG-Jaya, and ERao-2 algorithms, respectively. The MOPSO algorithm has the best IGD value. However, the proposed modified algorithms have achieved better mean values than the other algorithms. It indicates that the proposed algorithms are consistent in finding the optimal Pareto-front.

Similarly, the comparison of the statistical results achieved by the proposed modified algorithms with other algorithms in solving the ZDT-2 problem is presented in Table 4.12. Here, an observation can be made that the proposed modified algorithms and basic Jaya and Rao algorithms have outperformed the MOGOA, NSGA-II, MODA, and MOALO algorithms in terms of IGD values. The performances of the proposed algorithms are competitive with the MOPSO algorithm. The SAP-Rao algorithm has achieved the least mean spacing value. The MOPSO algorithm has achieved the least mean IGD value. The next best mean IGD values are achieved by ERao-3, ERao-2, SAP-Rao, and AMTPG-Jaya algorithms, respectively.

Table 4.13 presents the comparison of the statistical results achieved by the proposed modified algorithms with other algorithms in solving the ZDT-3 problem. Here, an observation can be made that the proposed modified algorithms and basic Jaya and Rao algorithms have outperformed the other algorithms from the literature in terms of IGD values. The ERao-3 algorithm has achieved the least mean IGD and spacing values. The next best mean IGD values are achieved by SAP-Rao, ERao-2, ERao-2, and AMTPG-Jaya algorithms.

Similarly, the comparison of the statistical results achieved by the proposed modified algorithms with other algorithms in solving the ZDT-1L problem is presented in Table 4.14. For this problem also, the proposed modified algorithms and basic Jaya and Rao algorithms have outperformed the other algorithms compared in terms of mean IGD values. The SAP-Rao algorithm has achieved the least mean spacing value. The ERao-2 algorithm has achieved the least mean IGD value. The next best mean IGD values are achieved by ERao-1, AMTPG-Jaya, ERao-3, and SAP-Rao algorithms, respectively. Similarly, Table 4.15 presents the comparison of the statistical results achieved by the proposed modified algorithms with other algorithms in solving the ZDT2-3O problem. In this problem also, the proposed modified algorithms and basic Jaya and Rao algorithms have outperformed the other algorithms from the literature in terms of IGD values. The SAP-Rao algorithm has achieved the least mean IGD value. The next best mean IGD values are achieved by AMTPG-Jaya, ERao-3, ERao-2, and Jaya algorithms, respectively. The ERao-1 algorithm has achieved the least mean spacing value.

In addition, for qualitative assessment of the algorithms' performances, the Pareto-fronts achieved by the proposed modified algorithms, along with Jaya and Rao algorithms, are presented in Figs. 4.17, 4.18, 4.19, 4.20 and 4.21.

Table 4.11 Comparison of the results achieved by various algorithms in multi-objective optimization of the ZDT-1 problem

Algorithm	IGD					Spacing				
	Best	Worst	M	Median	SD	Best	Worst	M	Median	SD
MOGOA	0.0028	0.0822	0.0121	0.0046	0.0247	–	–	–	–	–
MOPSO	**0.0015**	0.0101	0.00422	0.0037	0.003103	–	–	–	–	–
NSGA-II	0.0546	0.0702	0.05988	0.0574	0.005436	–	–	–	–	–
MODA	0.0024	0.0096	0.00612	0.0072	0.002863	–	–	–	–	–
MOALO	0.0061	0.0209	0.01524	0.0166	0.005022	–	–	–	–	–
Jaya	0.0022848	0.0036293	0.0028700	0.0027842	0.0004846	0.0030862	0.0097681	0.0054556	0.0045533	0.0023200
Rao-1	0.0022577	0.0031300	0.0026479	0.0025347	0.0003014	0.0030372	0.0095579	0.0049653	0.0047373	0.0020980
Rao-2	0.0023985	0.0040621	0.0032697	0.0032697	0.0006065	0.0037399	0.0086020	0.0055581	0.0049339	0.0017361
Rao-3	0.0022729	0.0025932	0.0024277	0.0024228	0.0001149	0.0030360	0.0037921	0.0033705	0.0033788	0.0002450
ERao-1	0.0022155	0.0025364	0.0023957	0.0023957	0.0000959	0.0032215	0.0037840	0.0034621	0.0034621	0.0001688
ERao-2	0.0022931	0.0025271	0.0024221	0.0024221	**0.0000846**	**0.0028235**	0.0038067	0.0032328	0.0031770	0.0003324
ERao-3	0.0022214	0.0024956	0.0023560	0.0023916	0.0000957	0.0029571	0.0036882	0.0033311	0.0033311	0.0002319
AMTPG-Jaya	0.0023440	0.0026824	0.0024228	0.0024110	0.0001099	0.0031013	0.0039536	0.0034919	0.0034595	0.0002897
SAP-Rao	0.0021807	**0.0024752**	**0.0023407**	**0.0023407**	0.0000869	0.0028501	**0.0032986**	**0.0030606**	**0.0030606**	**0.0001578**

Source MOGOA, MOPSO, NSGA-II: Mirjalili et al. (2018); MODA: Mirjalili (2016); MOALO: Mirjalili et al. (2017)
Result in boldface indicates better values

Table 4.12 Comparison of the results achieved by various algorithms in multi-objective optimization of the ZDT-2 problem

Algorithm	IGD					Spacing				
	Best	Worst	M	Median	SD	Best	Worst	M	Median	SD
MOGOA	0.0016	0.0273	0.007	0.0049	0.009	–	–	–	–	–
MOPSO	**0.0013**	**0.0017**	**0.00156**	**0.0017**	0.000174	–	–	–	–	–
NSGA-II	0.1148	0.1834	0.13972	0.1258	0.026263	–	–	–	–	–
MODA	0.0023	0.006	0.00398	0.0033	0.001604244	–	–	–	–	–
MOALO	0.005	0.0377	0.01751	0.0165	0.010977	–	–	–	–	–
Jaya	0.0030770	0.0038014	0.0033411	0.0031579	0.0002801	0.0038416	0.0052066	0.0044104	0.0042834	0.0004967
Rao-1	0.0024935	0.0038934	0.0029074	0.0027294	0.0004580	0.0031799	0.0041586	0.0035683	0.0035279	0.0002996
Rao-2	0.0023552	0.0041087	0.0029883	0.0027617	0.0006216	0.0031329	0.0064620	0.0042046	0.0039475	0.0011133
Rao-3	0.0030140	0.0036245	0.0032858	0.0032330	0.0002315	0.0035532	0.0049834	0.0042331	0.0042331	0.0004602
ERao-1	0.0024032	0.0032612	0.0026477	0.0025724	0.0002764	0.0029941	0.0051762	0.0036644	0.0035615	0.0006950
ERao-2	0.0022341	0.0026368	0.0024325	0.0024291	0.0001173	0.0028565	**0.0035270**	0.0031859	0.0031859	0.0002121
ERao-3	0.0022758	0.0026128	0.0024245	0.0024093	**0.0001046**	**0.0026497**	0.0035915	0.0032201	0.0032441	0.0003372
AMTPG-Jaya	0.0023580	0.0029221	0.0025997	0.0025184	0.0001928	0.0032547	0.0036704	0.0034484	0.0034484	**0.0001430**
SAP-Rao	0.0023364	0.0029362	0.0025754	0.0025504	0.0002138	0.0028400	0.0037287	**0.0031847**	**0.0031474**	0.0002785

Source MOGOA, MOPSO, NSGA-II: Mirjalili et al. (2018); MODA: Mirjalili (2016); MOALO: Mirjalili et al. (2017)
Result in boldface indicates better values

Table 4.13 Comparison of the results achieved by various algorithms in multi-objective optimization of the ZDT-3 problem

Algorithm	IGD					Spacing				
	Best	Worst	M	Median	SD	Best	Worst	M	Median	SD
MOGOA	0.0224	0.0345	0.0306	0.0313	0.0034	–	–	–	–	–
MOPSO	0.0308	0.0497	0.03782	0.0362	0.006297	–	–	–	–	–
NSGA-II	0.0315	0.0557	0.04166	0.0403	0.008073	–	–	–	–	–
MODA	0.02	0.0304	0.02794	0.0302	0.004021	–	–	–	–	–
MOALO	0.0303	0.033	0.03032	0.0323	0.000969	–	–	–	–	–
Jaya	0.0051872	0.0057729	0.0053927	0.0052673	0.0002199	0.0044896	0.0064756	0.0053651	0.0052044	0.0006667
Rao-1	0.0050920	0.0058165	0.0053540	0.0053084	0.0002343	0.0040658	0.0063537	0.0048799	0.0045751	0.0007317
Rao-2	0.0048858	0.0055722	0.0051537	0.0051575	0.0002144	0.0037246	0.0050765	0.0044082	0.0044386	0.0004175
Rao-3	0.0050084	0.0054534	0.0052065	0.0052065	0.0001564	0.0039955	0.0053726	0.0046838	0.0046838	0.0004642
ERao-1	0.0025187	0.0027268	0.0026053	0.0026053	0.0000626	0.0019391	0.0024701	0.0021656	0.0021563	0.0001654
ERao-2	0.0025197	0.0027046	0.0025945	0.0025803	**0.0000609**	0.0019289	0.0025008	0.0021647	0.0020792	0.0002345
ERao-3	**0.0024503**	**0.0026772**	**0.0025445**	**0.0025275**	0.0000752	0.0018835	**0.0021931**	**0.0020180**	0.0020180	**0.0001020**
AMTPG-Jaya	0.0029888	0.0033379	0.0031541	0.0031573	0.0001151	0.0022910	0.0027968	0.0025368	0.0025343	0.0001531
SAP-Rao	0.0024656	0.0027676	0.0025747	0.0025347	0.0001038	**0.0018454**	0.0023592	0.0020464	**0.0019970**	0.0001700

Source MOGOA, MOPSO, NSGA-II: Mirjalili et al. (2018); MODA: Mirjalili (2016); MOALO: Mirjalili et al. (2017)
Result in boldface indicates better values

Table 4.14 Comparison of the results achieved by various algorithms in multi-objective optimization of the ZDT-1L problem

Algorithm	IGD					Spacing				
	Best	Worst	M	Median	SD	Best	Worst	M	Median	SD
MOGOA	0.0017	0.0498	0.0091	0.0023	0.0148	–	–	–	–	–
MOPSO	**0.0012**	0.0165	0.00922	0.0098	0.005531	–	–	–	–	–
NSGA-II	0.0773	0.0924	0.08274	0.0804	0.005422	–	–	–	–	–
MODA	0.0022	0.0163	0.00616	0.0038	0.005186	–	–	–	–	–
MOALO	0.0106	0.033	0.01982	0.0196	0.007545	–	–	–	–	–
Jaya	0.0029703	0.0033261	0.0031069	0.0030627	0.0001278	0.0039527	0.0064543	0.0046406	0.0044064	0.0008148
Rao-1	0.0028968	0.0033283	0.0030612	0.0030361	0.0001454	0.0035196	0.0051419	0.0045532	0.0046276	0.0005193
Rao-2	0.0028631	0.0034530	0.0030898	0.0030741	0.0001802	0.0037438	0.0044214	0.0041032	0.0041032	0.0002069
Rao-3	0.0027155	0.0033786	0.0028957	0.0027991	0.0002204	0.0033409	0.0047274	0.0040933	0.0041095	0.0004013
ERao-1	0.0021659	0.0024248	0.0023073	0.0023155	0.0000824	0.0030625	0.0036670	0.0033255	0.0033232	0.0002051
ERao-2	0.0022049	0.0024593	**0.0023063**	0.0023063	0.0000821	**0.0027937**	**0.0034583**	0.0031784	0.0031795	0.0002059
ERao-3	0.0021698	0.0025456	0.0023266	**0.0022982**	0.0001205	0.0032572	0.0036294	0.0034229	0.0034302	**0.0001192**
AMTPG-Jaya	0.0022559	**0.0023868**	0.0023089	0.0023089	**0.0000445**	0.0028441	0.0040103	0.0033502	0.0032842	0.0003506
SAP-Rao	0.0022487	0.0024537	0.0023403	0.0023392	0.0000711	0.0028657	0.0035153	**0.0030986**	**0.0030265**	0.0002279

Source MOGOA, MOPSO, NSGA-II: Mirjalili et al. (2018); MODA: Mirjalili (2016); MOALO: Mirjalili et al. (2017)
Result in boldface indicates better values

Table 4.15 Comparison of the results achieved by various algorithms in multi-objective optimization of the ZDT-3O problem

Algorithm	IGD					Spacing				
	Best	Worst	M	Median	SD	Best	Worst	M	Median	SD
MOGOA	0.0068	0.0289	0.0114	0.0081	0.0079	–	–	–	–	–
MOPSO	0.0189	0.0225	0.02032	0.0203	0.001278	–	–	–	–	–
NSGA-II	0.0371	0.0847	0.0626	0.0584	0.017888	–	–	–	–	–
MODA	0.0048	0.0191	0.00916	0.0063	0.005372	–	–	–	–	–
MOALO	0.0191	0.0315	0.02629	0.0288	0.004451	–	–	–	–	–
Jaya	0.0020173	0.0022966	0.0021641	0.0021641	0.0000889	0.0456603	0.0595403	0.0528256	0.0519847	0.0044812
Rao-1	0.0020965	0.0033595	0.0024693	0.0022880	0.0004120	**0.0406990**	0.0658425	0.0540621	0.0546690	0.0083277
Rao-2	0.0020227	0.0027209	0.0023362	0.0023137	0.0002245	0.0475266	0.0609942	0.0535145	0.0535145	0.0048116
Rao-3	0.0022324	0.0033942	0.0026145	0.0024506	0.0003970	0.0462863	0.0692203	0.0574161	0.0561472	0.0075947
ERao-1	0.0020711	0.0024332	0.0022214	0.0022214	0.0001195	0.0444766	**0.0558539**	**0.0512404**	**0.0512518**	**0.0036667**
ERao-2	0.0020921	0.0022527	0.0021572	0.0021572	**0.0000617**	0.0482222	0.0743055	0.0570622	0.0538116	0.0093994
ERao-3	0.0019836	0.0024472	0.0021456	0.0020888	0.0001500	0.0438590	0.0629687	0.0527477	0.0530921	0.0060305
AMTPG-Jaya	0.0020546	0.0022480	0.0021270	0.0021115	0.0000646	0.0452289	0.0587363	0.0539866	0.0549457	0.0044246
SAP-Rao	**0.0018341**	**0.0021865**	**0.0020453**	**0.0020763**	0.0001176	0.0446194	0.0604126	0.0521410	0.0525578	0.0053660

Source MOGOA, MOPSO, NSGA-II: Mirjalili et al. (2018); MODA: Mirjalili (2016); MOALO: Mirjalili et al. (2017)
Result in boldface indicates better values

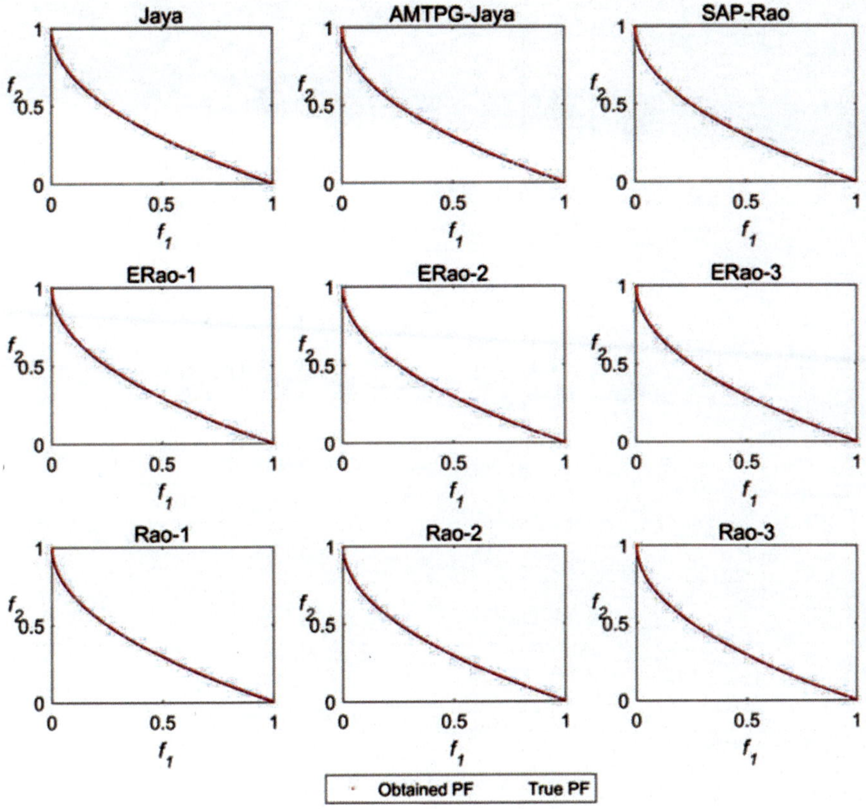

Fig. 4.17 True Pareto-front (PF) and the obtained Pareto-fronts in the objective space of the ZDT-1 problem

Furthermore, to judge the performance of the proposed modified algorithms, the Friedman statistical test is conducted based on the results achieved for all the problems. For ranking the algorithms, the mean values are given priority, and then, the priorities are given to the best, median, worst, and standard deviation values, respectively. Here, 14 algorithms are compared in this test with 13° of freedom for the χ^2 distribution. The average ranks, χ^2 value, and *p*-value achieved in the Friedman test are presented in Table 4.16. The ranks of these algorithms and their average rank values are plotted in Fig. 4.22.

The *p*-value of the Friedman test is 5.12859E−7, which indicates that the results are highly significant. The ERao-3 algorithm has achieved the least average rank of 2.4. The ranking of these algorithms in solving the ZDT test functions is ERao-3, SAP-Rao, ERao-2, ERao-1, AMTPG-Jaya, Rao-3, Rao-1, Rao-2, Jaya, MOPSO, MODA, MOGOA, MOALO, and NSGA-II. Here, an observation can be made that the proposed algorithms have achieved better average ranks compared to those of the

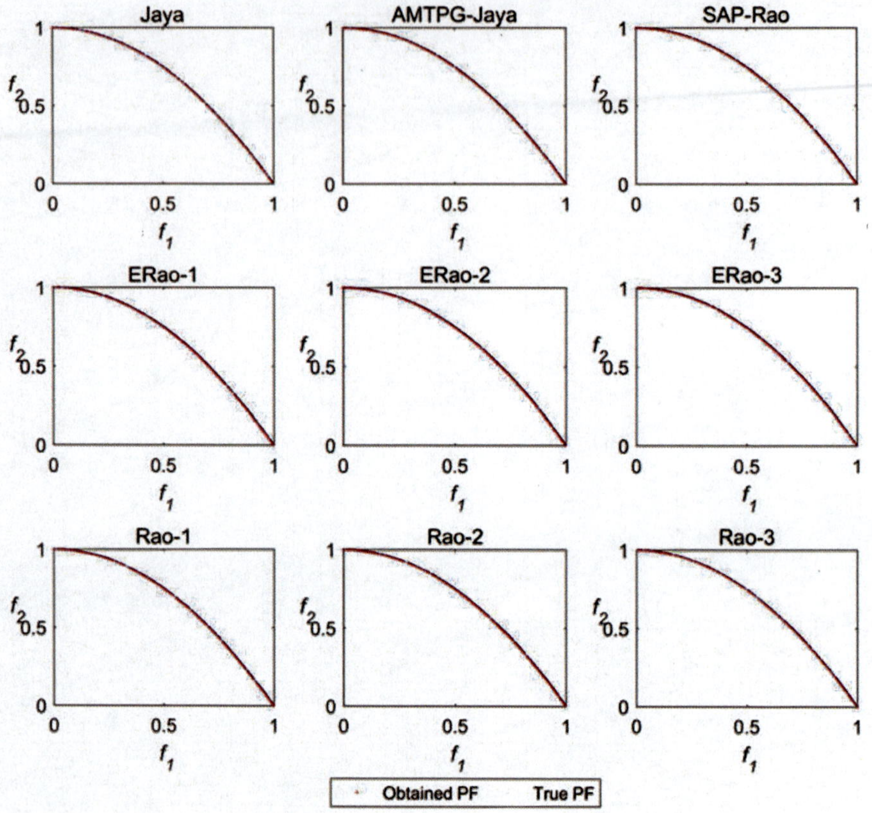

Fig. 4.18 True Pareto-front (PF) and the obtained Pareto-fronts in the objective space of the ZDT-2 problem

MOPSO, MODA, MOGOA, MOALO, and NSGA-II algorithms. From the computational results, it can be observed that the performances of the proposed modified algorithms are superior or competitive to that of the other algorithms compared.

Furthermore, mean IGD values are relatively closer to the best values. It indicates that the proposed algorithms are consistent in finding the optimal Pareto-front. Also, the proposed algorithms Pareto-fronts are consistent and along with the true Pareto-fronts. The performances of the proposed modified algorithms are superior or competitive to those of the other algorithms compared.

From the computational results in single-objective and multi-objective optimization benchmark test problems, it can be observed that the performance of the AMTPG-Jaya algorithm is superior to other proposed modified algorithms when the function evaluations are higher. The performances of the SAP-Rao and elitist Rao algorithms are superior to that of the AMTPG-Jaya algorithm for the problems where fewer function evaluations are considered as a termination criterion.

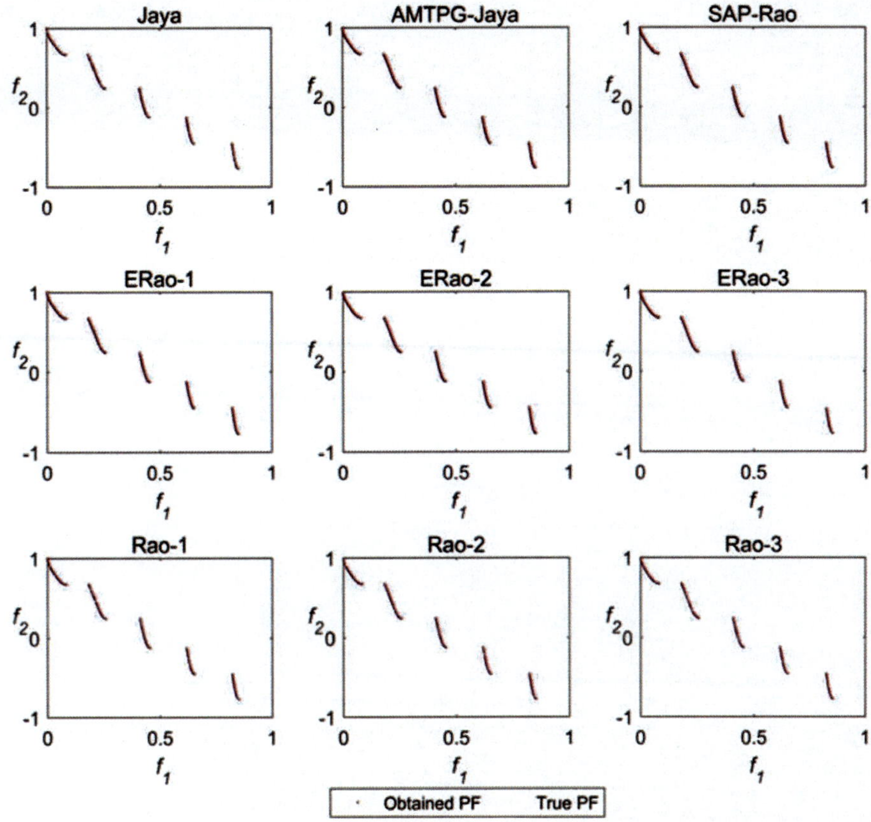

Fig. 4.19 True Pareto-front (PF) and the obtained Pareto-fronts in the objective space of the ZDT-3 problem

However, in both scenarios, the performances of the proposed algorithms are superior or competitive to those of the other algorithms compared. Hence, the modified versions of the Jaya and Rao algorithms can be considered as the improved versions of the Jaya and Rao algorithms, respectively.

The next chapter presents the multi-attribute decision-making methods and their implementation in multi-objective optimization scenarios.

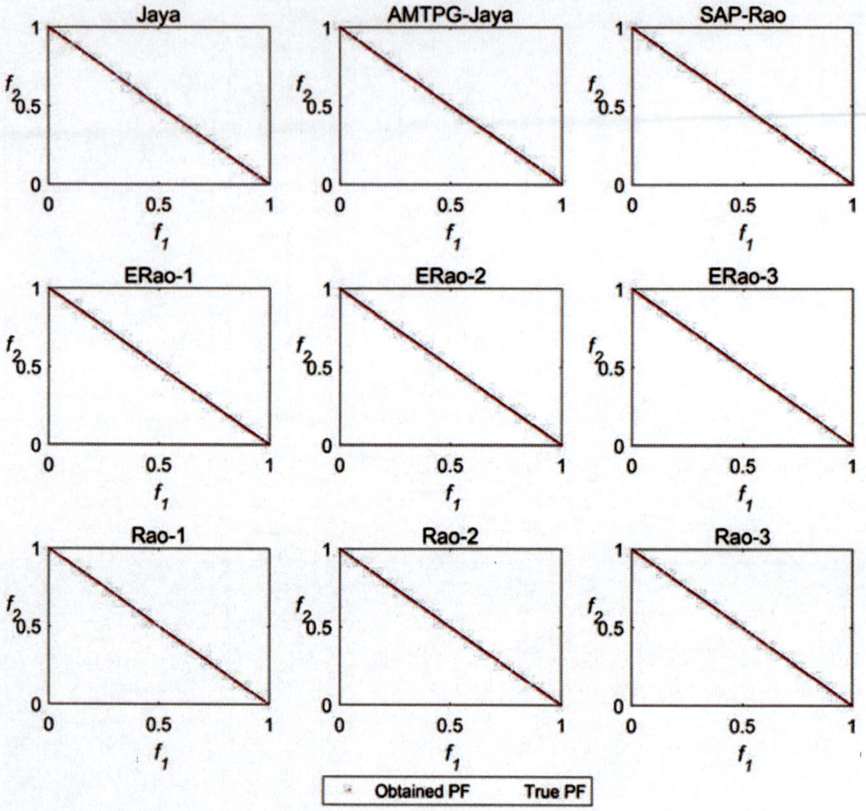

Fig. 4.20 True Pareto-front (PF) and the obtained Pareto-fronts in the objective space of the ZDT-1L problem

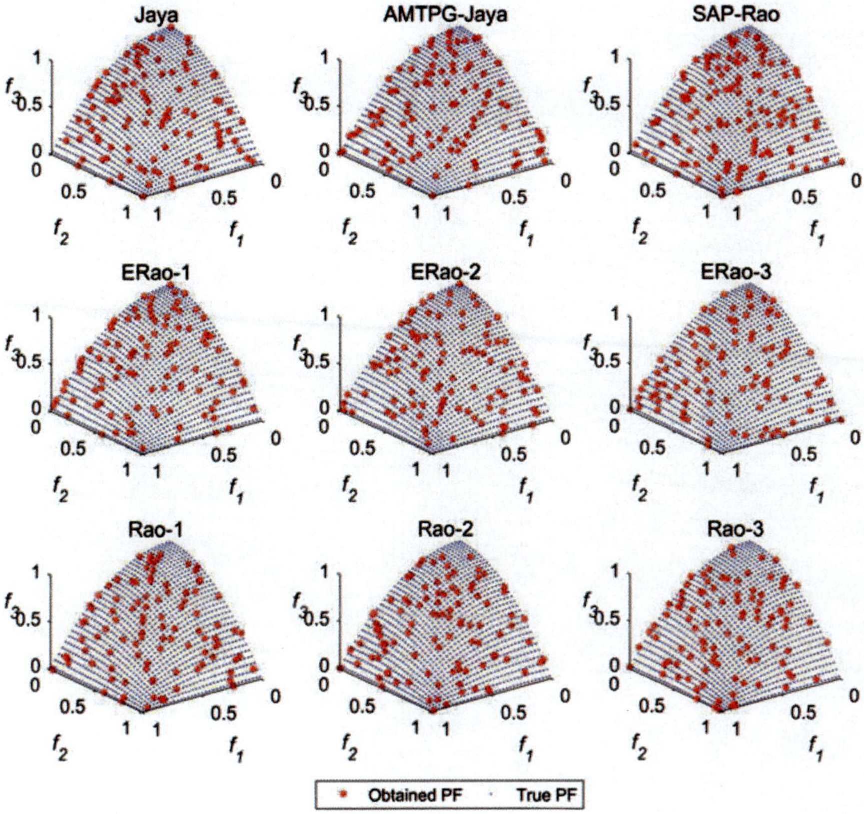

Fig. 4.21 True Pareto-front (PF) and the obtained Pareto-fronts in the objective space of the ZDT2-3O problem

Table 4.16 Friedman statistical test results for the results presented in Tables 4.11, 4.12, 4.13, 4.14 and 4.15 corresponding to the unconstrained multi-objective benchmark problems

Algorithm	MOGOA	MOPSO	NSGA-II	MODA	MOALO	Jaya	Rao-1	Rao-2	Rao-3	ERao-1	ERao-2	ERao-3	AMTPG-Jaya	SAP-Rao
Avg. Rank	11.6	9.6	14	10.4	12.6	8.2	7.4	7.6	7.2	4.2	3	2.4	4.2	2.6
χ^2	54.44													
p-value	The *p*-value is **5.12859E−7**. The result is significant at $p < 0.01$													

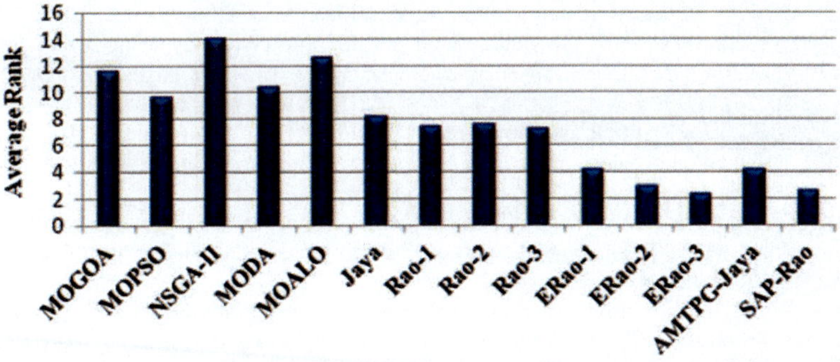

Fig. 4.22 Average ranks of various algorithms in the Friedman statistical test for the ZDT test functions results

References

Deb, K., Pratap, A., Agarwal, S., & Meyarivan, T. (2002). A fast and elitist multiobjective genetic algorithm: NSGA-II. *IEEE Transactions on Evolutionary Computation, 6*(2), 182–197. https://doi.org/10.1109/4235.996017

Eberhart, R., & Kennedy, J. (1995). A new optimizer using particle swarm theory. *MHS'95. Proceedings of the Sixth International Symposium on Micro Machine and Human Science*, 39–43. https://doi.org/10.1109/MHS.1995.494215

Haupt, R. L., & Haupt, S. E. (2003). *Practical Genetic Algorithms*. 2004 John Wiley & Sons, Inc. https://doi.org/10.1002/0471671746

Karaboga, D., & Akay, B. (2009). A comparative study of artificial bee colony algorithm. *Applied Mathematics and Computation, 214*(1), 108–132. https://doi.org/10.1016/j.amc.2009.03.090

Liang, J. J., Qin, A. K., Suganthan, P. N., & Baskar, S. (2006). Comprehensive learning particle swarm optimizer for global optimization of multimodal functions. *IEEE Transactions on Evolutionary Computation, 10*(3), 281–295. https://doi.org/10.1109/TEVC.2005.857610

Mendes, R., Kennedy, J., & Neves, J. (2004). The fully informed particle swarm: simpler, maybe better. *IEEE Transactions on Evolutionary Computation, 8*(3), 204–210. https://doi.org/10.1109/TEVC.2004.826074

Mirjalili, S. (2016). Dragonfly algorithm: a new meta-heuristic optimization technique for solving single-objective, discrete, and multi-objective problems. *Neural Computing and Applications, 27*(4), 1053–1073. https://doi.org/10.1007/s00521-015-1920-1

Mirjalili, S., Jangir, P., & Saremi, S. (2017). Multi-objective ant lion optimizer: a multi-objective optimization algorithm for solving engineering problems. *Applied Intelligence, 46*(1), 79–95. https://doi.org/10.1007/s10489-016-0825-8

Mirjalili, S. Z., Mirjalili, S., Saremi, S., Faris, H., & Aljarah, I. (2018). Grasshopper optimization algorithm for multi-objective optimization problems. *Applied Intelligence, 48*(4), 805–820. https://doi.org/10.1007/s10489-017-1019-8

Montes de Oca, M. A., Stutzle, T., Birattari, M., & Dorigo, M. (2009). Frankenstein's PSO: a composite particle swarm optimization algorithm. *IEEE Transactions on Evolutionary Computation, 13*(5), 1120–1132. https://doi.org/10.1109/TEVC.2009.2021465

Ngo, T. T., Sadollah, A., & Kim, J. H. (2016). A cooperative particle swarm optimizer with stochastic movements for computationally expensive numerical optimization problems. *Journal of Computational Science, 13*, 68–82. https://doi.org/10.1016/j.jocs.2016.01.004

Nickabadi, A., Ebadzadeh, M. M., & Safabakhsh, R. (2011). A novel particle swarm optimization algorithm with adaptive inertia weight. *Applied Soft Computing, 11*(4), 3658–3670. https://doi.org/10.1016/j.asoc.2011.01.037

Ocłoń, P., Cisek, P., Rerak, M., Taler, D., Rao, R. V., Vallati, A., & Pilarczyk, M. (2018). Thermal performance optimization of the underground power cable system by using a modified Jaya algorithm. *International Journal of Thermal Sciences, 123*, 162–180. https://doi.org/10.1016/j.ijthermalsci.2017.09.015

Rao, R. V. (2016). Jaya: A simple and new optimization algorithm for solving constrained and unconstrained optimization problems. *International Journal of Industrial Engineering Computations, 7*(1), 19–34. https://doi.org/10.5267/j.ijiec.2015.8.004

Rao, R. V. (2019). Jaya: An Advanced Optimization Algorithm And Its Engineering Applications. Springer International Publishing AG, part of Springer Nature. (2019). Springer. *Cham.* https://doi.org/10.1007/978-3-319-78922-4

Rao, R. V. (2020). Rao algorithms: Three metaphor-less simple algorithms for solving optimization problems. *International Journal of Industrial Engineering Computations, 11*, 107–130. https://doi.org/10.5267/j.ijiec.2019.6.002

Rao, R. V., & Keesari, H. S. (2018). Multi-team perturbation guiding Jaya algorithm for optimization of wind farm layout. *Applied Soft Computing Journal, 71*, 800–815. https://doi.org/10.1016/j.asoc.2018.07.036

Rao, R. V., & Keesari, H. S. (2019). Solar assisted heat engine systems: multi-objective optimisation and decision making. *International Journal of Ambient Energy, 0*(0), 1–27.https://doi.org/10.1080/01430750.2019.1636870

Rao, R. V., & Keesari, H. S. (2020). Rao algorithms for multi-objective optimization of selected thermodynamic cycles. *Engineering with Computers.* https://doi.org/10.1007/s00366-020-01008-9

Rao, R. V., & Keesari, H. S. (2021). A self-adaptive population Rao algorithm for optimization of selected bio-energy systems. *Journal of Computational Design and Engineering, 8*(1), 69–96. https://doi.org/10.1093/jcde/qwaa063

Rao, R. V., Keesari, H. S., Oclon, P., & Taler, J. (2019). Improved multi-objective Jaya optimization algorithm for a solar dish Stirling engine. *Journal of Renewable and Sustainable Energy, 11*(2), 25903. https://doi.org/10.1063/1.5083142

Rao, R. V., Keesari, H. S., Oclon, P., & Taler, J. (2020). An adaptive multi-team perturbation-guiding Jaya algorithm for optimization and its applications. *Engineering with Computers, 36*(1), 391–419. https://doi.org/10.1007/s00366-019-00706-3

Rao, R. V., Keesari, H. S., Taler, D., Taler, J., & Ocłoń, P. (2020b). Multi-objective optimization of a solar-assisted Stirling heat engine system using elitist Rao algorithms. In B. Samojeden (Ed.), *Energy Fuels Environment 2020, Kraków (Poland), 1–4 December 2020 : book of abstracts* (p. 77). Kraków: Faculty of Energy and Fuels, AGH University of Science and Technology.

Rao, R. V., & Patel, V. (2013). Comparative performance of an elitist teaching-learning-based optimization algorithm for solving unconstrained optimization problems. *International Journal of Industrial Engineering Computations, 4*(1), 29–50. https://doi.org/10.5267/j.ijiec.2012.09.001

Rao, R. V., & Rai, D. P. (2017). Optimisation of welding processes using quasi-oppositional-based Jaya algorithm. *Journal of Experimental & Theoretical Artificial Intelligence, 29*(5), 1099–1117. https://doi.org/10.1080/0952813x.2017.1309692

Rao, R. V., Rai, D. P., & Balic, J. (2017). A multi-objective algorithm for optimization of modern machining processes. *Engineering Applications of Artificial Intelligence, 61*(August 2015), 103–125. https://doi.org/10.1016/j.engappai.2017.03.001

Rao, R. V., & Saroj, A. (2017). A self-adaptive multi-population based Jaya algorithm for engineering optimization. *Swarm and Evolutionary Computation, 37*, 1–26. https://doi.org/10.1016/j.swevo.2017.04.008

Rashedi, E., Nezamabadi-pour, H., & Saryazdi, S. (2009). GSA: A gravitational search algorithm. *Information Sciences, 179*(13), 2232–2248. https://doi.org/10.1016/j.ins.2009.03.004

Schott, J. R. (1995). *Fault Tolerant Design Using Single and Multicriteria Genetic Algorithm Optimization.* Thesis, Massachusetts Institute of Technology, Boston, MA.

vanden Bergh, F., & Engelbrecht, A. P. (2004). A cooperative approach to particle swarm optimization. *IEEE Transactions on Evolutionary Computation, 8*(3), 225–239. https://doi.org/10.1109/TEVC.2004.826069

Yu, K., Liang, J. J., Qu, B. Y., Chen, X., & Wang, H. (2017). Parameters identification of photovoltaic models using an improved JAYA optimization algorithm. *Energy Conversion and Management, 150*, 742–753. https://doi.org/10.1016/j.enconman.2017.08.063

Zhou, A., Zhang, Q., & Jin, Y. (2009). Approximating the set of pareto-optimal solutions in both the decision and objective spaces by an estimation of distribution algorithm. *IEEE Transactions on Evolutionary Computation, 13*(5), 1167–1189. https://doi.org/10.1109/TEVC.2009.2021467

Zitzler, E., Deb, K., & Thiele, L. (2000). Comparison of multiobjective evolutionary algorithms: Empirical Results. *Evolutionary Computation, 8*(2), 173–195. https://doi.org/10.1162/106365600568202

Chapter 5
Multi-attribute Decision-Making Methods and Their Implementation in Energy Systems

Check for updates

Abstract This chapter presents multi-attribute decision-making methods and their implementation in energy systems' optimization for identifying the best compromise solution from the Pareto-fronts.

Decision-making in the presence of multiple conflicting criteria can be considered as multiple criteria decision-making (MCDM). Based on the problem type, the MCDM problems can be divided into multiple attribute decision-making (MADM) problems and multiple objective decision-making (MODM) problems. The MODM problems involve finding the best-suited solution from an infinitive or a large number of choices, which satisfies problem constraints and priorities. The MADM problems involve selecting the best choice from a limited number of alternatives. A MADM method specifies how the attribute information is to be processed in order to arrive at a choice. MADM methods require both inter- and intra-attribute comparisons and involve appropriate explicit trade-offs. This chapter presents the multi-attribute decision-making methods and their implementation in multi-objective optimization.

Identifying the best Pareto-optimal solution from a Pareto-front is treated as a multiple attribute decision-making problem (Rao & Keesari, 2019) in this book. MADM methods are a class of MCDM methods which are employed when a limited number of alternatives are available. These methods suggest the best alternative using the attribute information of various alternatives. The set of Pareto-optimal solutions are regarded as alternatives (A_i ∀ alternatives $i = 1, 2 \ldots N$), and the objective functions are regarded as attributes (B_j ∀ attributes $j = 1, 2 \ldots M$). The objective value of a non-dominated solution i corresponding to an objective or attribute j can be considered as the performance measure (m_{ij}). The decision-maker can appraise the relative importance or weights (w_j) of the objectives such that the summation of weights of all the attributes together is equal to 1. The objective functions considered in this book have different units. Thus, all the objective values must be normalized (nm_{ij}) to the same units before applying the MADM methods. Based on the potential to find the best alternative, the following methods have been considered in this book (Rao, 2007, 2013).

V. R. Ravipudi et al., *Design Optimization of Renewable Energy Systems Using Advanced Optimization Algorithms*, Green Energy and Technology,
https://doi.org/10.1007/978-3-030-95589-2_5

5.1 Simple Additive Weighing (SAW)

Fishburn (1967) developed the SAW method. For the given weights of the attributes, the cumulative performance (P_i) of alternative i is given by the following equation:

$$P_i = \sum_{j=1}^{M} W_j \left(\mathrm{nm}_{\mathrm{ij}}\right) \tag{5.1}$$

where $\mathrm{nm}_{\mathrm{ij}}$ is the normalized measure of performance, and the best alternative is the one with the highest P_i value.

5.2 Weighted Product Method (WPM)

Miller and Starr (1969) developed the WPM method. For the given weights of the attributes, the cumulative performance (P_i) of alternative i is given by the following equation:

$$P_i = \prod_{j=1}^{M} \left(\mathrm{nm}_{\mathrm{ij}}\right)^{w_j} \tag{5.2}$$

where $\mathrm{nm}_{\mathrm{ij}}$ is the normalized measure of performance, and the best alternative is the one with the highest P_i value.

5.3 Preference Ranking Organization Method for Enrichment Evaluation (PROMETHEE)

Brans et al. (1986) developed the PROMETHEE method, which belongs to the class of outranking methods. In this method, for each objective function, the Pareto-optimal solutions are compared pair-wise to determine the strength of a solution a_i over the solution a_k. For the given weights of the objectives, in pair-wise comparison of the solutions usual preference function ($P_{j,\ \mathrm{aiak}}$) is considered in this book. The usual preference function is the difference between the values of solutions a_i and a_k for an objective b_j. Now, the objective preference (Π_{aiak}) index is given by:

$$\Pi_{\mathrm{ajak}} = \sum_{j=1}^{M} w_j P_{j,\,\mathrm{aiak}} \tag{5.3}$$

The value of Π_{ajak} varies between zero and 1, and it denotes the intensity of the inclination toward the solution a_i over a_k, when concurrently all the objectives are compared. Now, for a solution a_i, the outranking relations are calculated using the following equations:

$$\text{Leaving flow}\left(\varnothing_i^+\right) = \sum_{x \in A} \Pi_{a_x a_i} \tag{5.4}$$

$$\text{Entering flow}\left(\varnothing_i^-\right) = \sum_{x \in A} \Pi_{a_i a_x} \tag{5.5}$$

$$\text{Net flow } \left(\varnothing(a_i)\right) = \left(\varnothing_i^+\right) - \left(\varnothing_i^-\right) \tag{5.6}$$

The leaving flow denotes the supremacy of solution a_i over all other solutions. The entering flow denotes the degree to which all other solutions are dominating the solution a_i. The net flow denotes the outranking relationship of the solutions, and higher values of the net flow mean the best Pareto-optimal solution. For more details about the PROMETHEE method, the readers may refer to Rao (2013).

5.4 Technique for Order Preference by Similarity to Ideal Solution (TOPSIS)

The TOPSIS method developed by Hwang and Yoon (1981) provides the relative similarity index (SI) near to an ideal and non-ideal solution. The ideal and non-ideal solutions represent the hypothetical best and worst scenarios of all the objectives, respectively. The normalized values (nm_{ij}) of the objective function values are calculated by using the following equation:

$$\text{nm}_{\text{ij}} = m_{\text{ij}} / \sqrt{\left(\sum_{j=1}^{M} m_{\text{ij}}^2\right)} \tag{5.7}$$

Let R_{ij} be the matrix of normalized values and w_j be the weights of the objective functions, and then the weighted normalized matrix can be calculated by using the following equation:

$$V_{\text{ij}} = w_j R_{\text{ij}} \tag{5.8}$$

Now, identify the ideal (O_j^{ideal}) and non-ideal ($O_j^{\text{non - ideal}}$) solutions with respect to all objective functions using the following criteria, respectively. For a maximization objective function (j), the solution with maximum V_{ij} value will be considered as an ideal solution and the solution with minimum V_{ij} value will be considered as the non-ideal solution. Similarly, for minimization objective function (j), the solution with minimum V_{ij} value will be considered as an ideal solution and the solution with maximum V_{ij} value will be considered as the non-ideal solution. Then, calculate the Pareto-optimal solution distances (d_i) from the ideal and non-ideal solution using the following equations respectively:

$$d_i^{\text{ideal}} = \sqrt{\sum_{j=1}^{M} \left(V_{ij} - O_j^{\text{ideal}}\right)^2} \tag{5.9}$$

$$d_i^{\text{non - ideal}} = \sqrt{\sum_{j=1}^{M} \left(V_{ij} - O_j^{\text{non - ideal}}\right)^2} \tag{5.10}$$

where $i = 1, 2 \ldots N$ (*i.e.*, number of Pareto-optimal candidate solutions or alternatives in a Pareto frontier), $j = 1, 2 \ldots M$ (*i.e.*, the number of objectives or attributes of the Pareto frontier). Now, the similarity index can be calculated using the following equation:

$$\text{SI} = \frac{d_i^{\text{non - ideal}}}{d_i^{\text{ideal}} + d_i^{\text{non - ideal}}} \tag{5.11}$$

If the similarity index of a Pareto-optimal solution is higher than that of all other solutions in the Pareto frontier, then it is the nearest solution to the ideal solution. Similarly, if the similarity index of a Pareto-optimal solution is lower than that of other solutions in the Pareto frontier, then it is the nearest solution to a non-ideal solution.

5.5 Modified TOPSIS (MTOPSIS)

Deng et al. (2000) proposed a modification to the TOPSIS method by suggesting weighted Euclidean distances. In this modified version, the ideal and non-ideal solutions are selected from the normalized matrix (R_{ij}) and the weighted Euclidean distances are calculated from the ideal and non-ideal solutions. According to the modified TOPSIS (MTOPSIS) method, the Pareto-optimal solution distances (d_i) from the ideal and non-ideal solution are calculated by the following equations:

$$d_i^{\text{ideal}} = \sqrt{\sum_{j=1}^{M} w_j \left(\text{nm}_{ij} - \text{nm}_j^{\text{ideal}}\right)^2} \tag{5.12}$$

$$d_i^{\text{non - ideal}} = \sqrt{\sum_{j=1}^{M} w_j \left(\text{nm}_{\text{ij}} - \text{nm}_j^{\text{non - ideal}}\right)^2} \quad (5.13)$$

Now, the similarity index is calculated using Eq. (5.11). The solution with the highest similarity index will be considered as the best solution, and the solution with the lowest similarity index will be considered as the relatively worst solution.

5.6 Compromise Ranking Method (VIKOR)

The idea of compromise ranking was introduced by Yu (1973) and Zeleny (1982) and later advanced by Opricovic and Tzeng (2002, 2003). This method is also called as VIšekriterijumsko KOmpromisno Rangiranje (VIKOR). The procedure for VIKOR is described below:

Step 1: Identify the ideal ($m_{\text{ij}}^{\text{ideal}}$) and non-ideal ($m_{\text{ij}}^{\text{non - ideal}}$) solutions for all the attributes based on the objective function values of the solutions for respective attributes using the following criteria. For a maximization objective function (*j*), the solution with maximum m_{ij} value will be considered as the ideal solution and the solution with minimum m_{ij} value will be considered as the non-ideal solution. Similarly, for minimization objective function (*j*), the solution with minimum m_{ij} value will be considered as the ideal solution and the solution with maximum m_{ij} value will be considered as the non-ideal solution.

Step 2: Calculate E_i and F_i values for all Pareto-optimal solutions in the Pareto-front.

$$E_i = \sum_{j=1}^{M} w_j \left[m_{\text{ij}}^{\text{ideal}} - m_{\text{ij}}/m_{\text{ij}}^{\text{ideal}} - m_{\text{ij}}^{\text{non - ideal}}\right] \quad (5.14)$$

For Pareto-optimal solution *i*,

$$F_i = \text{maximum of}\left\{w_j \left[m_{\text{ij}}^{\text{ideal}} - m_{\text{ij}}/m_{\text{ij}}^{\text{ideal}} - m_{\text{ij}}^{\text{non - ideal}}\right] | j = 1, 2 \ldots M\right\} \quad (5.15)$$

Step 3: Calculate the performance index (P_i) using the following equation:

$$P_i = V\left[\left(E_i - E_i^-\right)/\left(E_i^+ - E_i^-\right)\right] + (1 - V)\left[\left(F_i - F_i^-\right)/\left(F_i^+ - F_i^-\right)\right] \quad (5.16)$$

where E_i^+ and E_i^- are the maximum and minimum values of E_i, respectively. Similarly, F_i^+ and F_i^- are the maximum and minimum values of F_i, respectively. V is taken as 0.5.

Step 4: The solution with the least P_i will be considered as the best alternative if it satisfies the following conditions (Tzeng et al., 2005).

1. Condition-1 (Acceptable advantage): $P_{i\,\min} - P_{i\,\text{next}\min} \geq (1/(N-1))$.
2. Condition-2 (Acceptable stability in decision-making): The solution with least P_i must also be the best solution according to the E_i and/or F_i values. The solution with least E_i values is considered best according to E_i values, and the solution with least F_i value is considered as best according to F_i values.
3. Condition-3: If these conditions failed, then a set of compromise solutions is proposed based on the following criteria.
   - The solutions $P_{i\,\min}, P_{i\,\text{next}\min} \cdots P_{\text{ik}}$ are compromise solutions if condition-1 failed, where $P_{\text{ik}} - P_{i\,\text{next}\min} \approx (1/(N-1))$.
   - The solutions $P_{i\,\min}$ and $P_{i\,\text{next}\min}$ are compromise solutions if only condition-2 fails.

5.7 Complex Proportional Assessment (COPRAS)

The complex proportional assessment method was introduced by Edmundas Kazimieras Zavadskas in 1994 (Hajiagha et al., 2013). This method is simple to use and stepwise procedure for ranking and evaluating the alternatives. The stepwise procedure of COPRAS method is as follows:

The normalized values (nm_{ij}) of the objective function values are calculated by using the following equation:

$$\text{nm}_{\text{ij}} = m_{\text{ij}} / \left(\sum_{j=1}^{N} m_{\text{ij}} \right) \tag{5.17}$$

Let R_{ij} be the matrix of normalized values and w_j be the weights of the objective functions. Then, the weighted normalized matrix can be calculated by using the following equation:

$$V_{\text{ij}} = w_j R_{\text{ij}} \tag{5.18}$$

Now, calculate the summation P_i, only for the maximization objectives using the following equation:

$$P_i = \sum_{j=1}^{k} V_{\text{ij}} \tag{5.19}$$

where k is the number of maximization objectives. Now, calculate summation R_i, only for minimization objectives using the following equation:

$$R_i = \sum_{j=k+1}^{M} V_{ij} \tag{5.20}$$

Now, calculate the relative weight of each solution (Q_i) using the following equation:

$$Q_i = P_i + \left[\left(\sum_{i}^{M} R_i\right) \Big/ \left(R_i \sum_{i}^{M} (1/R_i)\right)\right] \tag{5.21}$$

Now, calculate the degree of utility (U_i) for each solution using the following equation:

$$U_i = (Q_i / Q_{i\,\max}) \times 100\,\% \tag{5.22}$$

The solution with a high degree of utility will be considered as the best solution.

5.8 Gray Relational Analysis (GRA)

Gray relational analysis is one of the derived evaluation methods based on the concept of gray relational space (GRS). The normalized values ($\mathrm{nm_{ij}}$) of the objective function values are calculated by using the following equations:

$$\mathrm{nm_{ij}} = \frac{\mathrm{m_{ij}} - \min\{\mathrm{m_{ij}}\}}{\max\{\mathrm{m_{ij}}\} - \min\{\mathrm{m_{ij}}\}} \text{For maximization} \tag{5.23}$$

$$\mathrm{nm_{ij}} = \frac{\max\{\mathrm{m_{ij}}\} - m_{\mathrm{ij}}}{\max\{m_{\mathrm{ij}}\} - \min\{m_{\mathrm{ij}}\}} \text{For minimization} \tag{5.24}$$

Now, calculate the gray relational coefficient (*GR*) using the following equation:

$$\mathrm{GR}\left(\mathrm{nm}_{0j}, \mathrm{nm_{ij}}\right) = \frac{\Delta_{\min} - \xi\,\Delta_{\max}}{\Delta_{ij} - \xi\,\Delta_{\max}} \tag{5.25}$$

where

$\Delta_{\mathrm{ij}} = \left|\mathrm{nm}_{0j} - \mathrm{nm_{ij}}\right|$

$\mathrm{nm}_{0j} \in$ reference sequence$\{\mathrm{nm}_{01}, \mathrm{nm}_{02}, \ldots, \mathrm{nm}_{0M}\}$

$\Delta_{\min} = \min\left\{\Delta_{\mathrm{ij}}, i = 1, 2, \ldots, N; j = 1, 2, \ldots, M\right\}$

$\Delta_{\max} = \max\left\{\Delta_{\mathrm{ij}}, i = 1, 2, \ldots, N; j = 1, 2, \ldots, M\right\}$

Distinguishing coefficient (ξ) $= 0.5$

Now, the gray relational grade can be calculated by the following equation:

$$\Gamma(\mathrm{nm}_0, \mathrm{nm}_i) = \sum_{j=1}^{M} w_j \mathrm{GR}(\mathrm{nm}_{0j}, \mathrm{nm}_{\mathrm{ij}}) \quad \text{for } i = 1, 2, \ldots, N \tag{5.26}$$

If a solution has the highest gray relational grade with the reference sequence, it means that the comparability sequence is most similar to the reference sequence, and that solution would be the best choice.

In the implementation of the MADM methods, the case studies' objective functions are considered as attributes and Pareto-optimal solutions as alternatives. In all the multi-objective optimization case studies, equal weights are taken for all the objectives. The MADM methods use different characteristics to rank the alternatives, which indicate that these methods rank the solutions differently. Therefore, to identify the final best solution, average ranks have been calculated. The average rank for a solution is nothing but the average value of the ranks given by different MADM methods.

Furthermore, to assess the correlation and strength of the ranking given by the MADM methods, Spearman's correlation coefficient is calculated between different pairs of methods. The Spearman correlation coefficient is 1 when the ranking given by two MADM methods is identical and -1 when the ranking given by two MADM methods is the opposite. If the Spearman correlation coefficient between two MADM methods is near 1, the ranking of those two methods is similar to each other. However, in this book, the ranking given by different MADM methods is considered for calculating average ranks based on the following conditions:

1. If a MADM method has a positive Spearman correlation coefficient greater than or equal to 0.5 with all other methods, then it is considered for calculating average ranks.
2. If a pair of MADM methods have the Spearman correlation coefficient value lesser than 0.5 (including negative values), then the MADM method, which has a better Spearman correlation coefficient value with more methods, is considered for the average ranking. For example, the SAW-WPM pair has a Spearman correlation less than 0.5. The SAW method has better correlation values (greater than or equal to 0.5) with TOPSIS, MTOPSIS, VIKOR, PROMETHEE, and GRA methods. The WPM method has better correlation values (greater than or equal to 0.5) with TOPSIS, MTOPSIS, VIKOR, and COPRAS methods. Then, the ranks given by the SAW method will be considered for calculating the average rank, and the ranks given by the WPM method will not be considered for calculating the average rank because the SAW method has better correlation values with five methods, whereas the WPM method has a better correlation value with four methods. Furthermore, if both methods have better correlation values (greater than or equal to 0.5) with the same number of methods, then the correlation values are compared, and the method with better correlation values will be considered.

The best Pareto-optimal solution identified based on the average rank is then similarly compared with the solutions identified for the other algorithms. The next chapter presents the applications of the Jaya and Rao algorithms, along with their improved versions, to wind farm layout optimization problems.

References

Brans, J. P., Vincke, P., & Mareschal, B. (1986). How to select and how to rank projects: The Promethee method. *European Journal of Operational Research, 24*, 228–238. https://doi.org/10.1016/0377-2217(86)90044-5

Deng, H., Yeh, C. H., & Willis, R. J. (2000). Inter-company comparison using modified TOPSIS with objective weights. *Computers and Operations Research, 27*, 963–973. https://doi.org/10.1016/S0305-0548(99)00069-6

Fishburn, P. C. (1967). Additive utilities with incomplete product sets: Application to priorities and assignments. *Operations Research, 15*(3), 537–542. https://doi.org/10.1287/opre.15.3.537

Hajiagha, S. H. R., Hashemi, S. S., & Zavadskas, E. K. (2013). A complex proportional assessment method for group decision making in an interval-valued intuitionistic fuzzy environment. *Technological and Economic Development of Economy, 19*(1), 22–37. https://doi.org/10.3846/20294913.2012.762953

Hwang, C.-L., & Yoon, K. (1981). *Multiple Attribute Decision Making: Methods and Applications A State-of-the-Art Survey.* https://doi.org/10.1007/978-3-642-48318-9

Miller, D. W., & Starr, M. K. (1969). *Executive decisions with operations research.* Prentice Hall, Englewood Cliffs.

Opricovic, S., & Tzeng, G. H. (2002). Multicriteria planning of post-earthquake sustainable reconstruction. *Computer-Aided Civil and Infrastructure Engineering, 17*, 211–220. https://doi.org/10.1111/1467-8667.00269

Opricovic, S., & Tzeng, G.-H. (2003). Fuzzy multicriteria model for postearthquake land-use planning. *Natural Hazards Review, 4*, 59–64. https://doi.org/10.1061/(asce)1527-6988(2003)4:2(59)

Rao, R. V. (2007). Decision making in the manufacturing environment using graph theory and fuzzy multiple attribute decision making methods. *Springer Series in Advanced Manufacturing, Springer-Verlag, London, UK.* https://doi.org/10.1007/978-1-84628-819-7

Rao, R. V. (2013). Decision making in the manufacturing environment using graph theory and fuzzy multiple attribute decision making methods (Volume 2). *Springer Series in Advanced Manufacturing, Springer-Verlag, London, UK.* https://doi.org/10.1007/978-1-4471-4375-8_1

Rao, R. V., & Keesari, H. S. (2019). Solar assisted heat engine systems: multi-objective optimisation and decision making. *International Journal of Ambient Energy, 0*(0), 1–27. https://doi.org/10.1080/01430750.2019.1636870.

Tzeng, G. H., Lin, C. W., & Opricovic, S. (2005). Multi-criteria analysis of alternative-fuel buses for public transportation. *Energy Policy, 33*, 1373–1383. https://doi.org/10.1016/j.enpol.2003.12.014

Yu, P. L. (1973). A class of solutions for group decision problems. *Management Science, 19*, 936–946. https://doi.org/10.1287/mnsc.19.8.936

Zeleny, M. (1982). *Multiple criteria decision making.* McGraw Hill.

Chapter 6
Optimization of Wind Farm Layouts

Abstract This chapter presents the applications of different versions of Jaya and Rao algorithms to wind farm layout optimization. Three scenarios of wind farm layout optimization (WFLO) problem are considered, which are fixed wind speed and direction (case-I); fixed wind speed with changing wind direction (case-II); changing wind speed with changing direction (case-III). The computational results of the proposed algorithms are compared with those of the GA and its variant, TLBO and its variants, ABC algorithm, and other algorithms from the literature.

6.1 Problem Definition and Wind Scenarios of the Wind Farm Layout Optimization

The optimal placing of wind turbines results in better mean incident velocity for a wind farm; thus, more power is produced with the same investment. In the optimization of the WFLO problem, each wind turbine location is taken as an independent variable for a specified farmland size and number of turbines. The characteristics of the wind turbines considered in this book are shown in Table 6.1.

The present work considers a square region of $2 \times 2\ \mathrm{km}^2$ in which a wind turbine can be placed at any position in the specified area, maintaining the minimum distance ($5d = 200$ m) between two adjacent turbines. The surface roughness of the region is assumed as 0.3 m. The C_T is considered constant throughout the processes (Grady et al., 2005). In optimizing the WFLO problem, three different cases of WFLO are considered, which are: fixed wind speed and fixed direction (case-I), fixed wind speed and variable direction (case-II), and variable wind speed and variable direction (case-III). In each case, optimal layouts are obtained by minimizing the cost per unit power. Along with the minimum value of the objective function, total power (KW) and farm efficiency are also calculated in computational experiments.

Computational experiments are performed by the Jaya algorithm and its modified versions (MTPG-Jaya and AMTPG-Jaya) and Rao algorithms and their modified versions (elitist Rao and SAP-Rao algorithms). In the implementation of the Jaya algorithm, the wind farm area considered is divided into 100 possible turbine locations as a square grid. A wind turbine can be placed at the center of every cell in

V. R. Ravipudi et al., *Design Optimization of Renewable Energy Systems Using Advanced Optimization Algorithms*, Green Energy and Technology,
https://doi.org/10.1007/978-3-030-95589-2_6

Table 6.1 Wind turbine properties

Property	Value
Hub height (*Z*)	60 m
Rotor diameter (*d*)	40 m
Thrust coefficient (C_T)	0.88
Ground roughness (Z_0)	0.3 m
Wind velocity (u_0)	12 m/s
Axial induction factor (*a*)	0.33
Entrainment constant (*k*)	0.094

the square grid. In the implementation of the remaining algorithms, the turbines are placed in the wind farm, restricting the minimum distance between the adjacent turbines to 200 m.

The population size is varied between 10 and 100 for different algorithms in the computational experiments. For the MTPG-Jaya and AMTPG-Jaya algorithms after performing few trials with different configurations, the probability of accepting the worst solution and maximum movement iterations is chosen as 0.35 and 10, respectively. Four teams are taken for the MTPG-Jaya algorithm. The elite size for the elitist Rao algorithms is varied from 10 to 40%. The number of teams and the population size are selected suitable for each case. Each candidate solution gives one layout for the WFLO problem. The computational results are compared with the well-established algorithms such as GA, lazy greedy algorithm (LGA), ABC, TLBO, TLBO enhanced with ABC (AL), and TLBO enhanced with ABC and PSO (PAL). The GA results are taken from Grady et al. (2005), and the results of LGA are taken from Changshui et al. (2011). The TLBO results in case-I are taken from Patel et al. (2015). In case-II and case-III, results for the TLBO, ABC, AL, and PAL are taken from Patel et al. (2017). In the next sub-section, the analysis of computational results related to WFLO case-I is presented.

6.2 Case-I: Fixed Wind Speed and Fixed Direction

It is the case of unidirectional wind at a steady speed. The incident wind speed is taken as 12 m/s in the normal direction. The wind speed inside the wake effect of a turbine is reduced. Thus, the incident wind speed for a turbine, which is placed behind an upstream turbine, will be modified. The incident wind speed of a turbine will not be affected by the wake effect of its adjacent wind turbine. In this book, optimization is carried out with the 30 turbines configuration. The comparative performance results are presented in the form of the total power (kW), farm efficiency, and objective value (cost per unit power generated). The TLBO algorithm results are presented for 9000 function evaluations, which are least when compared to GA (180,000). Thus, Jaya and Rao algorithms, including their modified versions, are tested with 9000

Table 6.2 Results obtained by the various algorithms in case-I with 30 turbines

Algorithm	Total power (kW/year)	Farm efficiency	Objective value
GA	14,310	92.015	0.0015436
LGA	14,310	92.01	0.0015436
TLBO	14,310	92.01	0.0015436
Jaya	14,310	92.01	0.0015436
MTPG-Jaya	14,601.0174	93.88514275	0.001512825
AMTPG-Jaya	14,606.10621	93.91786401	0.001512298
Rao-1	14,503.42196	93.25760006	0.001523005
Rao-2	14,470.30126	93.04463259	0.001526491
Rao-3	14,514.64737	93.32977991	0.001521828
ERao-1	14,528.11154	93.41635505	0.001520417
ERao-2	14,544.37374	93.52092171	0.001518717
ERao-3	14,549.54668	93.55418388	0.001518177
SAP-Rao	14,689.3961	94.45342116	0.001503724

Source GA-Grady et al. (2005), LGA-Changshui et al. (2011), TLBO- Patel et al. (2015)
Result in boldface indicates a better performing algorithm

function evaluations. Computational results of various algorithms for the case-I with 30 turbines are summarized in Table 6.2.

Here, an observation can be made that the Jaya algorithm and its modified versions and Rao algorithms and their modified versions have obtained better layouts compared to the layouts achieved by the GA, LGA, and TLBO algorithms. The graphical representation of the total power output of the solutions achieved by various algorithms in this case study is presented in Fig. 6.1. The SAP-Rao algorithm has achieved the most efficient layout, producing 14,689.39 kW power with 94.45%

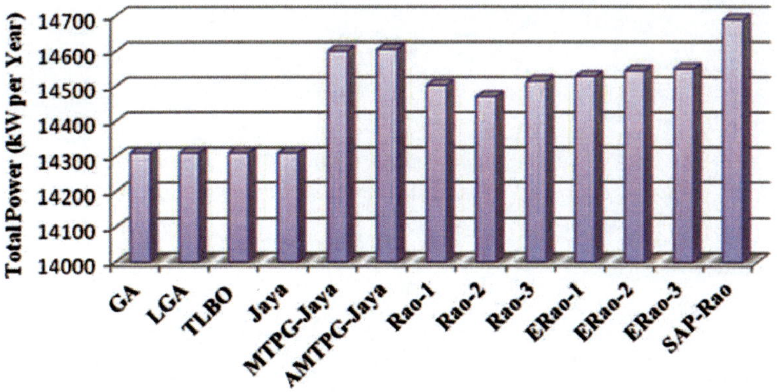

Fig. 6.1 Total power produced by the layouts achieved by different algorithms in case-I with 30 turbines

efficiency. Furthermore, the layout achieved by the SAP-Rao algorithm has resulted in a 2.7% increment in total power when compared to that of the solutions of the GA, LGA, TLBO, and Jaya algorithms. Similarly, the total power of the SAP-Rao algorithm is 1.3%, 1.5%, 1.2%, 1.1%, 1%, and 1% higher when compared to that of the Rao-1, Rao-2, Rao-3, ERao-1, ERao-2, and ERao-3 algorithms, respectively. The solutions of the AMTPG-Jaya, MTPG-Jaya, ERao-3, ERao-2, ERao-1, Rao-3, Rao-1, and Rao-2 algorithms are the next best solutions. The GA, LGA, TLBO, and Jaya algorithms have achieved identical layouts.

Also, to draw attention to the computational cost of the Jaya and Rao algorithms along with their modified versions, convergence plots are presented in Fig. 6.2. In the graph, curves represent the variation of the objective function with respect to the function evaluations presented. It can be observed that the SAP-Rao algorithm is converged faster than the other algorithms. Also, the convergence of the modified versions is faster than the basic algorithms.

The next sub-section presents the analysis of the computational results related to WFLO case-II.

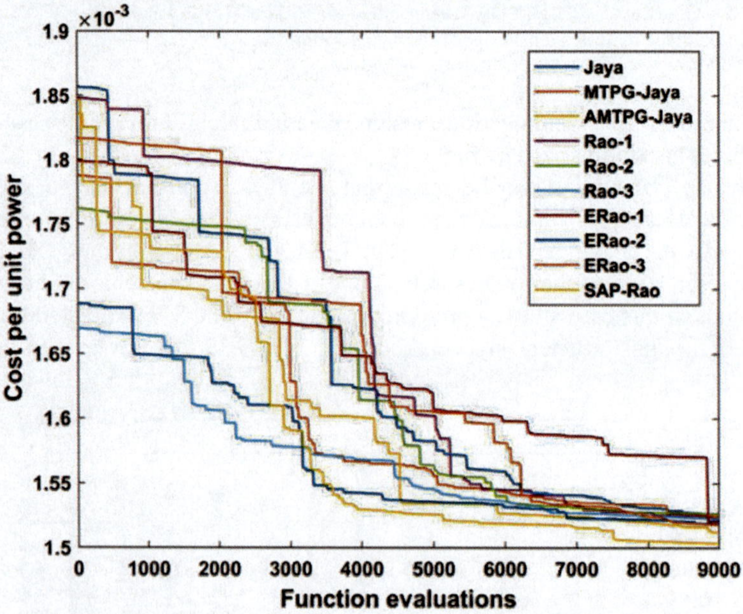

Fig. 6.2 Convergence plots of the basic Jaya and Rao algorithms along with their modified versions in case-I with 30 turbines

6.3 Case-II: Fixed Wind Speed and Variable Wind Direction

It is the case of multi-directional wind with a fixed speed. In case-II, unlike case-I, wind direction is varied from 0 to 360° with equal probability of occurrence. Figure 6.3 shows the wind profile for the case-II.

A total of 36 directions are considered with 10° intervals. The incident wind speed of the wind farm is taken as 12 m/s in all directions. The performances of proposed algorithms are tested for 39 turbines configuration in case-II. Similar to the case-I, in this case, also 9000 function evaluations are taken as termination criterion for all the algorithms. In this case, the computational performances of the proposed algorithms are compared with GA, LGA, ABC, TLBO, and TLBO variants (AL and PAL).

The computational results of various algorithms for the case-II with 39 turbines are summarized in Table 6.3. In this case also, the Jaya algorithm and its modified versions, Rao algorithms and their modified versions have obtained better layouts in comparison to the layouts achieved by the GA, LGA, TLBO, ABC, AL, and PAL algorithms. The AMTPG-Jaya algorithm has achieved the most efficient layout, which produces 18,552.48 kW power with 91.76% efficiency. The SAP-Rao, ERao-3, MTPG-Jaya, ERao-1, ERao-2, Rao-1, Rao-2, Rao-3, and Jaya algorithms have achieved the next best solutions, respectively. The graphical representation of the total power output of the solutions achieved by various algorithms in this case is presented in Fig. 6.4. Here, an observation can be made that the layout achieved by the AMTPG-Jaya algorithm has resulted in 7.7%, 5.4%, 1%, 2.7%, and 1.4% increment in total power when compared to that of the solutions of the GA, LGA, TLBO, ABC, and AL algorithms, respectively. In addition, the convergence plots of

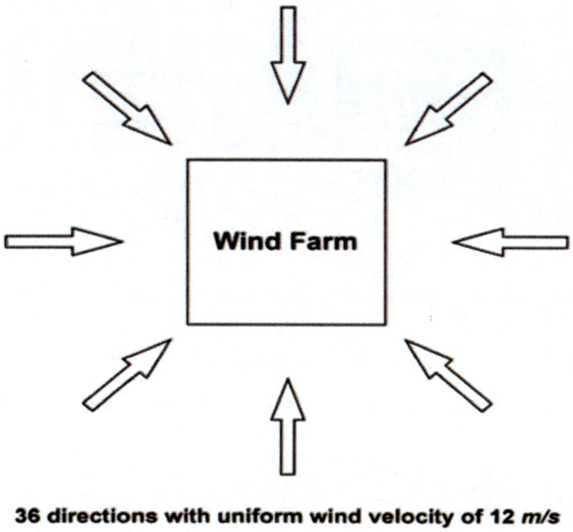

Fig. 6.3 Wind profile for WFLO problem case-II

Table 6.3 Results obtained by the various algorithms in case-II with 39 turbines

Algorithm	Total power (kW/year)	Farm efficiency	Objective value
GA	17,220	85.17	0.001567
LGA	17,611	87.11	0.0015318
TLBO	18,401	91.01	0.001463
ABC	18,062	89.34	0.00149
AL	18,305	90.53	0.001466
PAL	18,441	91.21	0.001459
Jaya	18,448.3	91.24	0.0014589
MTPG-Jaya	18,478.89	91.40	0.001456902
AMTPG-Jaya	18,552.48	91.764	0.001451108
Rao-1	18,463.98	91.326	0.001468063
Rao-2	18,460.92	91.311	0.001468305
Rao-3	18,450.64	91.260	0.001459117
ERao-1	18,468.07	91.347	0.00145774
ERao-2	18,466.30	91.338	0.00145788
ERao-3	18,494.23	91.476	0.001455678
SAP-Rao	18,541.83	91.711	0.001451941

Source GA- Grady et al. (2005), LGA- Changshui et al. (2011), TLBO- Patel et al. (2015), ABC-, AL-, and PAL- Patel et al. (2017)
Result in boldface indicates a better performing algorithm

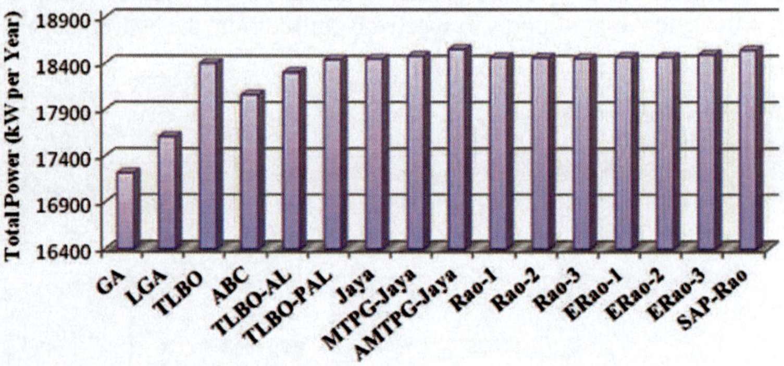

Fig. 6.4 Total power produced by the layouts achieved by different algorithms in case-II with 39 turbines

the Jaya and Rao algorithms, along with their modified versions, are presented in Fig. 6.5.

It can be observed that the convergence of the modified versions is faster than the basic algorithms. The ERao-3 and AMTPG-Jaya algorithms have converged

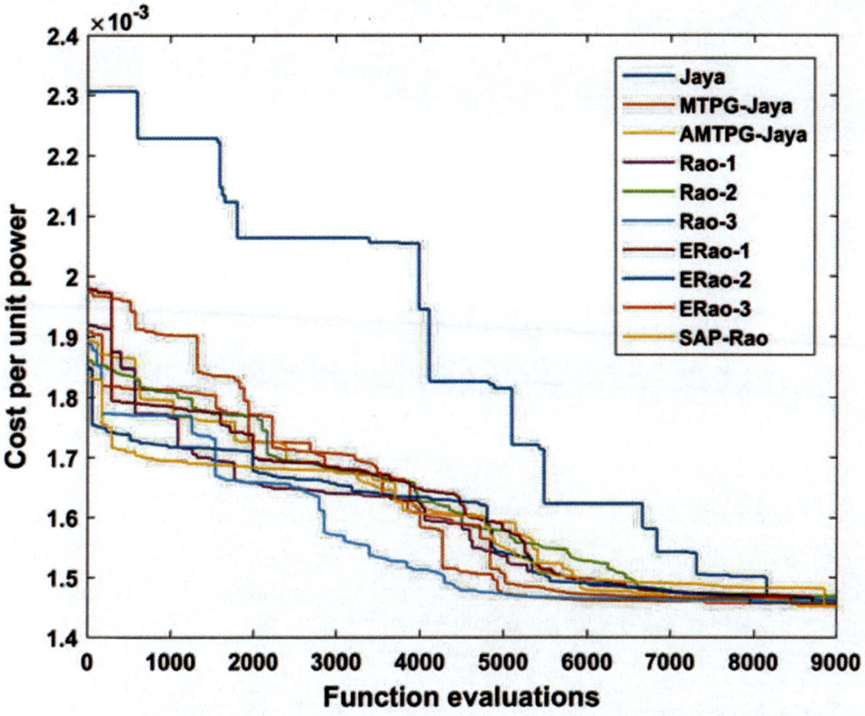

Fig. 6.5 Convergence plots of the basic Jaya and Rao algorithms along with their modified versions in case-II with 39 turbines

faster than the other algorithms. The next sub-section presents the analysis of the computational results related to the WFLO case-III problems.

6.4 Case-III: Fixed Wind Speed and Variable Wind Direction

Case-III is the case of multi-directional wind with varying wind speed. The variation of wind direction is the same as that in case-II. In this case, the wind flows from 36 rotational directions with an unequal probability of occurrence for all wind velocity in each direction. The incident wind speed of the wind farm is taken as 17, 12, and 8 m/s. The fraction of the occurrence of these velocities in 36 rotational directions is presented in Fig. 6.6. The probability distribution for various wind speeds in all considered directions is taken from Patel et al. (2017). In this case, 39 turbines wind farm configuration is considered. The layouts obtained by the proposed algorithms are compared with GA, LGA, ABC, and TLBO variants (AL and PAL). The proposed algorithms are tested with 50,000 function evaluations as the termination criterion,

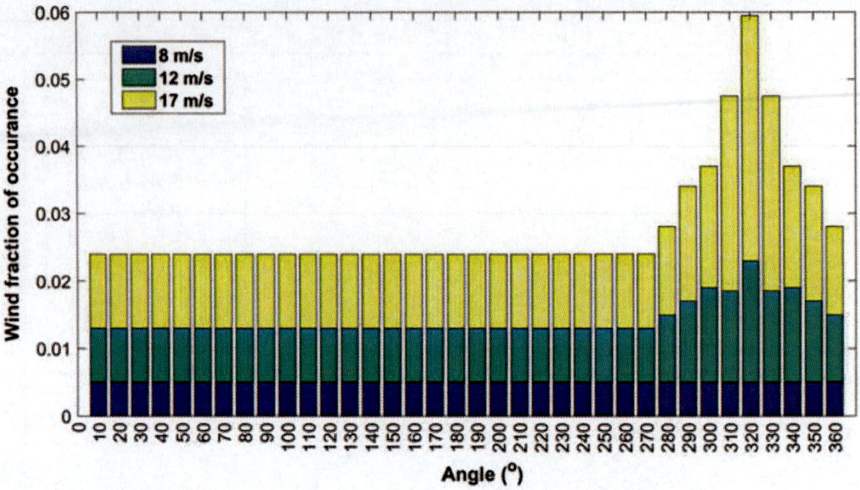

Fig. 6.6 Wind direction and velocity profile bar graph for WFLO case-III

which is the same as that of ABC, AL, and PAL algorithms. The computational results of various algorithms are summarized in Table 6.4.

Table 6.4 Results obtained by the various algorithms in case-III with 39 turbines

Algorithm	Total power (kW/year)	Farm efficiency	Objective value
GA	32,038	86.619	0.00080314
LGA	33,553	90.5	0.00080236
ABC	33,652	90.97	0.0008
AL	33,732	91.18	0.000798
PAL	33,810	91.4	0.000796
Jaya	34,147.92	92.3141	0.000788383
MTPG-Jaya	34,710.90	93.836	0.000775596
AMTPG-Jaya	34,703.54	93.816	0.000775761
Rao-1	34,089.86	92.157	0.000789726
Rao-2	34,093.57	92.167	0.00078964
Rao-3	34,108.13	92.207	0.000789303
ERao-1	34,614.30	93.575	0.000777761
ERao-2	34,708.46	93.829	0.000775651
ERao-3	34,647.53	93.665	0.000777015
SAP-Rao	34,665.86	93.714	0.000776604

Source GA- Grady et al. (2005), LGA- Changshui et al. (2011), TLBO- Patel et al. (2015), ABC-, AL-, and PAL- Patel et al. (2017)
Result in boldface indicates a better performing algorithm.

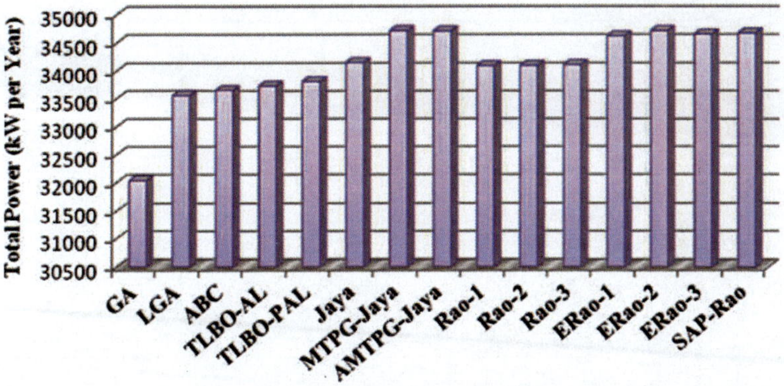

Fig. 6.7 Total power produced by the layouts achieved by different algorithms in case-III with 39 turbines

In this case also, the Jaya algorithm and its modified versions, Rao algorithms and their modified versions have obtained better layouts in comparison with the layouts achieved by the GA, LGA, TLBO, ABC, AL, and PAL algorithms. The graphical representation of the total power output of the solutions achieved by various algorithms is presented in Fig. 6.7. The MTPG-Jaya algorithm has achieved the most efficient layout, which produces 34,710.89 kW power with 93.836% efficiency. The ERao-2, AMTPG-Jaya, SAP-Rao, ERao-3, ERao-1, Jaya, Rao-3, Rao-2, and Rao-1 algorithms have achieved the next best solutions, respectively. Here, an observation can be made that the layout achieved by the MTPG-Jaya algorithm has resulted in 8.34%, 3.5%, 3.14%, 2.9%, 2.7%, 1.65%, 1.82%, 1.81%, and 1.77% increment in total power when compared to that of the solutions achieved by the GA, LGA, ABC, AL, PAL, Jaya, Rao-1, Rao-2, and Rao-3 algorithms, respectively.

In addition, the convergence plots of the Jaya and Rao algorithms, along with their modified versions, in this case, are presented in Fig. 6.8. For this problem also, the convergence of the modified versions is faster than the basic algorithms. The ERao-2, ERao-3, and MTPG-Jaya algorithms have converged relatively faster than the other algorithms.

The wind farm layouts achieved by the basic and proposed modified algorithms in case-I with 30 turbines configuration are presented in Figs. 6.9 and 6.10. Similarly, the wind farm layouts achieved by the basic and proposed modified algorithms in case-II with 39 turbines configuration are presented in Figs. 6.11 and 6.12. The wind farm layouts achieved by the basic Jaya and Rao algorithms and their modified versions in case-III with 39 turbines configuration are presented in Figs. 6.13 and 6.14. From the visual representations of the layouts obtained by Jaya and its modified versions, Rao and its modified versions, it can be observed that most of the turbines are placed at the outer edge by both proposed algorithms.

For the considered wind farm characteristics, it can be observed that the Jaya algorithm's optimal layout is the same as that obtained by GA, and TLBO algorithms

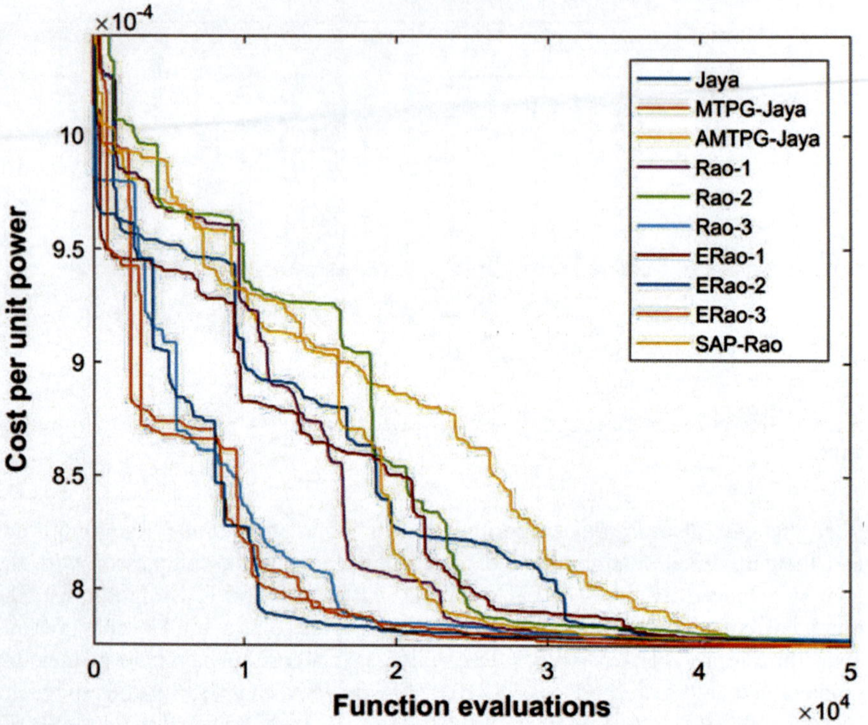

Fig. 6.8 Convergence plots of the basic Jaya and Rao algorithms along with their modified versions in case-III with 39 turbines

for case-I with 30 turbines problem. In the implementation of Jaya and TLBO algorithms to the WFLO problem, the wind farm area is divided into 100 discrete grid locations, and a turbine can be placed at the center of every cell in the square grid. However, in the implementation of modified versions of the Jaya algorithm, Rao algorithms, and modified versions of the Rao algorithms, it is not discrete. Instead, it is considered as continuous space. In this implementation, turbine locations are considered as two-dimensional Cartesian coordinates within the specified area, and these variables are considered as continuous variables.

In general, an upstream wind turbine wake effect will be gradually expanding behind the turbine. As the distance from the upstream turbine increases, the diameter of the wake effect will increase. Wake radius will be crossing 100 m approximately after moving 851 m behind the upstream turbine. Thus, there will be a greater number of alternative locations in case-I with continuous space layout when compared to a discrete grid layout, as in the case of TLBO and Jaya algorithms.

Here, it can be observed that, in all three cases, the performances of the proposed modified algorithms are superior or competitive to those of the basic Jaya and Rao algorithms. Hence, the modified versions of the Jaya and Rao algorithms can be considered as the improved versions of the Jaya and Rao algorithms respectively.

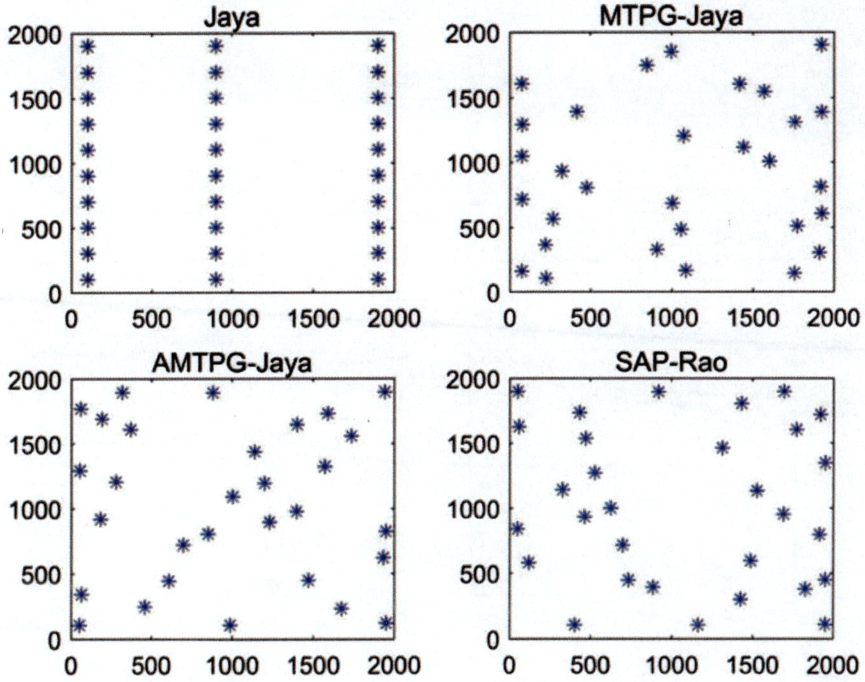

Fig. 6.9 Wind farm layouts achieved by the Jaya, MTPG-Jaya, AMTPG-Jaya, and SAP-Rao algorithms for case-I with 30 turbines

In all three cases, the results of the improved versions are superior to the results presented by other algorithms in terms of farm efficiency, the total power generated, and cost per unit power generated. In case-I with 30 turbines configuration, the layout achieved by the SAP-Rao algorithm has resulted in a 2.7% increment in total power when compared to that of the solutions of the GA, LGA, TLBO, and Jaya algorithms. Similarly, the total power of the SAP-Rao algorithm is 1.3%, 1.5%, 1.2%, 1.1%, 1%, and 1% higher when compared to that of the Rao-1, Rao-2, Rao-3, ERao-1, ERao-2, and ERao-3 algorithms, respectively. In case-II with 39 turbines configuration, the layout obtained by the AMTPG-Jaya has increased the total power by 7.7%, 5.4%, 1%, 2.7%, and 1.4% compared to that of GA, LGA, TLBO, ABC, and AL, respectively. In case-III with 39 turbines configuration, the layout obtained by the MTPG-Jaya has increased the total power by 8.34%, 3.5%, 3.14%, 2.9%, 2.7%, 1.65%, 1.82%, 1.81%, and 1.77% when compared to that of the GA, LGA, ABC, AL, PAL, Jaya, Rao-1, Rao-2, and Rao-3 algorithms, respectively.

The next chapter presents the Jaya algorithm and Rao algorithms' application along with their modified versions to the solar-assisted energy systems.

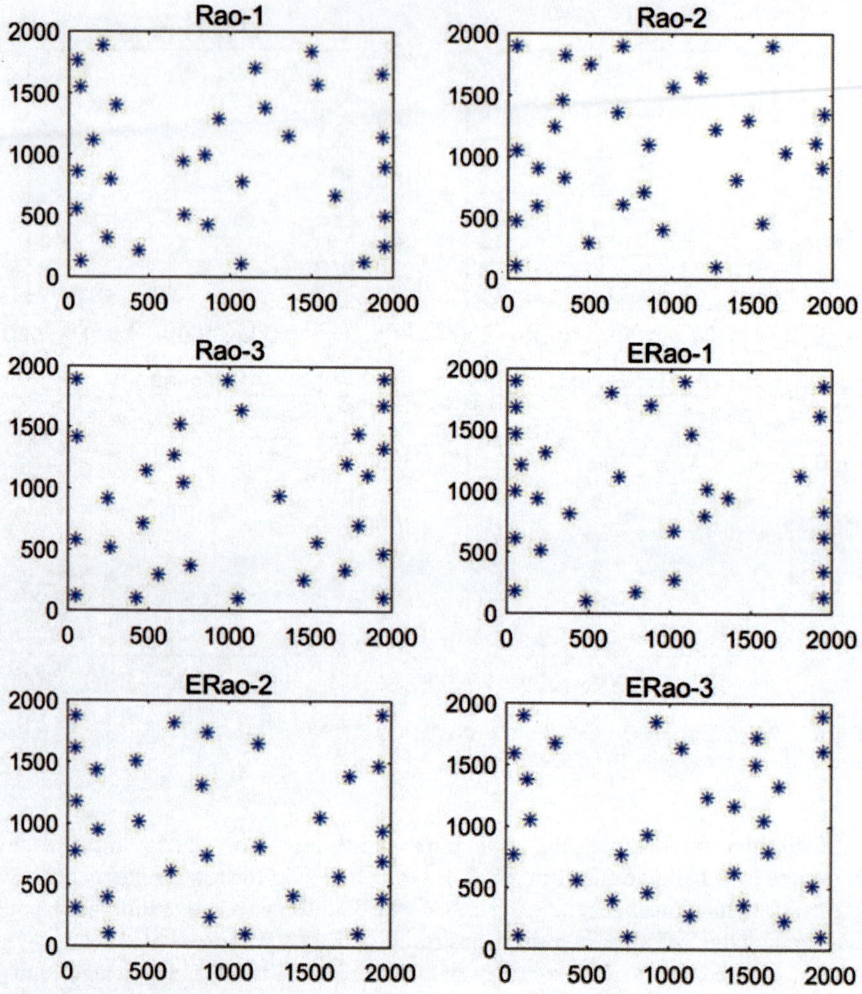

Fig. 6.10 Wind farm layouts achieved by the Rao and elitist Rao algorithms for case-I with 30 turbines

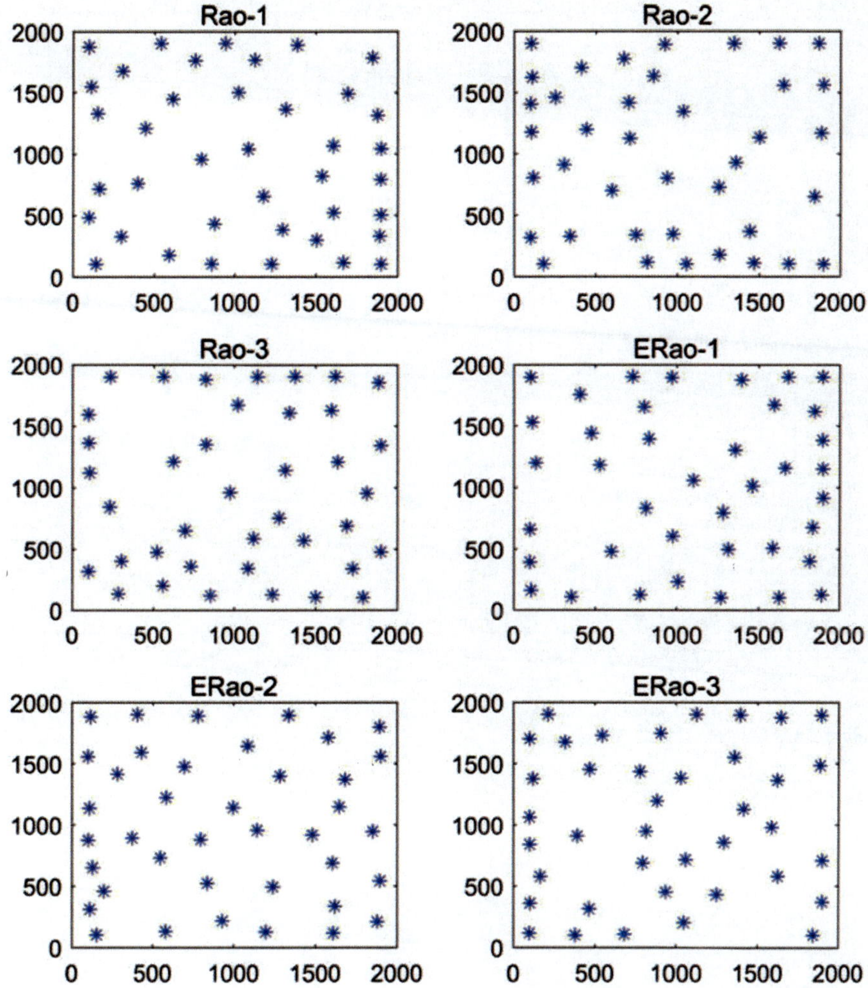

Fig. 6.11 The wind farm layouts achieved by the Rao and elitist Rao algorithms for case-II with 39 turbines

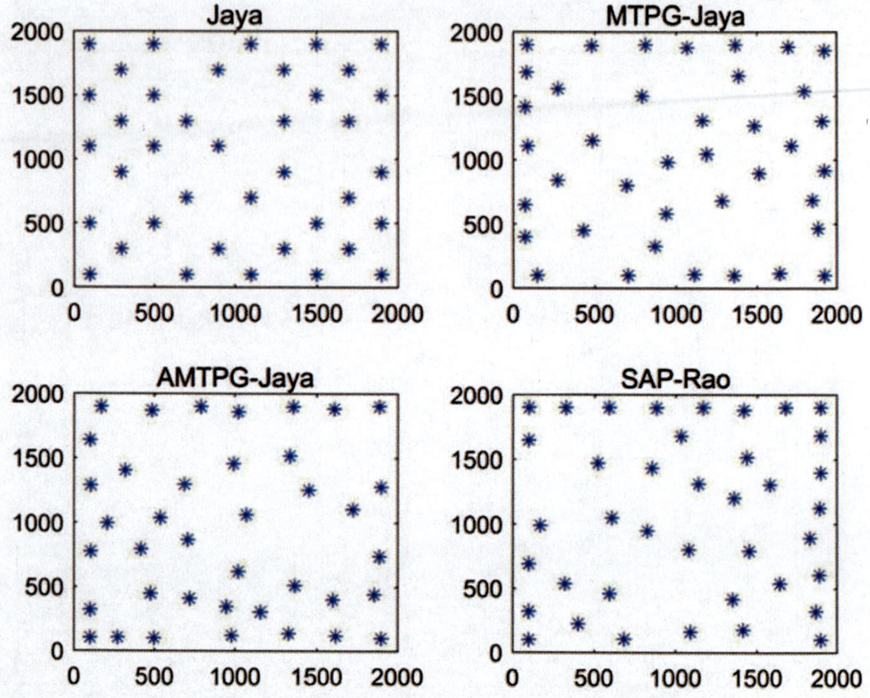

Fig. 6.12 Wind farm layouts achieved by the Jaya, MTPG-Jaya, AMTPG-Jaya, and SAP-Rao algorithms for case-II with 39 turbines

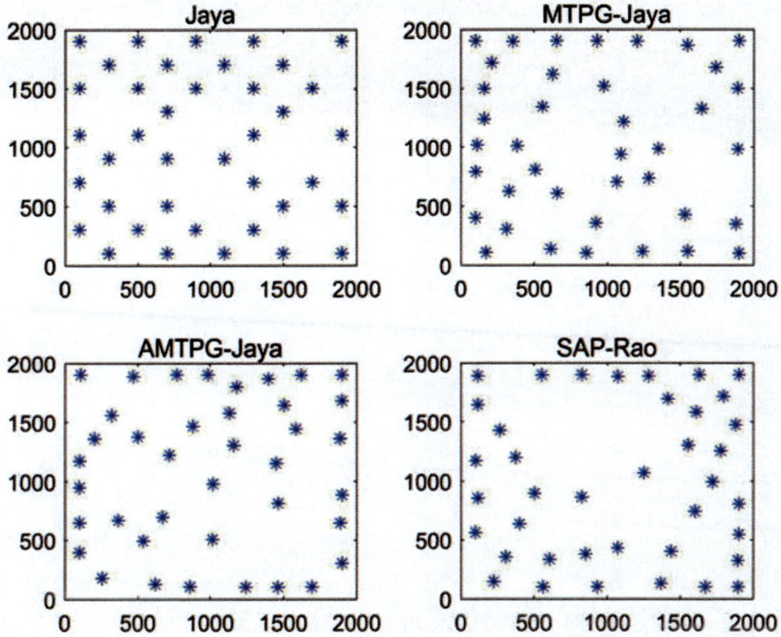

Fig. 6.13 Wind farm layouts achieved by the Jaya, MTPG-Jaya, AMTPG-Jaya, and SAP-Rao algorithms for case-III with 39 turbines

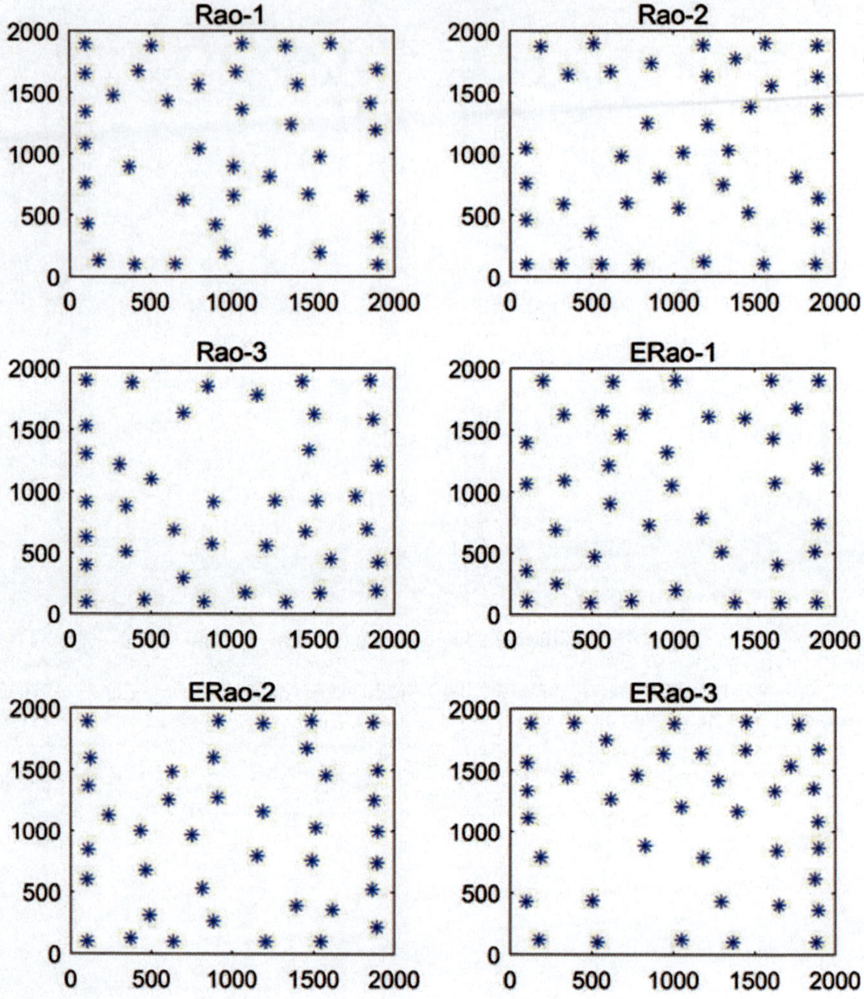

Fig. 6.14 Wind farm layouts achieved by the Rao and elitist Rao algorithms for case-III with 39 turbines

References

Changshui, Z., Guangdong, H., & Jun, W. (2011). A fast algorithm based on the submodular property for optimization of wind turbine positioning. *Renewable Energy, 36*(11), 2951–2958. https://doi.org/10.1016/j.renene.2011.03.045

Grady, S. A., Hussaini, M. Y., & Abdullah, M. M. (2005). Placement of wind turbines using genetic algorithms. *Renewable Energy, 30*(2), 259–270. https://doi.org/10.1016/j.renene.2004.05.007

Patel, J., Savsani, V., Patel, V., & Patel, R. (2017). Layout optimization of a wind farm to maximize the power output using enhanced teaching learning based optimization technique. *Journal of Cleaner Production, 158*, 81–94. https://doi.org/10.1016/j.jclepro.2017.04.132

Patel, J., Savsani, V., & Patel, R. (2015). Maximizing energy output of a wind farm using teaching–learning-based optimization. *Volume 2: Photovoltaics; Renewable-Non-Renewable Hybrid Power System; Smart Grid, Micro-Grid Concepts; Energy Storage; Solar Chemistry; Solar Heating and Cooling; Sustainable Cities and Communities, Transportation; Symposium on Integrated/Sustainable Built*. https://doi.org/10.1115/ES2015-49164

Chapter 7
Optimization of the Selected Solar-Assisted Energy Systems

Abstract This chapter presents the applications of different versions of Jaya and Rao algorithms to the problems of solar-assisted heat engine systems. In this chapter, five multi-objective optimization case studies of solar-assisted energy systems which includes a case study of solar-assisted Brayton heat engine system, three case studies of solar-assisted Stirling heat engine system, a case study of solar-assisted Carnot-like heat engine system, are considered for optimization. The optimization is carried out using the Jaya algorithm, Rao algorithms, and their modified versions. Furthermore, the solutions obtained in multi-objective optimization scenarios are non-dominated in nature due to the conflicting nature of the objectives. Hence, to identify the best solutions from the Pareto-fronts, the average rank method described in Chap. 5 is used. Computational results revealed that the performances of the modified Jaya and Rao algorithms are superior to those of the other algorithms. Also, the performances of the selected systems are improved by the solutions of the proposed algorithms.

7.1 Optimization of a Solar-Assisted Brayton Heat Engine System

The description of the selected solar-assisted Brayton heat engine system is presented in Sect. 2.2.1. The detailed description and thermodynamic analysis of the solar-assisted Brayton heat engine system considered in this book were presented by Li et al. (2015). The three objective functions considered in this case study are thermal efficiency (η_m), power output (P), and non-dimensional thermo-economic performance function (F) of the solar-assisted Brayton heat engine system. These three objective functions are maximization functions which are given in Eq. 2.11, Eq. 2.22, and Eq. 2.23. The decision variables of these objective functions are the temperature of the hot reservoir (T_H), the temperature of the cold reservoir (T_L), the temperature of the working fluid at state 1 of Brayton cycle (T_1), hot side heat exchange effectiveness (ε_H), cold side heat exchange effectiveness (ε_L), and regenerator effectiveness (ε_R). The lower and upper boundaries of the decision variables are as follows:

$$700\ \mathrm{K} \le T_H \le 1000\ \mathrm{K} \tag{7.1}$$

V. R. Ravipudi et al., *Design Optimization of Renewable Energy Systems Using Advanced Optimization Algorithms*, Green Energy and Technology,
https://doi.org/10.1007/978-3-030-95589-2_7

$$400\text{ K} \le T_L \le 500\text{ K} \tag{7.2}$$

$$T_L \le T_1 \le T_H \tag{7.3}$$

$$0.5 \le \varepsilon_H \le 0.7 \tag{7.4}$$

$$0.5 \le \varepsilon_L \le 0.7 \tag{7.5}$$

$$0.5 \le \varepsilon_R \le 0.8 \tag{7.6}$$

The characteristics of the solar-assisted Brayton heat engine system taken are as follows: $I = 1000\text{Wm}^{-2}$, $e_C = 0.9$, $R_C = 1300$, $\eta_0 = 0.85$, $k = 4$, $\xi = 0.02$, $\delta = 5.67 \times 10^{-8}\text{WK}^{-4}\text{m}^{-2}$, $h_C = 20\text{Wm}^{-2}\text{K}^{-1}$, $C_{wf} = 1050\text{ WK}^{-1}$, $T_0 = 300\text{ K}$, and $h_H = h_L = 2000\text{ WK}^{-1}\text{m}^{-2}$.

Li et al. (2015) had reported optimal solutions by employing the NSGA-II algorithm and reported three optimal solutions using the TOPSIS, Shannon's entropy, and LINMAP methods. The computations of the NSGA-II algorithms were performed by taking 125,000 function evaluations as the termination criterion. However, in this book, the proposed algorithms are executed using only 10,000 function evaluations as the termination criterion and reported the Pareto-optimal solutions using the average rank based on the multiple decision-making methods. Firstly, single-objective optimization is performed independently for each objective function, and then MOO is performed in this book. In all the computational experiments, the population size is maintained as 25 for all the algorithms. The elite size for the elitist Rao algorithms is taken as 20%.

Table 7.1 presents the single-objective optimization results of this case study. The Jaya algorithm, Rao algorithms, and their improved versions have obtained identical solutions in all the single-objective optimization scenarios of this case study. Hence, only one solution is presented here. Also, the solutions obtained by the improved algorithms in all the single-objective optimization scenarios are superior to those obtained by the NSGA-II algorithm. The maximum power output, thermal efficiency, and thermo-economic function obtained by the proposed algorithms are 71.60 kW, 23.77%, and 0.3144, respectively.

In the multi-objective optimization, to demonstrate the proposed average rank method for identifying the best compromise solution, the Pareto-front achieved by the SAP-Rao algorithm is considered. The best compromise solution of the SAP-Rao algorithm is found to be the best compromise solution among the solutions of the other algorithms compared. Hence, the SAP-Rao algorithm Pareto-front is considered for the demonstration of the proposed average rank method. Table 7.2 presents the Pareto-optimal solutions obtained by the SAP-Rao algorithm in multi-objective optimization of this case study. These solutions are non-dominated in nature, and all the solutions can be considered as equivalent. However, the compromise among the objectives

Table 7.1 Results obtained by the NSGA-II and proposed algorithms in single-objective optimization scenarios of the solar-assisted Brayton engine system case study

Objective	Algorithm	ε_H	ε_L	ε_R	T_H (K)	T_L (K)	T_1 (K)	P (kW)	η_m	F
Maximize-P	NSGA-II	0.69	0.69	0.8	999.83	400.19	604.92	71.42	0.2203	0.2932
	Proposed algorithms	0.7	0.7	0.8	1000	400	612.2243	71.6037	0.2276918	0.29949823
Maximize-η_m	NSGA-II	0.7	0.69	0.79	999.84	400.15	567.97	68.06	0.2376	0.3124
	Proposed algorithms	0.7	0.7	0.8	1000	400	566.3031	67.86533	0.23776	0.31442866
Maximize-F	NSGA-II	0.69	0.69	0.79	999.78	400.11	564.21	67.42	0.2372	0.3144
	Proposed algorithms	0.7	0.7	0.8	1000	400	563.2638	67.32262	0.2377065	0.3145

Source NSGA-II—Li et al. (2015)
Result in boldface indicates a better performing algorithm

Table 7.2 Pareto-optimal solutions obtained by the SAP-Rao algorithm for the solar-assisted Brayton engine system case study

Solution	ε_H	ε_L	ε_R	T_H (K)	T_L (K)	T_1 (K)	P (kW)	η_m	F
1	0.7	0.7	0.8	1000	400	612.224	71.60	0.22769	0.29950
2	0.7	0.7	0.8	1000	400	566.306	67.87	0.23776	0.31443
3	0.7	0.7	0.8	1000	400	563.262	67.32	0.23771	0.31450
4	0.7	0.7	0.8	1000	400	608.822	71.59	0.22901	0.30133
5	0.7	0.7	0.8	1000	400	596.782	71.21	0.23304	0.30700
6	0.7	0.7	0.8	1000	400	605.909	71.54	0.23008	0.30282
7	0.7	0.7	0.8	1000	400	610.018	71.60	0.22856	0.30070
8	0.7	0.7	0.8	1000	400	603.221	71.47	0.23101	0.30413
9	0.7	0.7	0.8	1000	400	601.597	71.42	0.23155	0.30489
10	0.7	0.7	0.8	1000	400	586.947	70.52	0.23550	0.31059
11	0.7	0.7	0.8	1000	400	589.335	70.72	0.23498	0.30981
12	0.7	0.7	0.8	1000	400	599.993	71.36	0.23207	0.30562
13	0.7	0.7	0.8	1000	400	594.434	71.08	0.23370	0.30795
14	0.7	0.7	0.8	1000	400	590.984	70.85	0.23459	0.30924
15	0.7	0.7	0.8	1000	400	570.475	68.54	0.23766	0.31411
16	0.7	0.7	0.8	1000	400	605.053	71.52	0.23038	0.30325
17	0.7	0.7	0.8	1000	400	567.980	68.15	0.23775	0.31433
18	0.7	0.7	0.8	1000	400	592.524	70.96	0.23420	0.30868
19	0.7	0.7	0.8	1000	400	598.676	71.30	0.23247	0.30620
20	0.7	0.7	0.8	1000	400	582.310	70.08	0.23637	0.31192
21	0.7	0.7	0.8	1000	400	583.643	70.21	0.23614	0.31156
22	0.7	0.7	0.8	1000	400	585.205	70.37	0.23585	0.31112
23	0.7	0.7	0.8	1000	400	580.282	69.85	0.23669	0.31242
24	0.7	0.7	0.8	1000	400	593.177	71.00	0.23403	0.30843
25	0.7	0.7	0.8	1000	400	571.973	68.77	0.23758	0.31393

is different for each solution. Moreover, the solution that has the best compromise among the objectives can be considered as the best solution. Hence, to identify the solution with the best compromise, MADM methods can be used.

In the multi-attribute decision-making scenario, the MADM methods are used to rank various alternatives based on their values of attributes. By considering each solution as an alternative, and each objective function as an attribute, these solutions are ranked using decision-making methods. Here, each method uses different principles to rank the solutions. Thus, eight MADM methods are considered for ranking these solutions instead of depending on a single method. Now, the average rank is calculated based on the ranks suggested by these methods to the solutions. The solution with the least average rank value is considered as the best solution. The ranks obtained by each solution using different decision-making methods are shown in

Table 7.3. Furthermore, to see if there is any correlation between the ranks suggested by different pairs of MADM methods, Spearman's correlation coefficients are calculated for different pairs of MADM methods. For calculating the average rank values, only the ranks of the methods with a correlation coefficient value greater than 0.5 with the other methods are considered. Spearman's correlation for different pairs of rankings given by decision-making methods is shown in Table 7.4.

Spearman's correlation coefficients for all the pairs of decision-making methods are positive, which indicates that there is some similarity among the ranks suggested by the MADM methods. Also, the pairs SAW-COPRAS and TOPSIS-MTOPSIS have Spearman's correlation coefficient equal to 1, which indicates that the ranking of these pairs is identical. It can be confirmed by the ranks shown in Table 7.3. The ranks suggested by the SAW and COPRAS methods are identical, and the ranks suggested by the TOPSIS and MTOPSIS methods are also identical. If we consider the SAW-WPM pair, Spearman's correlation value is 0.997. It indicates that these methods have a higher similarity in ranks suggested, and it can be observed in Table 7.3.

Similarly, for the PROMETHEE and GRA methods, Spearman's correlation coefficient values are lesser with other methods. It indicates that these methods have lesser similarities in the ranks suggested to the solutions. Table 7.3 shows that the ranks of these methods are different from those suggested by the other methods for the first 19 solutions, and for the bottom six solutions, the suggested ranks are the same as those given by other methods. If Spearman's correlation coefficient value is negative, then the ranks can be considered dissimilar and cannot be used to calculate the average ranks. However, in this book, the lower threshold value for Spearman's correlation is taken as 0.5. Spearman's correlation values for the PROMETHEE-TOPSIS, PROMETHEE-MTOPSIS, GRA-TOPSIS, and GRA-MTOPSIS pairs are less than 0.5. Also, for these four methods, Spearman's correlation value is greater than 0.5 with the remaining methods. If we compare the pair PROMETHEE-TOPSIS, Spearman's correlation values of TOPSIS method with SAW, WPM, VIKOR, MTOPSIS, COPRAS, and GRA methods are 0.893, 0.913, 1, 0.966, 0.893, and 0.357, respectively. Moreover, Spearman's correlation values of the PROMETHEE method with SAW, WPM, VIKOR, MTOPSIS, COPRAS, and GRA methods are 0.671, 0.64, 0.342, 0.541, 0.671, and 0.997, respectively. Here, the TOPSIS method has better correlation values with five (SAW, WPM, VIKOR, MTOPSIS, and COPRAS) methods, whereas the PROMETHEE method has a better correlation value with one (GRA) method. Hence, the ranks suggested by the TOPSIS method are considered for calculating average ranks and the ranks suggested by the PROMETHEE method will be unused. Similarly, by comparing the PROMETHEE-MTOPSIS, GRA-TOPSIS, and GRA-MTOPSIS pairs, the TOPSIS and MTOPSIS methods are selected for calculating the average ranks.

Now, the corrected average ranks are calculated, excluding the ranks suggested by the PROMETHEE and GRA methods, and presented in Table 7.3 as the corrected ranks. Solution 20, with an average rank of 2.33, can be considered as the best solution, which has the best compromise among the power output, system efficiency,

Table 7.3 Ranks suggested by the MADM methods for the SAP-Rao algorithm solutions for the solar-assisted Brayton engine system case study

Solution	SAW	WPM	TOPSIS	MTOPSIS	VIKOR	PROMETHEE	COPRAS	GRA	AR	CR	FR
20	1	1	5	5	1	7	1	7	3.5	2.33	1
21	3	2	3	3	1	8	3	8	3.875	2.5	2
22	4	4	2	2	1	9	4	9	4.375	2.83	3
10	5	5	1	1	1	10	5	10	4.75	3	4
23	2	3	7	7	6	6	2	6	4.875	4.5	5
11	6	6	4	4	1	11	6	11	6.125	4.5	5
14	7	7	6	6	7	12	7	12	8	6.67	7
18	9	8	8	8	8	13	9	13	9.5	8.33	8
24	10	10	9	9	9	14	10	14	10.625	9.5	9
25	8	9	14	14	11	5	8	4	9.125	10.67	10
13	12	12	10	10	10	15	12	15	12	11	11
5	14	13	11	11	12	16	14	16	13.375	12.5	12
15	11	11	16	16	13	4	11	3	10.625	13	13
19	16	15	12	12	14	17	16	17	14.875	14.17	14
17	13	14	18	18	15	2	13	1	11.75	15.17	15
12	17	17	13	13	16	18	17	18	16.125	15.5	16
9	18	18	15	15	17	19	18	19	17.375	16.83	17
2	15	16	21	21	18	1	15	2	13.625	17.67	18
8	20	20	17	17	19	20	20	20	19.125	18.83	19
16	21	21	19	19	21	21	21	21	20.5	20.33	20
3	19	19	23	23	20	3	19	5	16.375	20.5	21

(continued)

Table 7.3 (continued)

Solution	SAW	WPM	TOPSIS	MTOPSIS	VIKOR	PROMETHEE	COPRAS	GRA	AR	CR	FR
6	22	22	20	20	22	22	22	22	21.5	21.33	22
4	23	23	22	22	23	23	23	23	22.75	22.67	23
7	24	24	24	24	24	24	24	24	24	24	24
1	25	25	25	25	25	25	25	25	25	25	25

AR average rank, *CR* corrected rank, *FR* final rank

Table 7.4 Spearman's rank correlation coefficients between different pairs of MADM method's ranking for the SAP-Rao algorithm solutions

Method	SAW	WPM	TOPSIS	MTOPSIS	VIKOR	PROMETHEE	COPRAS	GRA
SAW	1	0.997	0.893	0.893	0.970	0.671	1	0.687
WPM	0.997	1	0.913	0.913	0.980	0.640	0.997	0.655
TOPSIS	0.893	0.913	1	1	0.966	0.342	0.893	0.357
MTOPSIS	0.893	0.913	1	1	0.966	0.342	0.893	0.357
VIKOR	0.970	0.980	0.966	0.966	1	0.541	0.970	0.556
PROMETHEE	0.671	0.640	0.342	0.342	0.541	1	0.671	0.997
COPRAS	1	0.997	0.893	0.893	0.970	0.671	1	0.687
GRA	0.687	0.655	0.357	0.357	0.556	0.997	0.687	1

and thermo-economic function values. In this way, the best solution from the Pareto-optimal solutions is identified for all the MOO scenarios in this report.

The Pareto-optimal solutions achieved by the Jaya algorithm, Rao algorithms, and their modified versions in this case study are presented in Tables 7.5, 7.6, 7.7, 7.8, 7.9, 7.10, 7.11 and 7.12. Here an observation can be made that the system thermal efficiency is varied between 0.2276 and 0.2378, power output is varied between 67.31 and 71.60 kW, and non-dimensional thermo-economic function is varied from 0.2994 to 0.315. Similar to the SAP-Rao algorithm's best solution, the best solutions to these algorithms are identified. Solution 18 of the Jaya algorithm, Solution 6 of the Rao-1 algorithm, Solution 16 of the Rao-2 algorithm, Solution 9 of the Rao-3 algorithm, Solution 5 of the AMTPG-Jaya algorithm, Solution 20 of the ERao-1 algorithm, Solution 10 of the ERao-2, and Solution 7 of the ERao-3 algorithm are identified as the best solutions from the respective algorithm Pareto-front. Now, these best solutions are compared with those of the NSGA-II algorithm in Table 7.13.

In Table 7.13, the solutions with the best average rank from the Pareto-optimal solutions of the various algorithms are compared with those reported by the NSGA-II algorithm. Figure 7.1 presents the Pareto-fronts achieved by the AMTPG-Jaya, SAP-Rao, and elitist Rao algorithms, including the optimal solutions reported for the NSGA-II, Jaya, and Rao algorithms. Here, an observation can be made that the NSGA-II algorithm solutions selected by employing TOPSIS, Shannon-entropy, and LINMAP methods belong to a lower level Pareto-front, which is dominated by the proposed algorithm's Pareto-fronts. The thermal efficiency, thermo-economic function, and power output of the proposed algorithms solutions are better than those reported for the NSGA-II algorithm. Furthermore, power output of the system for the ERao-2 algorithm solution is highest, which is 3%, 2.35%, 2.4%, 0.34%, 0.4%, 0.18%, 0.5%, 0.33%, 0.3%, 0.2%, and 0.14% higher than that of the NSGA-II (Shannon-entropy, LINMAP, and TOPSIS), Jaya, Rao-1, Rao-2, Rao-3, AMTPG-Jaya, SAP-Rao, ERao-1, and ERao-3 algorithms solutions, respectively.

Similarly, the thermal efficiency for the Rao-3 algorithm solution is best, which is 1.17%, 1.51%, 1.47%, 0.07%, 0.07%, 0.15%, 0.07%, 0.1%, 0.2%, 0.2%, and 0.2% higher when compared to that of the NSGA-II (Shannon-entropy, LINMAP,

Table 7.5 Pareto-optimal solutions obtained by the Jaya algorithm for the solar-assisted Brayton engine system case study

Solution	ε_H	ε_L	ε_R	T_H (K)	T_L (K)	T_1 (K)	P (kW)	η_m	F
1	0.7	0.7	0.8	1000	400	566.305	67.87	0.23776	0.31443
2	0.7	0.7	0.8	1000	400	563.267	67.32	0.23771	0.31450
3	0.7	0.7	0.8	1000	400	612.225	71.60	0.22769	0.29950
4	0.7	0.7	0.8	1000	400	596.314	71.19	0.23317	0.30719
5	0.7	0.7	0.8	1000	400	591.744	70.90	0.23440	0.30896
6	0.7	0.7	0.8	1000	400	580.194	69.84	0.23671	0.31244
7	0.7	0.7	0.8	1000	400	588.914	70.69	0.23507	0.30995
8	0.7	0.7	0.8	1000	400	606.486	71.55	0.22987	0.30253
9	0.7	0.7	0.8	1000	400	610.046	71.60	0.22855	0.30068
10	0.7	0.7	0.8	1000	400	564.353	67.52	0.23774	0.31449
11	0.7	0.7	0.8	1000	400	598.489	71.29	0.23253	0.30628
12	0.7	0.7	0.8	1000	400	583.930	70.24	0.23609	0.31148
13	0.7	0.7	0.8	1000	400	587.566	70.58	0.23537	0.31039
14	0.7	0.7	0.8	1000	400	602.978	71.46	0.23110	0.30425
15	0.7	0.7	0.8	1000	400	569.780	68.44	0.23769	0.31418
16	0.7	0.7	0.8	1000	400	586.322	70.47	0.23563	0.31078
17	0.7	0.7	0.8	1000	400	585.158	70.36	0.23586	0.31113
18	0.7	0.7	0.8	1000	400	581.822	70.02	0.23645	0.31204
19	0.7	0.7	0.8	1000	400	590.691	70.83	0.23466	0.30934
20	0.7	0.7	0.8	1000	400	599.285	71.33	0.23229	0.30593
21	0.7	0.7	0.8	1000	400	571.966	68.77	0.23758	0.31393
22	0.7	0.7	0.8	1000	400	568.977	68.31	0.23772	0.31425
23	0.7	0.7	0.8	1000	400	611.535	71.60	0.22797	0.29988
24	0.7	0.7	0.8	1000	400	576.678	69.42	0.23716	0.31318
25	0.7	0.7	0.8	1000	400	575.305	69.24	0.23731	0.31343

and TOPSIS), Jaya, Rao-1, Rao-2, AMTPG-Jaya, SAP-Rao, ERao-1, ERao-2, and ERao-3 algorithms solutions, respectively. In addition, thermo-economic function of the system for the Rao-3 algorithm solution is best, which is 1.1%, 1.49%, 1.46%, 0.08%, 0.08%, 0.2%, 0.09%, 0.12%, 0.2%, 0.3%, and 0.2% higher when compared to that of the NSGA-II (Shannon-entropy, LINMAP, and TOPSIS), Jaya, Rao-1, Rao-2, AMTPG-Jaya, SAP-Rao, ERao-1, ERao-2, and ERao-3 algorithms solutions, respectively. However, these solutions may not be the best as they have lesser fitness values in other objectives. Hence, to identify the best solution, which has the best compromise among the objectives, the MADM methods based average ranks are calculated and presented in Table 7.14.

The decision-making methods ranking of the Pareto-optimal solutions obtained by different algorithms is shown in Table 7.14. Spearman's correlation coefficients

Table 7.6 Pareto-optimal solutions obtained by the Rao-1 algorithm for the solar-assisted Brayton engine system case study

Solution	ε_H	ε_L	ε_R	T_H (K)	T_L (K)	T_1 (K)	P (kW)	η_m	F
1	0.7	0.7	0.8	1000	400	612.214	71.60	0.22770	0.29950
2	0.7	0.7	0.8	1000	400	566.303	67.87	0.23776	0.31443
3	0.7	0.7	0.8	1000	400	563.264	67.32	0.23771	0.31450
4	0.7	0.7	0.8	1000	400	585.056	70.35	0.23588	0.31116
5	0.7	0.7	0.8	1000	400	590.844	70.84	0.23462	0.30929
6	0.7	0.7	0.8	1000	400	581.727	70.01	0.23647	0.31206
7	0.7	0.7	0.8	1000	400	605.101	71.52	0.23037	0.30322
8	0.7	0.7	0.8	1000	400	583.929	70.24	0.23609	0.31148
9	0.7	0.7	0.8	1000	400	606.784	71.56	0.22977	0.30238
10	0.7	0.7	0.8	1000	400	568.159	68.18	0.23774	0.31432
11	0.7	0.7	0.8	1000	400	579.597	69.78	0.23679	0.31257
12	0.7	0.7	0.8	1000	400	591.817	70.91	0.23438	0.30894
13	0.7	0.7	0.8	1000	400	573.878	69.04	0.23744	0.31366
14	0.7	0.7	0.8	1000	400	593.790	71.04	0.23387	0.30820
15	0.7	0.7	0.8	1000	400	611.429	71.60	0.22801	0.29993
16	0.7	0.7	0.8	1000	400	600.333	71.37	0.23196	0.30547
17	0.7	0.7	0.8	1000	400	610.164	71.60	0.22850	0.30062
18	0.7	0.7	0.8	1000	400	572.734	68.88	0.23753	0.31383
19	0.7	0.7	0.8	1000	400	576.185	69.35	0.23722	0.31328
20	0.7	0.7	0.8	1000	400	607.992	71.57	0.22932	0.30176
21	0.7	0.7	0.8	1000	400	563.572	67.38	0.23772	0.31450
22	0.7	0.7	0.8	1000	400	594.373	71.07	0.23371	0.30797
23	0.7	0.7	0.8	1000	400	609.250	71.59	0.22885	0.30110
24	0.7	0.7	0.8	1000	400	570.771	68.59	0.23765	0.31408
25	0.7	0.7	0.8	1000	400	600.983	71.40	0.23175	0.30517

for different pairs of rankings given by decision-making methods for different algorithm's solutions are shown in Table 7.15. The ranks given by the COPRAS, WPM, and SAW methods are equal for each solution. Also, the TOPSIS and MTOPSIS methods ranks for each solution are identical. Furthermore, Spearman's correlation for all the pairs formed by all the MADM methods (except for the pairs GRA-TOPSIS, GRA-MTOPSIS, PROMETHEE-TOPSIS, and PROMETHEE-MTOPSIS) is greater than 0.5. Also, Spearman's correlation values for the GRA and PROMETHEE methods with other methods (SAW, WPM, and COPRAS) is lesser when compared to those of the TOPSIS and MTOPSIS methods. Thus, the ranks suggested by the GRA and PROMETHEE methods cannot be considered for calculating the average ranks. Hence, the corrected average ranks are calculated, excluding the ranks suggested by

Table 7.7 Pareto-optimal solutions obtained by the Rao-2 algorithm for the solar-assisted Brayton engine system case study

Solution	ε_H	ε_L	ε_R	T_H (K)	T_L (K)	T_1 (K)	P (kW)	η_m	F
1	0.7	0.7	0.8	1000	400	563.255	67.32	0.23771	0.31450
2	0.7	0.7	0.8	1000	400	566.297	67.86	0.23776	0.31443
3	0.7	0.7	0.8	1000	400	612.231	71.60	0.22769	0.29949
4	0.7	0.7	0.8	1000	400	602.434	71.45	0.23128	0.30450
5	0.7	0.7	0.8	1000	400	598.845	71.31	0.23242	0.30612
6	0.7	0.7	0.8	1000	400	603.701	71.49	0.23085	0.30390
7	0.7	0.7	0.8	1000	400	608.095	71.58	0.22928	0.30171
8	0.7	0.7	0.8	1000	400	592.338	70.94	0.23425	0.30875
9	0.7	0.7	0.8	1000	400	605.678	71.53	0.23016	0.30294
10	0.7	0.7	0.8	1000	400	606.290	71.55	0.22994	0.30263
11	0.7	0.7	0.8	1000	400	611.407	71.60	0.22802	0.29995
12	0.7	0.7	0.8	1000	400	610.149	71.60	0.22851	0.30063
13	0.7	0.7	0.8	1000	400	584.901	70.34	0.23591	0.31120
14	0.7	0.7	0.8	1000	400	609.331	71.59	0.22882	0.30106
15	0.7	0.7	0.8	1000	400	578.971	69.70	0.23688	0.31271
16	0.7	0.7	0.8	1000	400	582.905	70.14	0.23627	0.31176
17	0.7	0.7	0.8	1000	400	594.171	71.06	0.23377	0.30805
18	0.7	0.7	0.8	1000	400	570.213	68.50	0.23767	0.31414
19	0.7	0.7	0.8	1000	400	597.990	71.27	0.23268	0.30649
20	0.7	0.7	0.8	1000	400	573.696	69.02	0.23745	0.31369
21	0.7	0.7	0.8	1000	400	577.233	69.49	0.23710	0.31308
22	0.7	0.7	0.8	1000	400	586.318	70.47	0.23563	0.31078
23	0.7	0.7	0.8	1000	400	596.827	71.21	0.23302	0.30698
24	0.7	0.7	0.8	1000	400	589.028	70.70	0.23505	0.30991
25	0.7	0.7	0.8	1000	400	590.672	70.83	0.23466	0.30935

the GRA and PROMETHEE methods and presented in Table 7.14 as the corrected ranks.

The SAP-Rao and Rao-2 algorithms solutions have achieved the least average rank of 2. However, the SAP-Rao algorithm's solution ranked one by four (SAW, WPM, VIKOR, and COPRAS) methods, whereas the Rao-2 algorithm ranked one by three (TOPSIS, MTOPSIS, and VIKOR) methods. Hence, the SAP-Rao and Rao-2 algorithms solutions are assigned with the final ranks of 1 and 2, respectively. The AMTPG-Jaya, ERao-3, and ERao-1 algorithms solutions have the next best average ranks, which are 2.83, 3.17, and 3.33, respectively. The SAP-Rao algorithm solution ranked one by four methods (SAW, WPM, VIKOR, and COPRAS) and achieved a better average rank value. Furthermore, the power output of the system for the SAP-Rao algorithm solution is 2.72%, 2.1%, and 2.1% higher than that of

Table 7.8 Pareto-optimal solutions obtained by the Rao-3 algorithm for the solar-assisted Brayton engine system case study

Solution	ε_H	ε_L	ε_R	T_H (K)	T_L (K)	T_1 (K)	P (kW)	η_m	F
1	0.7	0.7	0.8	1000	400	563.259	67.32	0.23771	0.31450
2	0.7	0.7	0.8	1000	400	566.298	67.86	0.23776	0.31443
3	0.7	0.7	0.8	1000	400	612.229	71.60	0.22769	0.29950
4	0.7	0.7	0.8	1000	400	583.172	70.17	0.23623	0.31169
5	0.7	0.7	0.8	1000	400	586.873	70.52	0.23552	0.31061
6	0.7	0.7	0.8	1000	400	574.808	69.17	0.23736	0.31352
7	0.7	0.7	0.8	1000	400	578.664	69.67	0.23692	0.31278
8	0.7	0.7	0.8	1000	400	605.353	71.53	0.23028	0.30310
9	0.7	0.7	0.8	1000	400	580.778	69.91	0.23662	0.31230
10	0.7	0.7	0.8	1000	400	607.885	71.57	0.22936	0.30182
11	0.7	0.7	0.8	1000	400	569.338	68.37	0.23771	0.31422
12	0.7	0.7	0.8	1000	400	609.849	71.59	0.22862	0.30079
13	0.7	0.7	0.8	1000	400	611.078	71.60	0.22815	0.30013
14	0.7	0.7	0.8	1000	400	593.913	71.05	0.23384	0.30815
15	0.7	0.7	0.8	1000	400	592.007	70.92	0.23433	0.30887
16	0.7	0.7	0.8	1000	400	601.867	71.43	0.23146	0.30477
17	0.7	0.7	0.8	1000	400	594.656	71.09	0.23364	0.30786
18	0.7	0.7	0.8	1000	400	608.581	71.58	0.22910	0.30146
19	0.7	0.7	0.8	1000	400	572.332	68.82	0.23755	0.31389
20	0.7	0.7	0.8	1000	400	588.495	70.66	0.23517	0.31009
21	0.7	0.7	0.8	1000	400	604.496	71.51	0.23058	0.30352
22	0.7	0.7	0.8	1000	400	596.224	71.18	0.23320	0.30723
23	0.7	0.7	0.8	1000	400	571.704	68.73	0.23760	0.31397
24	0.7	0.7	0.8	1000	400	600.684	71.39	0.23185	0.30531
25	0.7	0.7	0.8	1000	400	565.190	67.67	0.23775	0.31447

the NSGA-II (Shannon-entropy, LINMAP, and TOPSIS), respectively. Similarly, the system's thermal efficiency for the SAP-Rao algorithm solution is 1.1%, 1.4%, and 1.34% higher than that of the NSGA-II algorithm (Shannon-entropy, LINMAP, and TOPSIS) solutions, respectively. The system's thermo-economic function for the SAP-Rao algorithm solution is 1%, 1.4%, and 1.34% higher when compared to that of the NSGA-II algorithm (Shannon-entropy, LINMAP, and TOPSIS) solutions, respectively.

Now, the performances of the Jaya algorithm, Rao algorithms, and their modified versions in the MOO scenario are evaluated based on the hypervolume, coverage, and spacing indicators. Table 7.16 presents the hypervolume and spacing values of the Pareto-fronts obtained by different algorithms in the MOO scenario. The Pareto-front obtained by the NSGA-II algorithm was not reported by Li et al. (2015). Thus, the

Table 7.9 Pareto-optimal solutions obtained by the AMTPG-Jaya algorithm for the solar-assisted Brayton engine system case study

Solution	ε_H	ε_L	ε_R	T_H (K)	T_L (K)	T_1 (K)	P (kW)	η_m	F
1	0.7	0.699999991	0.8	1000	400	563.262	67.32	0.23771	0.31450
2	0.7	0.7	0.8	1000	400	566.311	67.87	0.23776	0.31443
3	0.7	0.699999991	0.8	1000	400	612.244	71.60	0.22768	0.29949
4	0.7	0.699999993	0.8	1000	400	606.777	71.56	0.22977	0.30238
5	0.7	0.7	0.8	1000	400	581.877	70.03	0.23644	0.31203
6	0.7	0.699999994	0.8	1000	400	575.446	69.26	0.23729	0.31341
7	0.7	0.7	0.8	1000	400	608.815	71.59	0.22901	0.30133
8	0.7	0.699999999	0.8	1000	400	610.683	71.60	0.22830	0.30034
9	0.7	0.699999989	0.8	1000	400	600.180	71.37	0.23201	0.30553
10	0.7	0.699999998	0.8	1000	400	604.345	71.50	0.23063	0.30359
11	0.7	0.7	0.8	1000	400	597.888	71.27	0.23271	0.30654
12	0.7	0.699999993	0.8	1000	400	583.956	70.25	0.23609	0.31147
13	0.7	0.699999993	0.8	1000	400	573.810	69.03	0.23744	0.31367
14	0.7	0.7	0.8	1000	400	578.762	69.68	0.23691	0.31276
15	0.7	0.699999992	0.8	1000	400	586.778	70.51	0.23553	0.31064
16	0.7	0.7	0.8	1000	400	594.830	71.10	0.23359	0.30779
17	0.7	0.699999991	0.8	1000	400	596.870	71.21	0.23301	0.30696
18	0.7	0.699999973	0.8	1000	400	565.280	67.69	0.23776	0.31447
19	0.7	0.699999993	0.8	1000	400	593.262	71.01	0.23401	0.30840
20	0.7	0.69999999	0.8	1000	400	571.915	68.76	0.23758	0.31394
21	0.7	0.7	0.8	1000	400	578.820	69.68	0.23690	0.31275
22	0.7	0.699999992	0.8	1000	400	595.617	71.15	0.23337	0.30748
23	0.7	0.699999996	0.8	1000	400	588.576	70.66	0.23515	0.31006
24	0.7	0.699999992	0.8	1000	400	589.841	70.76	0.23486	0.30964
25	0.7	0.699999988	0.8	1000	400	603.016	71.47	0.23108	0.30423

performance metric values for the NSGA-II algorithm were not presented here. The SAP-Rao algorithm has a better spacing value, which is 49%, 6.2%, 17.64%, 10.32%, 28.43%, 43.44%, 18.34%, and 29.25% lesser value when compared to that of the Jaya, Rao-1, Rao-2, Rao-3, AMTPG-Jaya, ERao-1, ERao-2, and ERao-3 algorithms, respectively. The ERao-3 algorithm Pareto-front has a relatively higher hypervolume value. The ERao-1, ERao-2, and SAP-Rao algorithms have achieved the next better hypervolume values, respectively. Here, an observation can be made that the values of hypervolume achieved by the algorithms compared have relatively small differences with each other. It indicates that the Pareto-fronts of these algorithms are at the same level and spread over the maximum volume, which can be observed in Fig. 7.1. Besides, the distribution of the solutions along the Pareto-front is more uniform for the SAP-Rao algorithm.

Table 7.10 Pareto-optimal solutions obtained by the ERao-1 algorithm for the solar-assisted Brayton engine system case study

Solution	ε_H	ε_L	ε_R	T_H (K)	T_L (K)	T_1 (K)	P (kW)	η_m	F
1	0.7	0.7	0.8	1000	400	566.303	67.87	0.23776	0.31443
2	0.7	0.7	0.8	1000	400	612.242	71.60	0.22768	0.29949
3	0.7	0.7	0.8	1000	400	563.237	67.32	0.23771	0.31450
4	0.7	0.7	0.8	1000	400	601.768	71.43	0.23150	0.30481
5	0.7	0.7	0.8	1000	400	610.657	71.60	0.22831	0.30035
6	0.7	0.7	0.8	1000	400	606.756	71.56	0.22978	0.30240
7	0.7	0.7	0.8	1000	400	597.449	71.24	0.23284	0.30672
8	0.7	0.7	0.8	1000	400	604.745	71.51	0.23049	0.30340
9	0.7	0.7	0.8	1000	400	580.658	69.90	0.23664	0.31233
10	0.7	0.7	0.8	1000	400	594.270	71.07	0.23374	0.30801
11	0.7	0.7	0.8	1000	400	578.646	69.66	0.23692	0.31278
12	0.7	0.7	0.8	1000	400	596.705	71.21	0.23306	0.30703
13	0.7	0.7	0.8	1000	400	572.759	68.89	0.23753	0.31383
14	0.7	0.7	0.8	1000	400	592.204	70.94	0.23428	0.30880
15	0.7	0.7	0.8	1000	400	590.089	70.78	0.23480	0.30955
16	0.7	0.7	0.8	1000	400	585.153	70.36	0.23586	0.31113
17	0.7	0.7	0.8	1000	400	592.080	70.93	0.23431	0.30884
18	0.7	0.7	0.8	1000	400	588.618	70.67	0.23514	0.31005
19	0.7	0.7	0.8	1000	400	576.251	69.36	0.23721	0.31326
20	0.7	0.7	0.8	1000	400	583.024	70.15	0.23625	0.31173
21	0.7	0.7	0.8	1000	400	595.159	71.12	0.23350	0.30766
22	0.7	0.7	0.8	1000	400	583.121	70.16	0.23623	0.31170
23	0.7	0.7	0.8	1000	400	605.895	71.54	0.23008	0.30283
24	0.7	0.7	0.8	1000	400	587.024	70.53	0.23548	0.31056
25	0.7	0.7	0.8	1000	400	575.366	69.25	0.23730	0.31342

The coverage metric values of the Pareto-fronts obtained by the Jaya algorithm, Rao algorithms, and their modified versions are presented in Table 7.17. The coverage of all algorithms compared versus Rao-1 and Jaya algorithms is zero (column Jaya and Rao-1 in Table 7.17), which indicates that n the other methods dominated no solution of the Rao-1 and Jaya algorithms. In contrast, the coverage values of Rao-1 versus Rao-2, Rao-3, AMTPG-Jaya, SAP-Rao, ERao-1, ERao-2, and ERao-3 algorithms are 4, 4, 8, 4, 8, 4, and 8, respectively. It implies that the Rao-1 algorithm's solutions dominate 4% of the Rao-2 and Rao-3 algorithms solutions, 8% of AMTPG-Jaya algorithm solutions, 4% of the SAP-Rao algorithm solutions, 8% of ERao-1 algorithm solutions, 4% of ERao-2 algorithm solutions, and 8% of ERao-3 algorithm solutions. Similarly, the coverage values of all compared (except for Rao-1) algorithms versus the SAP-Rao algorithm are zero. The Jaya, Rao-1, and SAP-Rao

Table 7.11 Pareto-optimal solutions obtained by the ERao-2 algorithm for the solar-assisted Brayton engine system case study

Solution	ε_H	ε_L	ε_R	T_H (K)	T_L (K)	T_1 (K)	P (kW)	η_m	F
1	0.7	0.7	0.8	1000	400	563.245	67.32	0.23771	0.31450
2	0.7	0.7	0.8	1000	400	612.198	71.60	0.22770	0.29951
3	0.7	0.7	0.8	1000	400	566.302	67.87	0.23776	0.31443
4	0.7	0.7	0.8	1000	400	609.502	71.59	0.22875	0.30097
5	0.7	0.7	0.8	1000	400	605.098	71.52	0.23037	0.30322
6	0.7	0.7	0.8	1000	400	607.990	71.57	0.22932	0.30176
7	0.7	0.7	0.8	1000	400	593.544	71.02	0.23394	0.30829
8	0.7	0.7	0.8	1000	400	574.206	69.09	0.23741	0.31361
9	0.7	0.7	0.8	1000	400	594.693	71.09	0.23363	0.30785
10	0.7	0.7	0.8	1000	400	584.131	70.26	0.23605	0.31142
11	0.7	0.7	0.8	1000	400	598.798	71.31	0.23244	0.30614
12	0.7	0.7	0.8	1000	400	596.386	71.19	0.23315	0.30716
13	0.7	0.7	0.8	1000	400	586.394	70.48	0.23561	0.31076
14	0.7	0.7	0.8	1000	400	567.970	68.15	0.23775	0.31433
15	0.7	0.7	0.8	1000	400	602.189	71.44	0.23136	0.30462
16	0.7	0.7	0.8	1000	400	597.390	71.24	0.23286	0.30675
17	0.7	0.7	0.8	1000	400	579.851	69.80	0.23676	0.31252
18	0.7	0.7	0.8	1000	400	600.904	71.39	0.23178	0.30521
19	0.7	0.7	0.8	1000	400	581.666	70.01	0.23648	0.31208
20	0.7	0.7	0.8	1000	400	599.975	71.36	0.23207	0.30563
21	0.7	0.7	0.8	1000	400	606.915	71.56	0.22972	0.30231
22	0.7	0.7	0.8	1000	400	591.555	70.89	0.23445	0.30903
23	0.7	0.7	0.8	1000	400	588.627	70.67	0.23514	0.31005
24	0.7	0.7	0.8	1000	400	590.363	70.80	0.23474	0.30946
25	0.7	0.7	0.8	1000	400	576.956	69.45	0.23713	0.31313

algorithms have achieved better coverage values than the other algorithms compared. One out of 25 Pareto solutions of the SAP-Rao and ERao-2 algorithms are dominated by the solutions of the other algorithms compared. Similarly, two out of 25 Pareto solutions are dominated from the Pareto-fronts of the Rao-2, Rao-3, AMTPG-Jaya, ERao-1, and ERao-3 algorithms.

From the computational results of the solar-assisted Brayton engine case study, it can be observed that the performance of the considered system can be modified with the solutions achieved by the proposed algorithms. The proposed algorithms have achieved better solutions than those achieved by the NSGA-II algorithm in fewer function evaluations. The solution of the SAP-Rao and Rao-2 algorithm has better compromise among the power output, thermal efficiency, and non-dimensional thermo-economic function. By the SAP-Rao algorithm's solution, the power output

Table 7.12 Pareto-optimal solutions obtained by the ERao-3 algorithm for the solar-assisted Brayton engine system case study

Solution	ε_H	ε_L	ε_R	T_H (K)	T_L (K)	T_1 (K)	P (kW)	η_m	F
1	0.70000000	0.7	0.8	1000	400	612.237	71.60	0.22769	0.29949
2	0.69999993	0.7	0.8	999.9998262	400	566.308	67.87	0.23776	0.31443
3	0.69999994	0.7	0.8	999.9999246	400	563.211	67.31	0.23770	0.31450
4	0.69999998	0.7	0.8	999.99998	400	601.765	71.43	0.23150	0.30481
5	0.69999996	0.7	0.8	1000	400	575.366	69.25	0.23730	0.31342
6	0.70000000	0.7	0.8	999.9998695	400	605.294	71.53	0.23030	0.30313
7	0.69999998	0.7	0.8	999.9997704	400	583.180	70.17	0.23622	0.31169
8	0.70000000	0.7	0.8	1000	400	585.136	70.36	0.23586	0.31114
9	0.69999995	0.7	0.8	999.9999623	400	586.709	70.50	0.23555	0.31066
10	0.69999998	0.7	0.8	999.9999731	400	588.535	70.66	0.23516	0.31008
11	0.69999999	0.7	0.8	999.9999602	400	594.840	71.10	0.23359	0.30779
12	0.69999997	0.7	0.8	1000	400	572.018	68.78	0.23758	0.31393
13	0.70000000	0.7	0.8	999.9997307	400	610.956	71.60	0.22819	0.30019
14	0.69999993	0.7	0.8	999.9999285	400	608.760	71.58	0.22903	0.30136
15	0.69999988	0.7	0.8	999.9998717	400	580.594	69.89	0.23665	0.31234
16	0.69999991	0.7	0.8	1000	400	597.238	71.23	0.23290	0.30681
17	0.69999998	0.7	0.8	999.999813	400	592.112	70.93	0.23431	0.30883
18	0.69999995	0.7	0.8	1000	400	577.360	69.50	0.23708	0.31305
19	0.69999989	0.7	0.8	999.9999166	400	600.813	71.39	0.23181	0.30525
20	0.69999995	0.7	0.8	999.9997208	400	593.865	71.04	0.23385	0.30817
21	0.69999994	0.7	0.8	1000	400	607.084	71.56	0.22966	0.30223
22	0.69999985	0.7	0.8	999.9997966	400	567.982	68.15	0.23774	0.31433
23	0.69999981	0.7	0.8	999.9997094	400	578.461	69.64	0.23695	0.31282
24	0.69999983	0.7	0.8	999.9999883	400	606.634	71.55	0.22982	0.30246
25	0.69999992	0.7	0.8	999.9998098	400	608.556	71.58	0.22911	0.30147

is increased by 2.72%, the thermal efficiency is increased by 1.4%, and the thermo-economic function of the system is increased by 1.4% when compared to that of the NSGA-II (Shannon-entropy, LINMAP, and TOPSIS) algorithm solutions. Furthermore, the spacing value achieved by the SAP-Rao algorithm is much better than that achieved by other algorithms. Also, the SAP-Rao algorithm's performance in terms of hypervolume and coverage metrics is better or competitive to that of the other algorithms. In addition, the performances of the modified versions in this case study are better or competitive to those of the basic algorithms as well as the NSGA-II algorithm. The next section presents the application of the Jaya and Rao algorithms along with their modified versions, to the solar-assisted Stirling heat engine case studies.

Table 7.13 Best solutions obtained by various algorithms in MOO scenario of the solar-assisted Brayton engine system case study

Algorithm		ε_H	ε_L	ε_R	T_H (K)	T_L (K)	T_1 (K)	P (kW)	η_m	F
Entropy	NSGA-II	0.69	0.7	0.79	999.99	400.06	582.33	68.22	0.23390	0.30890
LINMAP		0.69	0.69	0.79	999.99	400.01	586.87	68.65	0.23310	0.30770
TOPSIS		0.7	0.7	0.8	999.99	400.01	586.69	68.64	0.23320	0.30780
Jaya		0.7	0.7	0.8	1000	400	581.8217	70.02	0.23645	0.31204
Rao-1		0.7	0.7	0.8	1000	400	581.7274	70.01	0.23647	0.31206
Rao-2		0.7	0.7	0.8	1000	400	582.9045	70.14	0.23627	0.31176
Rao-3		0.7	0.7	0.8	1000	400	580.7785	69.91	0.23662	0.31230
AMTPG-Jaya		0.7	0.7	0.8	1000	400	581.8773	70.03	0.23644	0.31203
SAP-Rao		0.7	0.7	0.8	1000	400	582.3098	70.08	0.23637	0.31192
ERao-1		0.7	0.7	0.8	1000	400	583.0244	70.15	0.23625	0.31173
ERao-2		0.7	0.7	0.8	1000	400	584.1309	70.26	0.23605	0.31142
ERao-3		0.699999976	0.7	0.8	999.99	400	583.1799	70.17	0.23622	0.31169

Result in boldface indicates better values

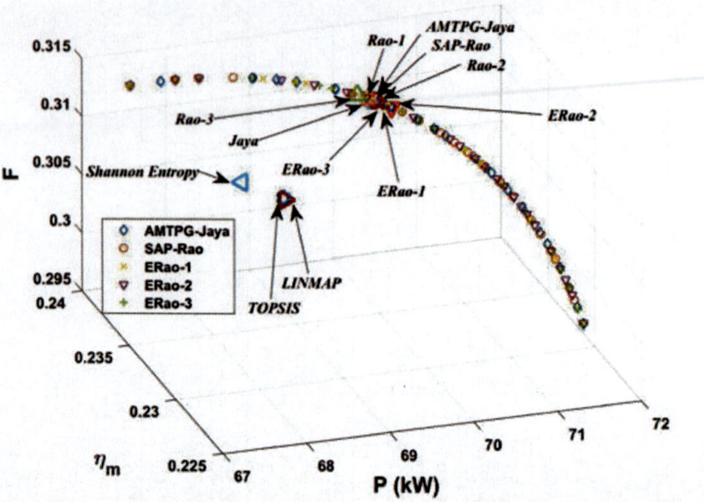

Fig. 7.1 Plot of Pareto-fronts of the proposed algorithms and solutions of the methods compared in solar-assisted Brayton engine system case study

7.2 Optimization of a Solar-Assisted Stirling Heat Engine System

The description of the solar-assisted Stirling heat engine system is presented in Sect. 2.2.2. Three case studies of the solar-assisted heat engine system are considered for optimization in this book. These case studies are presented in the subsequent sections.

7.2.1 *Case Study-1*

The description of this case study is presented in Sect. 2.2.2.1. The detailed description and thermodynamic analysis of the solar-assisted Stirling heat engine system considered in this case study were presented by Ahmadi et al. (2013a). The performance criteria of this case study are to maximize output power (P) and thermal efficiency (η_m), and minimize the rate of entropy generation (σ). These are given in Eqs. 2.24–2.26. The decision variables and their boundaries are presented in Table 7.18. The characteristics of the solar-assisted Stirling heat engine system taken are as follows: $\lambda = 2$, $\varepsilon = 0.9$, $I = 1000\text{Wm}^{-2}$, $K_0 = 2.5\text{WK}^{-1}$, $C_v = 15\text{Jmol}^{-1}\text{k}^{-1}$, $\zeta = 2\times10^{-10}$, $R = 4.3\text{Jmol}^{-1}\text{K}^{-1}$, $\delta = 5.67\times10^{-8}\text{WK}^{-4}\text{m}^{-2}$, $h = 20\text{Wm}^{-2}\text{K}^{-1}$, $\eta_0 = 0.9$ $(1/M_1 + 1/M_2) = 2 \times 10^{-5}\text{sK}^{-1}$, $T_{H1} = 1300\text{K}$, $T_{L1} = 290\text{K}$, $T_0 = 288\text{K}$, $C = 1300$, and $n = 1\text{mol}$.

Table 7.14 Ranks suggested by the MADM methods for different algorithm solutions presented in Table 7.13

Algorithm	SAW	WPM	TOPSIS	MTOPSIS	VIKOR	PROMETHEE	COPRAS	GRA	AR	CR	FR
Entropy	10	10	12	12	1	10	10	10	9.4	9.17	10
LINMAP	12	12	10	10	12	12	12	12	11.5	11.3	12
TOPSIS	11	11	11	11	11	11	11	11	11	11	11
Jaya	7	7	7	7	1	3	7	3	5.3	6	7
Rao-1	6	6	8	8	1	2	6	2	4.9	5.83	6
Rao-2	3	3	1	1	1	6	3	6	3	2	2
Rao-3	8	8	9	9	1	1	8	1	5.6	7.17	9
AMTPG-Jaya	2	2	5	5	1	4	2	4	3.1	2.83	3
SAP-Rao	1	1	4	4	1	5	1	5	2.8	2	1
ERao-1	5	5	2	2	1	7	5	7	4.3	3.33	5
ERao-2	9	9	6	6	1	9	9	9	7.3	6.67	8
ERao-3	4	4	3	3	1	8	4	8	4.4	3.17	4

AR average rank, *CR* corrected rank, *FR* final rank

Table 7.15 Spearman's rank correlation coefficients between different pairs of MADM method's rankings presented in Table 7.14

Method	SAW	WPM	TOPSIS	MTOPSIS	VIKOR	PROMETHEE	COPRAS	GRA
SAW	1	1	0.811	0.811	0.650	0.545	1	0.545
WPM	1	1	0.811	0.811	0.650	0.545	1	0.545
TOPSIS	0.811	0.811	1	1	0.511	0.266	0.811	0.266
MTOPSIS	0.811	0.811	1	1	0.511	0.266	0.811	0.266
VIKOR	0.650	0.650	0.511	0.511	1	0.650	0.650	0.650
PROMETHEE	0.545	0.545	0.266	0.266	0.650	1	0.545	1
COPRAS	1	1	0.811	0.811	0.650	0.545	1	0.545
GRA	0.545	0.545	0.266	0.266	0.650	1	0.545	1

Table 7.16 Hypervolume and spacing values of the Pareto-fronts obtained by the proposed algorithms in MOO for the solar-assisted Brayton engine case study

Algorithm	Hypervolume	Spacing
Jaya	0.0265084	0.0562112
Rao-1	0.0265094	0.0305907
Rao-2	0.0265131	0.0348516
Rao-3	0.0265120	0.0320051
AMTPG-Jaya	0.0265110	0.0401055
SAP-Rao	0.0265148	**0.0287033**
ERao-1	0.026521	0.050744
ERao-2	0.026517	0.035148
ERao-3	**0.026533**	0.040567

Results in the bold figure indicate better values

Table 7.17 Coverage (%) values of the Pareto-fronts obtained by the proposed algorithms in MOO for the solar-assisted Brayton engine case study

Algorithm	Jaya	Rao-1	Rao-2	Rao-3	AMTPG-Jaya	SAP-Rao	ERao-1	ERao-2	ERao-3
Jaya	–	0	8	8	8	0	8	4	8
Rao-1	0	–	4	4	8	4	8	4	8
Rao-2	0	0	–	0	4	0	8	4	8
Rao-3	0	0	8	–	4	0	8	4	8
AMTPG-Jaya	0	0	0	0	–	0	4	4	4
SAP-Rao	0	0	8	8	8	–	8	4	8
ERao-1	0	0	0	0	4	0	–	0	4
ERao-2	0	0	0	0	4	0	4	–	4
ERao-3	0	0	0	0	4	0	4	0	–

Table 7.18 Decision variables and their boundaries of the Stirling heat engine case study-1

Decision variables	Boundaries
Regenerator effectiveness (ε_R)	$0.4 \leq \varepsilon_R \leq 0.9$
High-temperature heat exchanger effectiveness (ε_H)	$0.4 \leq \varepsilon_H \leq 0.8$
Low-temperature heat exchanger effectiveness (ε_L)	$0.4 \leq \varepsilon_L \leq 0.8$
Heat capacitance rate of the heat source (C_H)	$300 \leq C_H \leq 1800$
Heat capacitance rate of the heat sink (C_L)	$300 \leq C_L \leq 1800$
Working fluid temperature during the isothermal expansion (T_h)	$800 \leq T_h \leq 1000$
Working fluid temperature during isothermal compression (T_c)	$400 \leq T_c \leq 510$

Ahmadi et al. (2013a) solved this multi-objective optimization problem using the NSGA-II algorithm and reported three Pareto-optimal solutions. These three solutions were documented based on the similarity to the ideal solutions calculated by employing TOPSIS, LINMAP, and Fuzzy Bellman–Zadeh decision-making methods. The number of function evaluations taken by the NSGA-II is 400,000. However, in this book, the Jaya algorithm, Rao algorithms, and their modified versions are tested by taking 10,000 function evaluations as the termination criterion. The best Pareto-optimal solutions are reported using the average rank based on multiple decision-making methods. Similar to the previous case studies, in this case study also, first single-objective optimization is performed for each objective function independently, and then multi-objective optimization is performed by the proposed algorithms. In all the computational experiments, the population size is maintained as 25 for all the algorithms. The elite size for the elitist Rao algorithms is taken as 20%.

Table 7.19 presents the single-objective optimization results of the Stirling heat engine case study-1. The Jaya algorithm, Rao algorithms, and their modified versions have obtained identical solutions in all the single-objective optimization scenarios,

Table 7.19 Results obtained by the proposed algorithms for single-objective optimization scenarios of the solar-assisted Stirling heat engine case study-1

Objective	ε_R	ε_H	ε_L	C_H	C_L	T_h	T_c	P (kW)	η_m (%)	σ (W/K)
Maximize-P	0.9	0.8	0.8	1800	1800	1000	510	70.3417	31.5795	139.2852
Maximize- η_m	0.9	0.8	0.8	1800	1800	1000	400	61.1952	36.5151	106.3605
Minimize- σ	0.9	0.4	0.4	300	300	1000	400	8.1778	32.2944	21.9193

which are presented in Table 7.19. The maximum power output and thermal efficiency obtained by the proposed algorithms are 70.3417 kW and 36.5151%, respectively. Similarly, the minimum entropy obtained by the proposed algorithms is 21.9193 W/K.

The Pareto-optimal solutions achieved by the Jaya algorithm, Rao algorithms, and their modified versions in this case study are presented in Tables 7.20, 7.21, 7.22, 7.23, 7.24, 7.25, 7.26, 7.27 and 7.28. The best solutions from the Pareto-fronts obtained by the proposed algorithms are identified based on the average rank method. Solution 13 of the Jaya algorithm, Solution 15 of the Rao-1 algorithm, Solution 17 of the Rao-2 algorithm, Solution 8 of the Rao-3 algorithm, Solution 7 of the AMTPG-Jaya algorithm, Solution 1 of the SAP-Rao algorithms, Solution 3 of the ERao-1 algorithm, Solution 18 of the ERao-2, and Solution 4 of the ERao-3 algorithm are identified as the best solutions from the respective algorithm Pareto-front and are compared in Table 7.29.

In Table 7.29, the solutions with the best average rank from the Pareto-optimal solutions of the proposed algorithms are compared with those reported by the NSGA-II algorithm. Figure 7.2 presents the Pareto-fronts achieved by the AMTPG-Jaya, SAP-Rao, and elitist Rao algorithms, including the optimal solutions reported for the NSGA-II, Jaya, and Rao algorithms. The conflicting nature of these objectives can be observed in Fig. 7.2. Any change in the design variables that will result in the increment of output power also leads to the increment of the generated entropy, which is not desirable. Also, the variation of thermal efficiency is relatively less when compared to the other objectives. It can also be observed from the Pareto-optimal solutions obtained by the algorithms. It indicates that the effect of the design variables on thermal efficiency is relatively less.

The decision-making methods ranking of the Pareto-optimal solutions obtained by different algorithms is shown in Table 7.30. Spearman's correlation coefficients for different pairs of rankings given by decision-making methods for different algorithm's solutions are shown in Table 7.31. The ranks given by the TOPSIS and MTOPSIS methods for each solution are identical. Furthermore, Spearman's correlation for all the pairs of decision-making methods (except with the VIKOR, PROMETHEE, and GRA methods) is greater than 0.5. Spearman's correlation values for the VIKOR-SAW, VIKOR-COPRAS, and VIKOR-GRA pairs are negative, and with other methods are less than 0.5. Similarly, the PROMETHEE method has Spearman's correlation values less than 0.5 with the SAW and COPRAS methods. The GRA method has Spearman's correlation values less than 0.5 with the WPM, TOPSIS, and MTOPSIS methods. Hence, the ranks suggested by the VIKOR, PROMETHEE, and GRA methods cannot be considered for calculating the average ranks. Now, the corrected average ranks are calculated, excluding the ranks suggested by the VIKOR, PROMETHEE, and GRA methods and presented in Table 7.30 as corrected ranks.

The ERao-3 algorithm solution has the least average rank, which is 1.4. Thus, it can be regarded as the best solution. The SAP-Rao, ERao-2, ERao-1, and AMTPG-Jaya algorithm's solutions have the next best average ranks, which are 2.4, 2.6, 4.8, and 6.2, respectively. The ERao-3 algorithm solution ranked one by three methods (SAW,

Table 7.20 Pareto-optimal solutions obtained by the Jaya algorithm for the solar-assisted Stirling heat engine case study-1

Solution	ε_R	ε_H	ε_L	C_H	C_L	T_h	T_c	P (kW)	η_m (%)	σ (W/K)
1	0.9	0.8	0.8	1800	1658.125	1000	400	58.9167	36.4942	102.5694
2	0.9	0.8	0.8	1800	1800	1000	508.8175	70.3137	31.6357	138.9775
3	0.899991	0.4	0.4	300	300	1000	400	8.1776	32.2937	21.9196
4	0.89996	0.8	0.8	1800	1800	1000	454.1751	67.6780	34.1603	124.2274
5	0.9	0.8	0.8	1800	1800	1000	482.4467	69.3978	32.8731	132.0330
6	0.899967	0.704188	0.8	1570.617	300	1000	440.3984	21.9561	33.7163	43.0688
7	0.9	0.8	0.8	1800	1285.157	1000	400	51.8134	36.4175	90.7505
8	0.89993	0.597117	0.698768	1744.565	300	1000	400	16.5111	34.6532	34.1212
9	0.9	0.77818	0.408339	1589.13	354.4436	1000	400.3019	12.0975	34.3312	28.2523
10	0.899946	0.792293	0.799236	1314.779	300	1000	400	18.4562	35.2864	35.3098
11	0.9	0.8	0.8	1800	1800	1000	503.9247	70.1866	31.8677	137.7023
12	0.899994	0.79626	0.799964	1271.048	768.6807	1000	400	36.7631	36.1514	65.7481
13	0.9	0.8	0.8	1183.395	1202.477	1000	400	47.2120	36.3557	83.0944
14	0.899995	0.8	0.8	657.2448	423.4564	1000	432.614	25.4262	34.4116	48.1493
15	0.9	0.8	0.8	1800	1800	1000	425.7529	64.9363	35.4189	115.5213
16	0.899995	0.787161	0.8	1319.502	1292.852	1000	406.9967	50.8080	36.0785	89.8207
17	0.899954	0.8	0.8	1123.41	678.7381	1000	400	33.4460	36.0686	60.2063
18	0.899981	0.523338	0.46452	468.7195	300	1000	400	10.3932	33.3658	25.3967
19	0.899996	0.607953	0.790123	1498.395	435.6892	1000	411.2081	25.5686	34.9166	49.0071
20	0.899996	0.795085	0.692617	1275.527	392.8277	1000	400.2389	20.4383	35.4386	40.0208
21	0.9	0.8	0.8	1800	1800	1000	418.5825	64.0346	35.7285	113.1250

(continued)

Table 7.20 (continued)

Solution	ε_R	ε_H	ε_L	C_H	C_L	T_h	T_c	P (kW)	η_m (%)	σ (W/K)
22	0.899981	0.556517	0.715937	1103.67	1381.123	1000	400	43.9929	35.7976	82.7098
23	0.899989	0.79149	0.798764	1080.825	540.97	1000	426.7059	31.5009	34.8672	58.4120
24	0.9	0.8	0.8	1433.868	510.697	1000	400.3588	28.1693	35.8777	51.4242
25	0.9	0.68388	0.67073	597.289	1800	1000	400	42.6730	36.0409	80.2348

Table 7.21 Pareto-optimal solutions obtained by the Rao-1 algorithm for the solar-assisted Stirling heat engine case study-1

Solution	ε_R	ε_H	ε_L	C_H	C_L	T_h	T_c	P (kW)	η_m (%)	σ (W/K)
1	0.9	0.7993391	0.4	300	300	1000	400	9.1893	33.6236	22.6777
2	0.9	0.7998926	0.8	1800	1800	1000	400	61.1940	36.5149	106.3603
3	0.9	0.7991472	0.8	1800	1800	1000	510	70.3278	31.5780	139.2756
4	0.9	0.7904755	0.8	1800	1800	1000	510	70.1851	31.5629	139.1747
5	0.8999992	0.7990978	0.4	300	300	1000	449.84906	11.1113	32.2664	28.5796
6	0.8999948	0.7977162	0.8	1800	723.36467	1000	445.46871	42.3253	34.2993	78.4398
7	0.9	0.7999288	0.8	1427.562	976.44811	1000	431.57393	47.4430	34.9905	86.0785
8	0.8999959	0.8	0.68864	497.71392	390.70304	1000	454.36008	21.9000	33.2993	45.3083
9	0.9	0.7954883	0.8	1800	1800	1000	479.9931	69.2135	32.9783	131.3343
10	0.8999999	0.7980945	0.8	1738.5758	1800	1000	427.71157	64.7946	35.3275	115.5526
11	0.8999987	0.7998035	0.7358987	300	300	1000	400	14.6285	34.8663	29.5143
12	0.8999959	0.7999697	0.5836188	497.71392	390.70304	1000	419.71686	17.6958	34.4209	37.5064
13	0.8999987	0.8	0.7613621	764.91225	649.6838	1000	400	29.8551	35.9538	54.9369
14	0.8999996	0.7999219	0.554752	1593.3967	628.1976	1000	400	25.1472	35.7367	50.4324
15	0.8999994	0.7999531	0.8	1014.3719	963.43731	1000	400	40.8272	36.2473	72.4717
16	0.9	0.794358	0.8	1800	1800	1000	492.82592	69.7394	32.3798	134.7348
17	0.8999986	0.8	0.7611377	1740.7027	740.78232	1000	400	35.8488	36.1345	65.0491
18	0.8999923	0.8	0.7461217	1516.9655	1239.4437	1000	417.19205	50.5072	35.6554	91.9435
19	0.8999995	0.7999323	0.8	1167.7503	1800	1000	400	56.9574	36.4747	99.3107
20	0.8999996	0.7999779	0.7615532	1593.3967	628.1976	1000	400	31.7681	36.0187	58.1590
21	0.9	0.7998574	0.8	1511.6418	1491.5292	1000	431.57393	58.9219	35.1131	105.8793

(continued)

Table 7.21 (continued)

Solution	ε_R	ε_H	ε_L	C_H	C_L	T_h	T_c	P (kW)	η_m (%)	σ (W/K)
22	0.8999998	0.7988243	0.7990539	1689.386	1367.0839	1000	409.67036	54.6618	36.0305	96.4075
23	0.8999951	0.7976211	0.5115517	1060.6435	440.2292	1000	430.38431	19.7992	34.1461	43.3318
24	0.8999925	0.7999699	0.7504792	1717.1978	1800	1000	417.19205	61.6812	35.7683	111.1097
25	0.8999991	0.7999573	0.7983882	784.03596	920.36155	1000	400	37.9666	36.1871	67.7496

Table 7.22 Pareto-optimal solutions obtained by the Rao-2 algorithm for the solar-assisted Stirling heat engine case study-1

Solution	ε_R	ε_H	ε_L	C_H	C_L	T_h	T_c	P (kW)	η_m (%)	σ (W/K)
1	0.9	0.8	0.4	300	300	1000	400	9.1902	33.6254	22.6773
2	0.9	0.8	0.8	1800	1800	1000	510	70.3417	31.5795	139.2852
3	0.9	0.8	0.8	1800	1800	1000	400	61.1952	36.5151	106.3605
4	0.89849	0.8	0.8	1800	300	1000	476.264	24.3999	32.3919	48.8891
5	0.9	0.8	0.54796	300	839.736	996.853	400	23.1091	35.5890	46.9788
6	0.89997	0.8	0.78054	833.376	1032.08	999.54	426.698	43.3582	35.1350	79.3279
7	0.89749	0.8	0.77452	434.719	300	1000	424.315	17.6961	34.0878	35.3926
8	0.89995	0.8	0.8	1800	1800	1000	475.228	69.0345	33.2031	130.1067
9	0.9	0.8	0.78212	1587.17	519.566	999.645	400	28.2197	35.8892	51.8265
10	0.9	0.8	0.8	374.233	1800	994.786	400	40.3039	36.1748	71.9599
11	0.9	0.8	0.69471	724.233	763.005	999.751	402.18	31.1674	35.9004	58.6238
12	0.8996	0.8	0.8	1800	675.108	1000	400	34.8627	36.0806	62.6978
13	0.9	0.8	0.55558	300	300	1000	400	11.9058	34.3738	26.4405
14	0.9	0.8	0.8	1800	1800	1000	447.668	67.1569	34.4544	122.3130
15	0.9	0.8	0.8	1800	733.798	999.725	400	36.9327	36.1602	66.0082
16	0.89986	0.8	0.8	1690.33	1800	1000	439.919	65.7667	34.7841	118.8811
17	0.9	0.8	0.76482	614.81	1800	999.994	400	47.5994	36.3597	84.7584
18	0.9	0.8	0.44915	702.996	437.89	999.811	400	15.0281	34.8825	33.2781
19	0.9	0.8	0.8	1800	1800	1000	462.356	68.2694	33.7938	126.5347
20	0.89934	0.8	0.611	300	322.502	1000	417.985	14.4749	34.0827	31.2396
21	0.9	0.8	0.8	1800	1800	999.874	505.109	70.2179	31.8092	138.0354

(continued)

Table 7.22 (continued)

Solution	ε_R	ε_H	ε_L	C_H	C_L	T_h	T_c	P (kW)	η_m (%)	σ (W/K)
22	0.89913	0.8	0.8	1800	1645.32	1000	422.99	62.0488	35.4543	110.8163
23	0.89998	0.8	0.8	1800	1800	999.892	428.497	65.2527	35.2963	116.4362
24	0.9	0.8	0.79721	1800	1676.84	999.639	427.318	63.1123	35.3308	112.7617
25	0.89903	0.8	0.77101	1089.26	1563.81	999.879	400	51.8502	36.3396	92.2784

Table 7.23 Pareto-optimal solutions obtained by the Rao-3 algorithm for the solar-assisted Stirling heat engine case study-1

Solution	ε_R	ε_H	ε_L	C_H	C_L	T_h	T_c	P (kW)	η_m (%)	σ (W/K)
1	0.9	0.8	0.8	1800	1800	1000	400	61.1952	36.5151	106.3605
2	0.9	0.8	0.8	1800	1800	1000	510	70.3417	31.5795	139.2852
3	0.9	0.759994	0.4	300	300	1000	400	9.1307	33.5151	22.6933
4	0.9	0.8	0.8	1800	1800	1000	480.7015	69.3158	32.9538	131.5648
5	0.9	0.8	0.8	1800	1800	999.8908	492.5472	69.8168	32.4009	134.7371
6	0.9	0.8	0.8	1029.872	1800	1000	400	55.4940	36.4596	96.8746
7	0.9	0.602877	0.739097	679.3916	830.7619	999.8761	420.6716	33.1765	34.7873	64.0671
8	0.9	0.8	0.764364	904.36	1800	1000	400	52.8852	36.4289	93.6753
9	0.9	0.78514	0.8	1800	788.0152	1000	400.8879	38.8023	36.1386	69.3042
10	0.9	0.737009	0.8	1800	940.6118	1000	430.7336	47.2919	34.9050	86.5230
11	0.9	0.8	0.8	1800	1800	999.8031	506.6204	70.2571	31.7361	138.4435
12	0.9	0.8	0.8	1800	1800	999.9968	418.4326	64.0147	35.7349	113.0741
13	0.9	0.796108	0.8	1800	1800	1000	449.0885	67.2234	34.3836	122.7064
14	0.9	0.8	0.542715	1800	1455.145	999.7456	400	44.5934	36.2996	86.0298
15	0.9	0.8	0.727704	1800	360.5263	1000	400	20.2042	35.4445	39.1078
16	0.9	0.740574	0.566445	1423.274	300	999.9941	400	13.8188	34.5943	30.0496
17	0.9	0.8	0.8	1771.044	1800	999.8016	433.0936	65.5975	35.0948	117.6151
18	0.9	0.71598	0.8	1800	809.9594	999.6809	421.3975	42.0913	35.1851	77.0150
19	0.9	0.8	0.669274	300	640.3507	1000	400	22.2237	35.5801	43.4088
20	0.9	0.587261	0.795635	1800	498.2299	1000	400	27.3234	35.4105	51.6205
21	0.9	0.8	0.8	1800	1800	1000	460.7346	68.1595	33.8673	126.0785

(continued)

Table 7.23 (continued)

Solution	ε_R	ε_H	ε_L	C_H	C_L	T_h	T_c	P (kW)	η_m (%)	σ (W/K)
22	0.9	0.771132	0.8	932.3202	1085.606	999.9201	407.9979	43.6879	35.9002	78.1181
23	0.9	0.782422	0.451933	1157.008	1042.876	999.9834	400	30.4433	35.9178	62.3189
24	0.9	0.727855	0.695886	1800	484.1864	1000	400	24.5437	35.5681	47.5053
25	0.9	0.787554	0.4	300	1763.795	1000	400	28.1942	35.8388	59.0871

Table 7.24 Pareto-optimal solutions obtained by the AMTPG-Jaya algorithm for the solar-assisted Stirling heat engine case study-1

Solution	ε_R	ε_H	ε_L	C_H	C_L	T_h	T_c	P (kW)	η_m (%)	σ (W/K)
1	0.9	0.7997667	0.4	300	300	1000	400	9.1899	33.6248	22.6774
2	0.9	0.8	0.8	1800	1800	1000	400	61.1952	36.5151	106.3605
3	0.9	0.8	0.8	1800	1800	1000	510	70.3417	31.5795	139.2852
4	0.9	0.8	0.7280096	300	300	1000	488.89549	17.9903	31.5698	38.8740
5	0.9	0.795675	0.7651346	1800	886.78117	1000	497.17225	50.2091	32.0413	100.6726
6	0.9	0.7998091	0.8	300	683.48999	1000	401.42904	25.3323	35.7071	46.7413
7	0.9	0.7992839	0.8	1800	1170.7675	1000	400	49.2301	36.3828	86.4613
8	0.9	0.8	0.8	1433.027	1732.6344	1000	408.77653	59.4577	36.1191	104.3307
9	0.9	0.7993774	0.7347561	635.03515	300	1000	401.32919	16.4163	35.0392	32.6046
10	0.9	0.7960857	0.7780951	1519.4813	853.20034	1000	424.0343	42.7386	35.2415	78.1069
11	0.9	0.7994338	0.4835576	1800	300	1000	400	12.1547	34.4122	27.5551
12	0.9	0.79957	0.8	1462.5721	1032.6008	1000	400	44.7435	36.3165	78.9921
13	0.9	0.8	0.8	1800	1454.501	1000	458.55906	62.3999	33.9292	115.4491
14	0.9	0.7995988	0.7500945	631.66397	391.31351	1000	400	20.2652	35.4505	38.9164
15	0.9	0.7992906	0.8	300	997.75294	1000	400	29.8357	35.9545	54.1883
16	0.9	0.8	0.8	1800	1800	1000	472.80059	68.9079	33.3176	129.4248
17	0.9	0.7998415	0.8	1505.7135	835.98003	1000	400	39.5294	36.2209	70.3132
18	0.9	0.8	0.8	1800	1800	1000	479.14725	69.2404	33.0256	131.1465
19	0.9	0.8	0.8	1800	1800	1000	453.86381	67.6589	34.1771	124.1197
20	0.9	0.7972386	0.7982886	707.17944	673.4566	1000	415.03194	32.6803	35.4109	59.7417
21	0.9	0.8	0.8	1800	1800	1000	496.58112	69.9606	32.2139	135.7796

(continued)

Table 7.24 (continued)

Solution	ε_R	ε_H	ε_L	C_H	C_L	T_h	T_c	P (kW)	η_m (%)	σ (W/K)
22	0.9	0.8	0.8	1648.3672	693.08294	1000	408.92034	36.6103	35.7783	65.9608
23	0.9	0.8	0.8	1293.1109	583.83794	1000	439.81491	35.1887	34.4156	65.4908
24	0.9	0.7991705	0.8	1800	1800	1000	441.50501	66.5952	34.7267	120.4671
25	0.9	0.790367	0.8	1800	1800	1000	427.31977	65.0008	35.3318	115.9869

Table 7.25 Pareto-optimal solutions obtained by the SAP-Rao algorithm for the solar-assisted Stirling heat engine case study-1

Solution	ε_R	ε_H	ε_L	C_H	C_L	T_h	T_c	P (kW)	η_m (%)	σ (W/K)
1	0.9	0.8	0.8	1800	1800	999.487	400	61.1910	36.5088	106.4076
2	0.9	0.8	0.4	300	300	999.693	400	9.1888	33.6223	22.6793
3	0.9	0.8	0.8	1800	1800	1000	510	70.3417	31.5795	139.2852
4	0.9	0.79144	0.8	1800	1800	1000	492.684	69.6880	32.3812	134.6675
5	0.9	0.8	0.8	1800	1800	999.884	501.222	70.1074	31.9931	137.0180
6	0.89977	0.79562	0.8	1588.27	326.146	999.883	404.088	20.4030	35.2752	38.6770
7	0.89988	0.8	0.8	1800	835.521	998.804	417.629	42.8593	35.5050	77.2393
8	0.9	0.79767	0.8	1800	895.226	999.682	419.679	44.9869	35.4664	80.9142
9	0.9	0.8	0.8	624.594	455.955	1000	400.36	23.6571	35.6631	43.9138
10	0.9	0.79767	0.79978	1469.99	864.425	999.761	400	40.2361	36.2282	71.5335
11	0.89856	0.79957	0.78421	953.614	407.592	999.874	433.086	25.6267	34.2974	49.2339
12	0.9	0.79641	0.8	1800	1800	999.911	439.244	66.3428	34.8200	119.7760
13	0.8992	0.79966	0.79938	1052.09	598.05	999.705	400	30.4068	35.9111	55.4246
14	0.89889	0.8	0.8	1560.48	1800	1000	415.379	62.0635	35.7697	110.1454
15	0.9	0.8	0.8	1800	1800	998.867	484.438	69.4788	32.7602	132.7598
16	0.89993	0.79996	0.49596	414.555	300	999.94	400	11.3677	34.2410	25.9941
17	0.89983	0.79479	0.68601	1456.87	300	999.612	400	16.3297	35.0491	33.0279
18	0.89997	0.8	0.74426	373.746	1208.09	999.92	400	34.0686	36.0832	62.4218
19	0.89998	0.79626	0.8	1800	1800	999.874	420.826	64.2788	35.6213	113.8938
20	0.9	0.8	0.4934	752.86	359.949	999.748	400	13.8936	34.7220	30.6847
21	0.9	0.8	0.8	1800	1634.38	999.621	444.846	64.3546	34.5572	117.0447

(continued)

Table 7.25 (continued)

Solution	ε_R	ε_H	ε_L	C_H	C_L	T_h	T_c	P (kW)	η_m (%)	σ (W/K)
22	0.9	0.8	0.8	1160.23	300	999.678	400	18.3419	35.2928	35.0680
23	0.89994	0.79788	0.8	1375.47	1245.01	999.631	401.416	49.3976	36.3125	86.9306
24	0.9	0.8	0.8	1800	993.484	999.571	439.187	50.1814	34.6822	91.5984
25	0.89959	0.8	0.8	593.161	925.798	998.95	405.246	36.3941	35.8853	65.6120

Table 7.26 Pareto-optimal solutions obtained by the ERao-1 algorithm for the solar-assisted Stirling heat engine case study-1

Solution	ε_R	ε_H	ε_L	C_H	C_L	T_h	T_c	P (kW)	η_m (%)	σ (W/K)
1	0.9	0.8	0.8	1800	1800	999.9999999	510	70.3417	31.5795	139.2852
2	0.9	0.8	0.8	1800	1800	999.9999999	510	70.3417	31.5795	139.2852
3	0.9	0.8	0.8	1800	1384.223	999.9999942	400	53.8803	36.4418	94.1895
4	0.9	0.8	0.4	300	300	1000	400	9.1902	33.6254	22.6773
5	0.9	0.8	0.8	1800	1800	999.9999978	402.7928	61.6738	36.3984	107.4368
6	0.9	0.8	0.8	1800	1800	999.9999975	464.3999	68.4037	33.7010	127.1065
7	0.9	0.8	0.8	1800	1560.526	999.9999	404.1432	57.9419	36.3058	101.3368
8	0.9	0.8	0.758302	300	484.8919	1000	403.7585	20.6328	35.3282	39.5468
9	0.9	0.8	0.8	1800	1800	1000	477.3006	69.1478	33.1108	130.6479
10	0.9	0.8	0.799987	1780.1	1800	1000	501.2853	69.9676	31.9917	136.7422
11	0.9	0.8	0.8	1800	1137.797	999.9998096	414.3926	50.7820	35.7835	90.2424
12	0.9	0.8	0.490663	1658.091	398.9937	1000	400	15.7543	34.9841	34.1529
13	0.9	0.8	0.8	1800	824.046	1000	446.8504	45.9118	34.2949	84.8385
14	0.9	0.8	0.8	1800	348.3517	1000	410.0981	22.3250	35.1873	41.9989
15	0.9	0.8	0.656474	1520.034	1001.658	1000	400	39.1493	36.2055	73.1958
16	0.9	0.8	0.8	1800	1800	1000	483.3937	69.4411	32.8292	132.2865
17	0.9	0.8	0.8	999.7795	693.4501	999.9999994	403.2892	33.8632	35.9459	61.0506
18	0.9	0.8	0.605649	300	300	1000	403.382	12.9052	34.4463	27.8750
19	0.9	0.8	0.671495	568.1433	574.2821	1000	400	24.1527	35.6942	46.7317
20	0.9	0.8	0.8	976.1455	1740.567	999.9999991	402.1646	54.3748	36.3542	95.2007
21	0.9	0.8	0.6999	300	959.0352	1000	406.9635	28.2064	35.5972	53.6635

(continued)

Table 7.26 (continued)

Solution	ε_R	ε_H	ε_L	C_H	C_L	T_h	T_c	P (kW)	η_m (%)	σ (W/K)
22	0.9	0.8	0.8	1800	1036.765	999.9998973	411.1758	47.7480	35.8833	84.8526
23	0.9	0.8	0.8	1656.925	454.2977	999.9999931	400	26.0326	35.7998	47.8548
24	0.9	0.8	0.452302	300	414.7224	999.9999939	400	13.0072	34.5703	29.4453
25	0.9	0.8	0.4	793.5581	300	1000	400	9.9568	33.8660	24.1422

Table 7.27 Pareto-optimal solutions obtained by the ERao-2 algorithm for the solar-assisted Stirling heat engine case study-1

Solution	ε_R	ε_H	ε_L	C_H	C_L	T_h	T_c	P (kW)	η_m (%)	σ (W/K)
1	0.89999	0.79967	0.4	300	302.674	1000	402.058	9.3562	33.6054	23.0647
2	0.9	0.79976	0.8	1800	1800	1000	400	61.1924	36.5144	106.3609
3	0.9	0.8	0.8	1800	1800	1000	509.812	70.3373	31.5884	139.2364
4	0.9	0.7984	0.78778	1800	1056.17	1000	413.622	48.1912	35.7799	86.2197
5	0.9	0.79728	0.8	313.946	949.652	1000	400	29.7390	35.9465	54.0450
6	0.9	0.8	0.7849	885.621	1708.63	1000	400	52.0730	36.4199	91.6585
7	0.9	0.8	0.78267	1098.45	729.57	1000	400	34.4642	36.0994	62.2540
8	0.9	0.8	0.4	549.054	444.381	1000	400	13.5309	34.6527	30.9718
9	0.9	0.8	0.74273	1056.76	300	1000	400	17.1622	35.1743	33.7460
10	0.9	0.79803	0.78403	1273.06	918.127	1000	418.425	43.1103	35.4948	78.0850
11	0.9	0.79939	0.8	1800	1800	1000	491.404	69.7651	32.4553	134.4097
12	0.9	0.7988	0.78215	1331.8	791.613	1000	400	37.1375	36.1646	66.7524
13	0.89999	0.79876	0.78903	628.371	1181.12	1000	400	40.7100	36.2416	72.5643
14	0.89999	0.7979	0.8	300	724.382	1000	400	25.9269	35.7898	47.6964
15	0.9	0.79796	0.46828	300	300	1000	410.919	10.9771	33.7392	25.8874
16	0.9	0.79649	0.8	1800	1800	1000	435.596	65.9810	34.9821	118.6415
17	0.9	0.79572	0.8	1800	1800	1000	454.517	67.6495	34.1396	124.2791
18	0.89999	0.79887	0.8	1161.01	1800	1000	400	56.8764	36.4712	99.1957
19	0.9	0.79679	0.8	1800	1800	1000	502.268	70.0878	31.9403	137.2358
20	0.9	0.8	0.8	1800	1800	1000	475.106	69.0333	33.2117	130.0530
21	0.89999	0.79946	0.7783	1474.38	378.29	1000	400	21.9403	35.5702	41.3556

(continued)

Table 7.27 (continued)

Solution	ε_R	ε_H	ε_L	C_H	C_L	T_h	T_c	P (kW)	η_m (%)	σ (W/K)
22	0.89999	0.79874	0.8	1800	1800	1000	414.186	63.4169	35.9137	111.6009
23	0.89999	0.79959	0.8	1178.3	685.771	1000	470.573	40.7407	33.1555	77.9010
24	0.9	0.8	0.8	1800	1800	1000	460.278	68.1280	33.8879	125.9497
25	0.89999	0.79684	0.78858	668.962	324.351	1000	423.058	20.2857	34.5139	39.1573

Table 7.28 Pareto-optimal solutions obtained by the ERao-3 algorithm for the solar-assisted Stirling heat engine case study-1

Solution	ε_R	ε_H	ε_L	C_H	C_L	T_h	T_c	P (kW)	η_m (%)	σ (W/K)
1	0.9	0.7993	0.8	1800	1800	1000	510	70.3303	31.5782	139.2776
2	0.9	0.8	0.40638	300	300	1000	400	9.3092	33.6663	22.8556
3	0.9	0.8	0.8	1800	1800	1000	510	70.3417	31.5795	139.2852
4	0.9	0.8	0.8	1800	1800	1000	400	61.1952	36.5151	106.3605
5	0.9	0.8	0.79984	1089.25	1377.47	1000	400	49.8168	36.3920	87.4332
6	0.89999	0.8	0.8	300	300	1000	479.001	18.8275	32.0670	38.4535
7	0.89999	0.79664	0.79917	1015.18	1618.7	1000	422.925	55.4980	35.4562	99.1495
8	0.9	0.8	0.8	1792.16	1800	1000	491.39	69.7198	32.4568	134.3101
9	0.9	0.79968	0.79993	1800	1800	1000	485.228	69.5157	32.7435	132.7746
10	0.89999	0.8	0.8	577.98	677.395	1000	400	30.0039	35.9612	54.4655
11	0.9	0.8	0.8	1322.66	657.441	1000	400	33.3474	36.0693	60.0257
12	0.9	0.79981	0.65757	300	300	1000	400	13.4957	34.6832	28.3266
13	0.9	0.79966	0.8	1800	980.22	1000	400.909	44.5633	36.2746	78.7565
14	0.9	0.8	0.54313	300	300	1000	400	11.7022	34.3286	26.1813
15	0.9	0.79963	0.8	1800	1800	1000	438.524	66.3147	34.8590	119.5639
16	0.89999	0.79731	0.79926	1728.64	300	1000	459.957	23.5262	33.1505	46.1309
17	0.9	0.8	0.8	1800	1800	1000	457.399	67.9233	34.0177	125.1345
18	0.89999	0.8	0.8	300	390.407	1000	400	18.4383	35.3045	35.2203
19	0.9	0.79961	0.8	321.168	1800	1000	400	37.4744	36.1750	66.8981
20	0.9	0.79824	0.79972	1800	641.843	1000	413.759	35.6716	35.5464	64.6867
21	0.89999	0.79717	0.79887	365.85	1800	1000	417.639	40.4816	35.4807	73.1211

(continued)

Table 7.28 (continued)

Solution	ε_R	ε_H	ε_L	C_H	C_L	T_h	T_c	P (kW)	η_m (%)	σ (W/K)
22	0.9	0.79997	0.7049	300	364.575	1000	400	16.2424	35.0690	32.6000
23	0.9	0.79897	0.8	1549.66	1632.02	1000	403.93	57.8917	36.3122	101.2500
24	0.89999	0.79952	0.8	1800	1662.64	1000	433.037	63.6254	35.0847	114.1846
25	0.9	0.8	0.49372	300	300	1000	400	10.8711	34.1284	25.0840

Table 7.29 Best solutions obtained by various algorithms in MOO scenario of the solar-assisted Stirling heat engine case study-1

Algorithm		ε_R	ε_H	ε_L	C_H	C_L	T_h	T_c	P (kW)	η_m (%)	σ (W/K)
TOPSIS	NSGA-II	0.9	0.8	0.8	1424	1252	996.9	400.5	49.64	**36.56**	89.47
LINMAP		0.9	0.8	0.8	1413	823	996.7	400.5	38.86	36.36	**69.50**
Bellman–Zadeh		0.9	0.8	0.8	1410	774	996.7	400.5	37.37	36.33	70.00
Jaya		0.9	0.8	0.8	1183.39	1202.477	1000	400	47.2120	36.3557	83.0944
Rao-1		0.89	0.79	0.8	1014.37	963.437	1000	400	40.8272	36.2473	72.4717
Rao-2		0.9	0.8	0.76482	614.81	1800	999.994	400	47.5994	36.3597	84.7584
Rao-3		0.9	0.8	0.76436	904.36	1800	1000	400	52.8852	36.4289	93.6753
AMTPG-Jaya		0.9	0.79	0.8	1800	1170.767	1000	400	49.2301	36.3828	86.4613
SAP-Rao		0.9	0.8	0.8	1800	1800	999.4868	400	61.1910	36.5088	106.4076
ERao-1		0.9	0.8	0.8	1800	1384.223	999.99999	400	53.8803	36.4418	94.1895
ERao-2		0.89	0.79	0.8	1161.010	1800	1000	400	56.8764	36.4712	99.1957
ERao-3		0.9	0.8	0.8	1800	1800	1000	400	**61.1952**	36.5151	106.3605

Source NSGA-II—Ahmadi et al. (2013a)
Result in boldface indicates better values

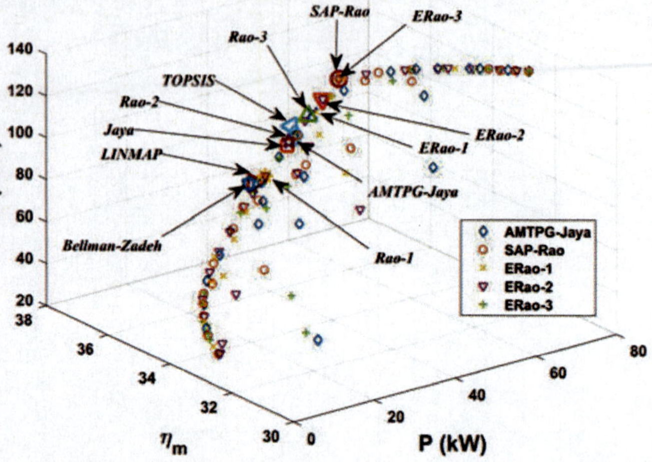

Fig. 7.2 Plot of Pareto-fronts of the proposed algorithms and solutions of the methods compared in solar-assisted Stirling heat engine system case study-1

WPM, and COPRAS) and ranked two by two methods (TOPSIS and MTOPSIS). Here an observation can be made that the ERao-3 algorithm solution has a higher power output, relatively similar thermal efficiency, and higher entropy generation. Even though the ERao-3 algorithm solution has higher entropy generation than all other solutions, it is ranked one because the relative improvement in the power output is higher than the relative deterioration in the entropy generation with the ERao-3 algorithm solution. Here, it can be noted that the MADM methods have ranked these solutions based on the relative importance of the solutions. The power output of the system for the ERao-3 algorithm solution is 23.4%, 57.5%, 63.8%, 29.6%, 49.9%, 28.6%, 15.7%, 24.3%, 13.6%, and 7.6% higher than that of the NSGA-II (TOPSIS, LINMAP, and Bellman–Zadeh), Jaya, Rao-1, Rao-2, Rao-3, AMTPG-Jaya, ERao-1, and ERao-2 algorithms solutions, respectively. Whereas, by the ERao-3 solution the entropy generation is deteriorated by 18.8%, 53%, 51.9%, 28%, 46.8%, 25.5%, 13.5%, 23%, 12.9%, and 7.2% when compared to that of the NSGA-II (TOPSIS, LINMAP, and Bellman–Zadeh), Jaya, Rao-1, Rao-2, Rao-3, AMTPG-Jaya, ERao-1, and ERao-2 algorithms solutions, respectively. Besides, the thermal efficiency is increased by the ERao-3 solution about 0.43%, 0.51%, 0.44%, 0.74%, 0.43%, 0.24%, 0.36%, 0.20%, and 0.12% when compared to that of the NSGA-II (LINMAP and Bellman–Zadeh), Jaya, Rao-1, Rao-2, Rao-3, AMTPG-Jaya, ERao-1, and ERao-2 algorithms solutions, respectively. Hence, the ERao-3 algorithm solution has achieved the least average rank.

Now, the performances of the Jaya algorithm, Rao algorithms, and their modified versions in the MOO scenario are evaluated based on the hypervolume, coverage, and spacing indicators. Table 7.32 presents the hypervolume and spacing values of

Table 7.30 Ranks suggested by the MADM methods for different algorithm solutions presented in Table 7.29

Algorithm	SAW	WPM	TOPSIS	MTOPSIS	VIKOR	PROMETHEE	COPRAS	GRA	AR	CR	FR
TOPSIS	12	11	9	9	1	1.5	12	3	7.3125	10.60	11
LINMAP	3	10	11	11	10	7.5	3	4	7.4375	7.60	9
Bellman–Zadeh	10	12	12	12	11	11.5	10	6	10.5625	11.2	12
Jaya	9	6	7	7	7	10	8	11	8.125	7.40	7
Rao-1	5	8	10	10	12	11.5	5	9	8.8125	7.60	8
Rao-2	11	9	8	8	6	9	11	12	9.25	9.40	10
Rao-3	8	7	5	5	4	6	9	8	6.5	6.80	6
AMTPG-Jaya	7	5	6	6	2	7.5	7	10	6.3125	6.20	5
SAP-Rao	2	2	3	3	9	3	2	2	3.25	2.40	2
ERao-1	6	4	4	4	3	5	6	7	4.875	4.80	4
ERao-2	4	3	1	1	5	4	4	5	3.375	2.60	3
ERao-3	1	1	2	2	8	1.5	1	1	2.1875	1.40	1

AR average rank, *CR* corrected rank, *FR* final rank

Table 7.31 Spearman's rank correlation coefficients between different pairs of MADM method's rankings presented in Table 7.30

Method	SAW	WPM	TOPSIS	MTOPSIS	VIKOR	PROMETHEE	COPRAS	GRA
SAW	1	0.699	0.510	0.510	−0.392	0.344	0.993	0.545
WPM	0.699	1	0.923	0.923	0.133	0.524	0.706	0.266
TOPSIS	0.510	0.923	1	1	0.371	0.668	0.5	0.280
MTOPSIS	0.510	0.923	1	1	0.371	0.668	0.5	0.280
VIKOR	−0.392	0.133	0.371	0.371	1	0.5	−0.413	−0.119
PROMETHEE	0.344	0.524	0.668	0.668	0.5	1	0.316	0.728
COPRAS	0.993	0.706	0.5	0.5	−0.413	0.316	1	0.524
GRA	0.545	0.266	0.280	0.280	−0.119	0.728	0.524	1

Table 7.32 Hypervolume and spacing values of the Pareto-fronts obtained by the proposed algorithms in MOO for the solar-assisted Stirling engine case study-1

Algorithm	Hypervolume	Spacing
Jaya	73,710.7	0.080901
Rao-1	73,044.9	0.090238
Rao-2	73,628.1	0.110388
Rao-3	73,055.7	0.086255
AMTPG-Jaya	74,402.4	0.146992
SAP-Rao	**74,487.9**	**0.079901**
ERao-1	73,906.6	0.080655
ERao-2	74,152.9	0.090039
ERao-3	74,388.8	0.087872

Results in the bold figure indicate better values

the Pareto-fronts obtained by different algorithms in the MOO scenario. The Pareto-front obtained by the NSGA-II algorithm was not reported by Ahmadi et al. (2013a). Thus, the performance metric values for the NSGA-II algorithm were not presented here. Here, an observation can be made that the SAP-Rao algorithm Pareto-front has higher hypervolume and the lower spacing value. The SAP-Rao algorithm has a better spacing, which is 1.2%, 11.5%, 27.61%, 7.4%, 45.6%, 1%, 11.3%, and 9.1% lesser value when compared to that of the Jaya, Rao-1, Rao-2, Rao-3, AMTPG-Jaya, ERao-1, ERao-2, and ERao-3 algorithms, respectively. Also, the SAP-Rao algorithm Pareto-front has a higher hypervolume value than the other algorithms. The AMTPG-Jaya, ERao-3, ERao-2, and ERao-1 algorithms have achieved the next better hypervolume values, respectively. The performances of the proposed modified algorithms are better or competitive to that of the basic algorithms in terms of the hypervolume and spacing values.

The coverage metric values of the Pareto-fronts obtained by the Jaya algorithm, Rao algorithms, and their modified versions are presented in Table 7.33. The ERao-1

Table 7.33 Coverage (%) values of the Pareto-fronts obtained by the proposed algorithms in MOO for the solar-assisted Stirling engine case study-1

Algorithm	Jaya	Rao-1	Rao-2	Rao-3	AMTPG-Jaya	SAP-Rao	ERao-1	ERao-2	ERao-3
Jaya	–	28	8	16	16	8	8	4	4
Rao-1	4	–	8	20	8	16	4	16	4
Rao-2	0	16	–	28	20	16	12	4	8
Rao-3	0	16	16	–	16	8	0	8	12
AMTPG-Jaya	12	24	24	32	–	16	8	8	8
SAP-Rao	0	24	20	16	16	–	8	16	12
ERao-1	16	28	8	16	8	12	–	0	0
ERao-2	12	28	16	16	12	16	8	–	8
ERao-3	12	12	20	20	24	12	8	4	–

and ERao-3 algorithms have achieved better coverage values than the other algorithms compared. From the coverage values, it can be observed that three solutions from the Pareto-fronts of the ERao-1 and ERao-3 algorithms are dominated. Similarly, four solutions are dominated from the Pareto-fronts achieved by the Jaya, SAP-Rao, and ERao-2 algorithms; five solutions are dominated from the Pareto-fronts achieved by the AMTPG-Jaya and Rao-2 algorithms; seven solutions are dominated from the Pareto-fronts achieved by the Rao-1 algorithm; eight solutions are dominated from the Pareto-fronts achieved by the Rao-3 algorithm. The performance of the proposed modified algorithms is better or competitive to that of the basic algorithms in terms of the coverage values.

From the computational results of the solar-assisted Stirling heat engine case study-1, it can be observed that the performance of the considered system can be improved with the solutions achieved by the proposed algorithms. The solution of the ERao-3 algorithm has better compromise among the power output, thermal efficiency, and entropy generation. By the ERao-3 algorithm's solution, the power output is increased by 63.8%, and the thermal efficiency is increased by 1% compared to that of the NSGA-II (TOPSIS, LINMAP, and Bellman–Zadeh) algorithm solutions. The SAP-Rao algorithm's performance in terms of the hypervolume and spacing values is much better than that achieved by other algorithms. The ERao-1 and ERao-3 algorithms have achieved better performance in terms of coverage values. In addition, the performances of the modified versions in this case study are better or competitive to those of the basic algorithms as well as the NSGA-II algorithm.

7.2.2 *Case Study-2*

The description of this case study is presented in Sect. 2.2.2.2. The detailed description and thermodynamic analysis of the solar-assisted Stirling heat engine system considered in this case study were presented by Ahmadi et al. (2013b). The objective functions of this case study are thermal efficiency of solar-assisted Stirling system (η_{th}), the thermo-economic objective function (F), and the dimensionless power output (P), which are given in Eqs. 2.29–2.32. The decision variables of these objective functions are internal irreversibility parameter (ϕ), heat transfer area ratio (A_r), temperature ratio (χ), the temperature of the heat source (T_H), the temperature of the working fluid in the high-temperature isothermal process (T_h).

All three objective functions considered are maximization functions. To maintain the uniformity and to facilitate a fair comparison of the computational results of the proposed algorithms with previous works, characteristics of the solar-assisted Stirling system taken are the same as given by the Ahmadi et al. (2013b). The characteristics of the solar-assisted Stirling system are as follows:

$h_h = h_c = 200\text{WK}^{-1}\text{m}^{-2}$, $f = 0.7$, $C = 1300$, $\delta = 5.67 \times 10^{-8}\text{Wm}^{-2}\text{K}^{-4}$, $T_L = 320$ K, $T_0 = 300$ K, $h = 20\text{WK}^{-1}\text{m}^{-2}$, $I = 1000\text{Wm}^{-2}$, $\left(\frac{1}{M_1} + \frac{1}{M_2}\right) = 2 \times 10^{-5}\text{sK}^{-1}$, $R = 4.3$ J $\text{mol}^{-1}\text{K}^{-1}$, $n = 1\text{mol}$, $\lambda = 2$, $\varepsilon = 0.9$, $K_0 = 2.5\text{WK}^{-1}$,

Table 7.34 Results obtained by the proposed algorithms in single-objective optimization scenarios of the solar-assisted Stirling heat engine case study-2

Objective	ϕ	χ	A_r	T_H	T_h	P	η_{th}	F
Maximize-P	1	0.4752	10	1400	850	0.615117604	0.358187	0.1163736
Maximize-η	1	0.45	10	1100	850	0.342704804	0.4292346	0.064836
Minimize-F	1	0.47809	1.63617	1400	989.9205124	0.352630625	0.3561972	0.2072814

$\eta_0 = 0.85$, $0.45 \le \chi \le 0.7$, $1100 \le T_H \le 1400$ K, $850 \le T_h \le 1000$ K, $0.25 \le A_r \le 10$, and $\phi \ge 1$.

Ahmadi et al. (2013b) solved this MOO problem using the NSGA-II algorithm and reported three Pareto-optimal solutions. These three solutions are documented based on the similarity to the ideal solutions calculated by employing TOPSIS, LINMAP, and Fuzzy Bellman–Zadeh decision-making methods. The number of function evaluations taken by the NSGA-II algorithm was not specified. However, in this work, the Jaya algorithm, Rao algorithms, and their modified versions are tested by taking 20,000 function evaluations as the termination criterion. The best Pareto-optimal solutions are reported using the average rank based on multiple decision-making methods. Similar to the previous case studies, in this case study also, first single-objective optimization is performed independently for each objective function, and then MOO is performed by the proposed algorithms. In all the computational experiments, the population size is maintained as 25 for all the algorithms. The elite size for the elitist Rao algorithms is taken as 20%.

The Jaya algorithm, Rao algorithms, and their modified versions have obtained identical solutions in all the single-objective optimization scenarios of this case study, and those solutions are presented in Table 7.34. The maximum non-dimensional power output, thermo-economic function, and thermal efficiency obtained by the proposed algorithms are 0.615117604, 0.2072814, and 0.4292346, respectively.

The Pareto-optimal solutions achieved by the Jaya algorithm, Rao algorithms, and their modified versions in MOO are presented in Tables 7.35, 7.36, 7.37, 7.38, 7.39, 7.40, 7.41, 7.42 and 7.43. Here an observation can be made that the non-dimensional power output is varied between 0.17014 and 0.61512, the thermal efficiency is varied between 0.292 and 0.4293, and non-dimensional thermo-economic function is varied from 0.0616 to 0.20725. The best solutions from the Pareto-fronts obtained by the proposed algorithms are identified based on the average rank method. Solution 9 of the Jaya algorithm, Solution 16 of the Rao-1 algorithm, Solution 9 of the Rao-2 algorithm, Solution 12 of the Rao-3 algorithm, Solution 8 of the AMTPG-Jaya algorithm, Solution 14 of the SAP-Rao algorithms, Solution 16 of the ERao-1 algorithm, Solution 9 of the ERao-2, and Solution 16 of the ERao-3 algorithm are identified as the best solutions from the respective algorithm Pareto-front. Now, these best solutions are compared with those of the NSGA-II algorithm in Table 7.44.

In Table 7.44, the solutions with the best average rank from the Pareto-optimal solutions of the proposed algorithms are compared with those reported by the NSGA-II algorithm. Figure 7.3 presents the Pareto-fronts achieved by the AMTPG-Jaya,

Table 7.35 Pareto-optimal solutions obtained by the Jaya algorithm for the solar-assisted Stirling heat engine case study-2

Solution	ϕ	χ	A_r	T_H	T_h	P	η_{th}	F
1	1	0.47243	1.66268	1400	989.501	0.3548535	0.3600603	0.2072045
2	1	0.47519	10	1400	850	0.6151176	0.3581945	0.1163736
3	1	0.45	9.8232	1100	850	0.3418216	0.4292344	0.0656095
4	1	0.57164	4.8373	1400	871.554	0.4805860	0.2923630	0.1563832
5	1	0.45	0.79922	1202.84	1000	0.1798537	0.4147325	0.1339669
6	1	0.53305	5.42186	1400	852.312	0.5195827	0.3187032	0.1563288
7	1	0.45	1.09691	1288.87	1000	0.2486357	0.3996464	0.1691279
8	1	0.5081	2.45764	1214.43	855.798	0.3167602	0.3692959	0.1542707
9	1	0.48666	2.51716	1400	901.694	0.4103274	0.3503541	0.1973881
10	1	0.45	2.2319	1100	857.86	0.2298330	0.4292051	0.1174698
11	1	0.51777	3.84643	1400	866.249	0.4739180	0.3291281	0.1789402
12	1	0.45	9.75773	1370.65	871.385	0.5850816	0.3823743	0.1129090
13	1	0.51106	6.36388	1400	860.526	0.5483205	0.3337116	0.1471062
14	1	0.45	1.10705	1349.82	989.018	0.2720724	0.3870409	0.1845249
15	1	0.50426	7.59815	1400	869.281	0.5714123	0.3383530	0.1342494
16	1	0.50777	5.69527	1362.65	863.752	0.5092648	0.3438474	0.1480064
17	1	0.45	7.62834	1238.33	856.033	0.4504540	0.4089154	0.1055103
18	1	0.45	10	1176.52	850	0.4214861	0.4188383	0.0797406
19	1	0.46591	1.28909	1400	1000	0.3170798	0.3645035	0.2042427
20	1	0.45	10	1138.71	850	0.3837377	0.4242221	0.0725990
21	1	0.45625	1.18381	1300.16	999.708	0.2633766	0.3929243	0.1747284
22	1	0.45	2.11063	1125.15	885.101	0.2392401	0.4260053	0.1256145
23	1	0.46839	5.08283	1291.51	863.271	0.4474201	0.3858162	0.1407708
24	1	0.48211	8.25916	1400	850	0.5893103	0.3534718	0.1298143
25	1	0.45	7.67138	1190.33	875.933	0.4028174	0.4167435	0.0939465

SAP-Rao, and elitist Rao algorithms, including the optimal solutions reported for the NSGA-II, Jaya, and Rao algorithms. The conflicting nature of these objectives can be observed in Fig. 7.3. Any change in the design variables, which will result in the improvement of one objective, is also leading to the deterioration of the other two objectives, which is not desirable.

Furthermore, the non-dimensional power output of the system for the Rao-2 algorithm solution is highest, which is 38.1%, 32.6%, 34%, 46.7%, 47.3%, 26.5%, 21%, 33%, 40.5%, 37%, and 38.5% higher than that of the NSGA-II (TOPSIS, LINMAP, and Fuzzy), Jaya, Rao-1, Rao-3, AMTPG-Jaya, SAP-Rao, ERao-1, ERao-2, and ERao-3 algorithms solutions, respectively. Similarly, the thermo-economic function of the system for the Rao-1 algorithm solution is highest, which is 7%, 12%, 26%,

Table 7.36 Pareto-optimal solutions obtained by the Rao-1 algorithm for the solar-assisted Stirling heat engine case study-2

Solution	ϕ	χ	A_r	T_H	T_h	P	η_{th}	F
1	1	0.47522	10	1400	850	0.6151176	0.3581761	0.1163736
2	1	0.48054	1.62447	1400	983.749	0.3515125	0.3545243	0.2072350
3	1	0.54076	4.32582	1400	850	0.4848592	0.3134404	0.1698922
4	1	0.45	10	1100	865.777	0.3298162	0.4292333	0.0623977
5	1	0.45917	1.26622	1100	899.987	0.1841288	0.4220305	0.1193575
6	1	0.48579	3.02882	1400	864.08	0.4285399	0.3509485	0.1864785
7	1	0.45	1.78525	1188.78	915.915	0.2567197	0.4169539	0.1454414
8	1	0.50046	5.86113	1400	861.614	0.5395252	0.3409488	0.1536272
9	1	0.46217	4.58789	1267.55	850	0.4133015	0.3947666	0.1393352
10	1	0.45092	9.1107	1162.74	868.512	0.3913541	0.4201571	0.0797935
11	1	0.45	2.77357	1347.49	996.766	0.3898552	0.3875787	0.1781242
12	1	0.45989	7.87454	1173.7	850.477	0.4014021	0.4117169	0.0917532
13	1	0.45189	2.14718	1100	919.42	0.2131402	0.4277327	0.1109979
14	1	0.45	1.82094	1276.04	951.23	0.3036086	0.4021049	0.1705281
15	1	0.45387	8.37204	1250.51	864.38	0.4698091	0.4039235	0.1023992
16	1	0.45	2.49222	1399.45	948.093	0.4088411	0.3755073	0.1976900
17	1	0.52486	4.71035	1400	850	0.5021918	0.3242925	0.1663590
18	1	0.53429	7.19616	1400	850	0.5551247	0.3178606	0.1359245
19	1	0.46721	10	1382.89	850	0.6014358	0.3676388	0.1137851
20	1	0.47044	2.26178	1399.45	941.75	0.3978422	0.3615525	0.2020185
21	1	0.46192	5.06017	1343.27	874.853	0.4792003	0.3800878	0.1512319
22	1	0.46381	2.30579	1347.94	1000	0.3674042	0.3777486	0.1847929
23	1	0.45	1.09509	1164.53	900.588	0.1930444	0.4205449	0.1313831
24	1	0.45655	3.36859	1382.78	921.189	0.4462364	0.3750051	0.1826083
25	1	0.45	8.36698	1100	850	0.3333641	0.4292328	0.0726941

0.2%, 74%, 21%, 17%, 8%, 0.6%, 4.8%, and 2% higher than that of the NSGA-II (TOPSIS, LINMAP, and Fuzzy), Jaya, Rao-2, Rao-3, AMTPG-Jaya, SAP-Rao, ERao-1, ERao-2, and ERao-3 algorithms solutions, respectively. The thermal efficiency of the system for the Fuzzy solution is higher than that of the other algorithm's solutions. As these solutions are non-dominated, to identify the best solution among the solutions of different algorithms, which has the best compromise among the objectives, the MADM methods-based average ranks are calculated and presented in Table 7.45.

The decision-making methods ranking of the Pareto-optimal solutions obtained by different algorithms is shown in Table 7.45. Spearman's correlation coefficients for different pairs of rankings given by decision-making methods for different algorithm's solutions are shown in Table 7.46. The ranks given by the TOPSIS and

Table 7.37 Pareto-optimal solutions obtained by the Rao-2 algorithm for the solar-assisted Stirling heat engine case study-2

Solution	ϕ	χ	A_r	T_H	T_h	P	η_{th}	F
1	1	0.47535	10	1400	850	0.6151174	0.3580891	0.1163736
2	1	0.48467	1.67085	1400	983.613	0.3555712	0.3517084	0.2072000
3	1	0.55429	4.48677	1400	850	0.4845037	0.3042074	0.1657613
4	1	0.45	10	1176.72	850	0.4216784	0.4188087	0.0797770
5	1	0.45	0.73455	1190.15	963.299	0.1701430	0.4167082	0.1294055
6	1	0.45	3.71136	1100	850	0.2756510	0.4292195	0.1064050
7	1	0.45141	1.387	1400	1000	0.3252330	0.3743961	0.2039807
8	1	0.45	10	1337.86	850	0.5607651	0.3896826	0.1060907
9	1	0.45208	10	1391.63	850	0.6020914	0.3759945	0.1139092
10	1	0.45	6.56738	1131.2	850	0.3473813	0.4252275	0.0910665
11	1	0.45	10	1214.81	850	0.4575556	0.4128630	0.0865646
12	1	0.49313	6.3795	1399.22	906.063	0.5444270	0.3461292	0.1457999
13	1	0.457	1.15377	1240.73	982.986	0.2344893	0.4032685	0.1569041
14	1	0.45041	2.80281	1352.93	998.311	0.3944172	0.3860868	0.1791825
15	1	0.53276	5.68522	1400	850	0.5262827	0.3188994	0.1531439
16	1	0.50965	1.91942	1400	954.327	0.3734701	0.3346602	0.2049099
17	1	0.45	1.86284	1100	868.388	0.2141476	0.4291996	0.1190795
18	1	0.53412	6.63805	1400	850	0.5455282	0.3179733	0.1418844
19	1	0.47145	5.95551	1252.29	850	0.4404524	0.3906165	0.1239885
20	1	0.45	1.01229	1142.15	926.471	0.1805630	0.4236977	0.1259297
21	1	0.45931	1.50596	1259.99	941.329	0.2750647	0.3982035	0.1671706
22	1	0.55156	4.19592	1400	850	0.4763056	0.3060659	0.1702154
23	1	0.53359	2.8402	1389.5	850	0.4151693	0.3204855	0.1872470
24	1	0.54203	2.30053	1398.38	963.251	0.3857805	0.3129021	0.1942558
25	1	0.52513	2.4346	1400	850	0.3925763	0.3240960	0.1921192

MTOPSIS methods for each solution are identical. Furthermore, Spearman's correlation for all the pairs formed by the SAW, WPM, TOPSIS, MTOPSIS, and COPRAS methods is greater than 0.5. Spearman's correlation values for all the pairs formed by the VIKOR, PROMETHEE, and GRA methods are either negative or very nearer to zero. Hence, the ranks suggested by the VIKOR, PROMETHEE, and GRA methods cannot be considered for calculating the average ranks. Now, the corrected average ranks are calculated, excluding the ranks suggested by the VIKOR, PROMETHEE, and GRA methods and presented in Table 7.45 as the corrected ranks.

The ERao-3 algorithm solution has the least average rank, which is 1.4. Thus, it can be regarded as the best solution. The AMTPG-Jaya, ERao-1, SAP-Rao, and ERao-2 algorithms' solutions have the next best average ranks, which are 2.6, 3.2, 3.8, and 4.4, respectively. The ERao-3 algorithm solution ranked 1 by three methods (SAW,

Table 7.38 Pareto-optimal solutions obtained by the Rao-3 algorithm for the solar-assisted Stirling heat engine case study-2

Solution	ϕ	χ	A_r	T_H	T_h	P	η_{th}	F
1	1	0.45	1.69862	1400	1000	0.3556207	0.3753673	0.2058012
2	1	0.45	10	1400	876.669	0.6116321	0.3753879	0.1157142
3	1	0.45	1.89605	1100	887.321	0.2151054	0.4292021	0.1186729
4	1	0.45	10	1100	850	0.3427048	0.4292346	0.0648360
5	1	0.45	4.60016	1325.77	944.944	0.4463408	0.3922457	0.1502074
6	1	0.45	2.45894	1380.49	966.273	0.3994052	0.3800659	0.1944683
7	1	0.45	4.37848	1222.09	1000	0.3075491	0.4116512	0.1069181
8	1	0.45	3.90207	1315.52	934.754	0.4225595	0.3943697	0.1581249
9	1	0.45	10	1387.47	850	0.5976537	0.3784222	0.1130696
10	1	0.45	2.55625	1342.04	946.331	0.3831792	0.3887658	0.1828552
11	1	0.45	1.89216	1138.01	901.326	0.2354851	0.4242865	0.1300357
12	1	0.45	4.45221	1366.51	925.322	0.4760553	0.3833191	0.1637003
13	1	0.45	8.35976	1188.52	885.494	0.4014555	0.4170229	0.0876014
14	1	0.45	7.04411	1396.61	914.073	0.5603855	0.3762152	0.1394374
15	1	0.45	8.76887	1178.92	911.711	0.3731480	0.4184761	0.0784240
16	1	0.45	6.11491	1241.44	997.089	0.3465512	0.4083705	0.0957145
17	1	0.45	2.05639	1275.46	989.029	0.3104563	0.4022167	0.1650214
18	1	0.45	5.92725	1289.27	909.757	0.4547534	0.3996025	0.1284524
19	1	0.45	4.04331	1256.3	976.134	0.3582437	0.4057442	0.1310881
20	1	0.45	9.38173	1155.99	906.705	0.3542621	0.4218193	0.0705597
21	1	0.45	1.86434	1298.75	918.927	0.3119856	0.3977249	0.1734213
22	1	0.45	5.97463	1348.65	980.338	0.4732221	0.3873317	0.1329069
23	1	0.45	6.48322	1400	850.588	0.5335651	0.3753805	0.1412100
24	1	0.45	1.22715	1361.97	958.164	0.2831489	0.3843213	0.1855593
25	1	0.45	4.39652	1360.78	906.79	0.4695661	0.3846194	0.1628050

WPM, and COPRAS) and ranked 2 by two methods (TOPSIS and MTOPSIS). Here an observation can be made that the ERao-3 algorithm solution has a better fitness value in none of the three objectives. However, it is ranked one because it has better compromise among the objectives. Here, it can be noted that the MADM methods have ranked these solutions based on the relative importance of the solutions. If the NSGA-II (TOPSIS) and ERao-3 algorithms solutions are compared, the NSGA-II (TOPSIS) solution has better power output and thermal efficiency, and the ERao-3 algorithm has a better thermo-economic function. However, the combined deterioration in the power output (0.3%) and thermal efficiency (2.25%) is relatively much lesser than the improvement in the thermo-economic function (4.9%) by the ERao-3 solution. Similarly, the ERao-3 algorithm solution has better relative importance when compared to the solutions of the other algorithms.

Table 7.39 Pareto-optimal solutions obtained by the AMTPG-Jaya algorithm for the solar-assisted Stirling heat engine case study-2

Solution	ϕ	χ	A_r	T_H	T_h	P	η_{th}	F
1	1	0.47523	10	1400	850	0.6151176	0.3581708	0.1163736
2	1	0.47429	1.70012	1400	984.776	0.3581183	0.3587884	0.2071696
3	1	0.54385	3.37925	1400	874.013	0.4479567	0.3113258	0.1829702
4	1	0.45	1.06142	1149.28	910.127	0.1866062	0.4227112	0.1282612
5	1	0.45	10	1100	870.232	0.3257729	0.4292329	0.0616327
6	1	0.4707	6.99366	1400	917.136	0.5592433	0.3612581	0.1399058
7	1	0.45	9.98302	1201.36	889.024	0.4233368	0.4150250	0.0802012
8	1	0.45377	4.54838	1393.86	923.845	0.4981830	0.3742949	0.1689154
9	1	0.53189	2.9567	1400	893.116	0.4322267	0.3194890	0.1906471
10	1	0.48782	5.0417	1400	891.645	0.5190372	0.3495748	0.1642143
11	1	0.51059	3.55161	1400	856.574	0.4595718	0.3340258	0.1822166
12	1	0.45	1.48514	1285.29	993.68	0.2814204	0.4003456	0.1719659
13	1	0.45473	10	1274.38	850	0.5112880	0.3989772	0.0967302
14	1	0.49373	3.94899	1400	933.421	0.4790397	0.3455400	0.1779214
15	1	0.49536	8.8077	1400	850	0.5960797	0.3444305	0.1248405
16	1	0.50232	6.37904	1400	879.452	0.5486068	0.3396782	0.1469270
17	1	0.45	1.35865	1244.47	1000	0.2478882	0.4078197	0.1566653
18	1	0.45	1.55317	1364.8	977.08	0.3245880	0.3836902	0.1948723
19	1	0.45	10	1157.6	917.761	0.3473153	0.4215906	0.0657083
20	1	0.52618	1.93654	1400	917.31	0.3705050	0.3233798	0.2024679
21	1	0.45	1.57438	1121.24	884.211	0.2100277	0.4265042	0.1254094
22	1	0.45848	6.52616	1195.05	850	0.4032398	0.4095936	0.1062017
23	1	0.47417	4.65899	1316.61	863.824	0.4530867	0.3768290	0.1511948
24	1	0.45	7.77134	1345.71	850	0.5306336	0.3879771	0.1225319
25	1	0.52424	5.6177	1400	850	0.5278638	0.3247167	0.1549084

In addition, the non-dimensional power output of the system for the ERao-3 algorithm solution is 6%, 6.4%, and 1.5% higher than that of the Jaya, Rao-1, and ERao-1 solutions, respectively. Also, the non-dimensional thermo-economic function of the system for the ERao-3 solution is 4.9%, 9.8%, 23.5%, 70.1%, 18.4%, 14.7%, 6%, and 2.8% higher than that of the NSGA-II (TOPSIS, LINMAP, and Bellman–Zadeh), Rao-2, Rao-3, AMTPG-Jaya, SAP-Rao, and ERao-2 algorithms solutions, respectively. In addition, the ERao-3 algorithm solution's thermal efficiency is 7.14%, 0.29%, and 1.62% higher when compared to that of the Jaya, AMTPG-Jaya, and ERao-1 algorithms solutions, respectively. Hence, the ERao-3 algorithm solution can be considered as the best solution.

Now, the performances of the Jaya algorithm, Rao algorithms, and their modified versions in the MOO scenario are evaluated based on the hypervolume, coverage,

Table 7.40 Pareto-optimal solutions obtained by the SAP-Rao algorithm for the solar-assisted Stirling heat engine case study-2

Solution	ϕ	χ	A_r	T_H	T_h	P	η_{th}	F
1	1	0.45	1.71204	1400	1000	0.3567995	0.3753675	0.2057987
2	1	0.45	10	1400	876.612	0.6116322	0.3753879	0.1157142
3	1	0.45	1.24252	1145.2	876.651	0.1935361	0.4232790	0.1262873
4	1	0.45	10	1100	850	0.3427048	0.4292346	0.0648360
5	1	0.45	1.2553	1300.33	992.743	0.2696118	0.3974069	0.1753017
6	1	0.45	10	1171	850	0.4161101	0.4196557	0.0787235
7	1	0.45	1.92037	1285.48	894.594	0.3019813	0.4003063	0.1656492
8	1	0.45	2.23873	1100	862.539	0.2305106	0.4292059	0.1176402
9	1	0.45	7.01104	1375.63	850	0.5330005	0.3812121	0.1330927
10	1	0.45	5.61806	1383.96	875.37	0.5138091	0.3792526	0.1507770
11	1	0.45	10	1260.89	878.674	0.4898969	0.4049275	0.0926832
12	1	0.45	2.9261	1311.57	937.723	0.3837980	0.3951712	0.1702709
13	1	0.45	1.80745	1356.36	973.487	0.3426676	0.3856023	0.1930934
14	1	0.45	3.44796	1393.93	981.638	0.4531617	0.3768601	0.1828964
15	1	0.45	2.05026	1261.23	988.123	0.3006076	0.4048442	0.1600097
16	1	0.45	2.42015	1194.09	884.873	0.2909584	0.4161412	0.1428223
17	1	0.45	7.5917	1221.14	850	0.4364477	0.4118194	0.1026070
18	1	0.45	2.91216	1204.13	859.136	0.3117243	0.4145640	0.1386632
19	1	0.45	8.95597	1242.3	908.541	0.4456168	0.4082301	0.0921025
20	1	0.45	10	1364.2	893.804	0.5790010	0.3838529	0.1095407
21	1	0.45	10	1149.17	850.615	0.3940691	0.4227820	0.0745536
22	1	0.45	10	1120.52	850	0.3647684	0.4266386	0.0690102
23	1	0.45	10	1233.54	850	0.4744574	0.4097393	0.0897622
24	1	0.45	3.9718	1326.75	868.76	0.4216985	0.3920336	0.1560574
25	1	0.45	5.68606	1225.44	906.469	0.3968878	0.4111016	0.1154790

and spacing indicators. Table 7.47 presents the hypervolume and spacing values of the Pareto-fronts obtained by different algorithms in the MOO scenario. The Pareto-front obtained by the NSGA-II algorithm was not reported by Ahmadi et al. (2013b). Thus, the performance metric values for the NSGA-II algorithm were not presented here. Here, an observation can be made that the SAP-Rao algorithm Pareto-front has the highest hypervolume value, and the AMTPG-Jaya algorithm Pareto-front has the least spacing value. The SAP-Rao algorithm has 5.72%, 5.8%, 4.74%, 6.1%, 4.6%, 4.7%, 1.6%, and 4.7% higher hypervolume value when compared to that of the Jaya, Rao-1, Rao-2, Rao-3, AMTPG-Jaya, ERao-1, ERao-2, and ERao-3 algorithms, respectively. The ERao-2, AMTPG-Jaya, ERao-3, and ERao-1 algorithms have achieved the next better hypervolume values, respectively. Similarly, the AMTPG-Jaya algorithm has 18%, 20.7%, 24.6%, 17.2%, 13.1%, 7%, 4.14%, and

Table 7.41 Pareto-optimal solutions obtained by the ERao-1 algorithm for the solar-assisted Stirling heat engine case study-2

Solution	ϕ	χ	A_r	T_H	T_h	P	η_{th}	F
1	1	0.47707	1.64488	1400	995.219	0.3533372	0.3568923	0.2072421
2	1	0.47437	10	1400	850	0.6151102	0.3587564	0.1163722
3	1	0.45	1.60377	1100	894.457	0.2007590	0.4291962	0.1189803
4	1	0.45	9.93364	1100	850	0.3423765	0.4292345	0.0651243
5	1	0.53359	3.73404	1400	893.423	0.4647414	0.3183302	0.1787259
6	1	0.52033	4.02608	1400	872.952	0.4807894	0.3273836	0.1764064
7	1	0.48953	5.69076	1399.8	860.51	0.5356768	0.3484499	0.1557699
8	1	0.45	1.58619	1219.23	895.985	0.2527946	0.4121014	0.1504912
9	1	0.45603	9.20774	1400	921.443	0.5910918	0.3712751	0.1195049
10	1	0.46265	3.05367	1321.07	887.254	0.3934555	0.3841716	0.1704217
11	1	0.4511	10	1211.41	850	0.4546602	0.4125888	0.0860168
12	1	0.4541	10	1283.77	850	0.5190538	0.3976800	0.0981994
13	1	0.45	1.39254	1175.48	906.515	0.2235552	0.4189527	0.1400017
14	1	0.45	4.34434	1128.28	883.008	0.3009289	0.4256097	0.1051514
15	1	0.48922	7.75176	1400	850	0.5802420	0.3486192	0.1342475
16	1	0.4588	2.75666	1400	969.261	0.4285094	0.3693724	0.1964355
17	1	0.45806	9.25184	1341.88	877.305	0.5555832	0.3831168	0.1118983
18	1	0.45226	1.60377	1297.87	894.457	0.2801840	0.3962575	0.1660518
19	1	0.45	9.25184	1149.88	877.305	0.3724772	0.4226819	0.0750195
20	1	0.45286	2.34635	1372.1	906.5	0.3738316	0.3800321	0.1863959
21	1	0.45	5.88401	1298.82	892.381	0.4653791	0.3977351	0.1321455
22	1	0.47194	1.45158	1400	985.773	0.3342937	0.3603890	0.2060865
23	1	0.46402	5.0195	1400	850	0.5002558	0.3658123	0.1587502
24	1	0.45	1.45158	1379.3	985.773	0.3215900	0.3803318	0.1982548
25	1	0.46983	4.34434	1156.37	883.008	0.3259623	0.4065528	0.1138987

38% less spacing value when compared to that of the Jaya, Rao-1, Rao-2, Rao-3, SAP-Rao, ERao-1, ERao-2, and ERao-3 algorithms, respectively. The performances of the proposed modified algorithms are better or competitive to that of the basic algorithms in terms of the hypervolume and spacing values.

The coverage metric values of the Pareto-fronts obtained by the Jaya algorithm, Rao algorithms, and their modified versions are presented in Table 7.48. The ERao-2 algorithm has achieved better coverage values than the other algorithms compared. From the coverage values, it can be observed that two solutions from the Pareto-front of the ERao-2 algorithm are dominated. Similarly, three solutions of the Pareto-fronts achieved by the SAP-Rao and ERao-1 algorithms are dominated, and four solutions of the Pareto-front achieved by the ERao-3 algorithm are dominated. Also, 7, 6, 8, 6, and 5 solutions from the Pareto-fronts achieved by the Jaya, Rao-1, Rao-2, Rao-3,

Table 7.42 Pareto-optimal solutions obtained by the ERao-2 algorithm for the solar-assisted Stirling heat engine case study-2

Solution	ϕ	χ	A_r	T_H	T_h	P	η_{th}	F
1	1	0.45	1.69724	1400	1000	0.3554984	0.3753673	0.2058011
2	1	0.45	10	1400	876.611	0.6116322	0.3753879	0.1157142
3	1	0.45	1.61005	1110.87	882.977	0.2071486	0.4278372	0.1225717
4	1	0.45	9.92067	1100	850	0.3423118	0.4292345	0.0651810
5	1	0.45	1.28658	1371.47	1000	0.3027167	0.3821551	0.1951261
6	1	0.45	10	1317.06	867.94	0.5439979	0.3940618	0.1029185
7	1	0.45	10	1161.31	850	0.4065615	0.4210646	0.0769170
8	1	0.45	3.19021	1355.88	945.794	0.4223584	0.3857184	0.1784187
9	1	0.45	3.1075	1400	1000	0.4396704	0.3753790	0.1885551
10	1	0.45	10	1365	867.672	0.5836720	0.3836699	0.1104244
11	1	0.45	4.65624	1390.89	1000	0.4778360	0.3775984	0.1595163
12	1	0.45	3.63427	1333.79	973.544	0.4172122	0.3905473	0.1631300
13	1	0.45	5.85474	1350.76	912.226	0.5018472	0.3868653	0.1430101
14	1	0.45	10	1124.11	850	0.3685645	0.4261692	0.0697284
15	1	0.45	2.84316	1262.88	850	0.3274019	0.4045388	0.1475783
16	1	0.45	1.00523	1229.68	997.307	0.2128587	0.4103494	0.1487679
17	1	0.45	10	1242.35	850	0.4822558	0.4082209	0.0912376
18	1	0.45	7.5421	1382	866.634	0.5565623	0.3797197	0.1315026
19	1	0.45	10	1171.6	850	0.4166949	0.4195676	0.0788342
20	1	0.45	2.28606	1291.95	982.339	0.3345231	0.3990693	0.1689734
21	1	0.45	1.65632	1205.04	1000	0.2366704	0.4144083	0.1384157
22	1	0.45	10	1279.03	850	0.5136592	0.4015648	0.0971788
23	1	0.45	3.37096	1303.44	850	0.3706979	0.3968012	0.1516335
24	1	0.45	1.61418	1243.68	907.907	0.2676593	0.4079549	0.1582108
25	1	0.45	3.55997	1100	850	0.2720516	0.4292185	0.1077132

and AMTPG-Jaya algorithms are dominated, respectively. The performance of the proposed modified algorithms is better or competitive to that of the basic algorithms in terms of the coverage values.

From the computational results of the solar-assisted Stirling heat engine case study-2, it can be observed that the performance of the considered system can be modified with the solutions achieved by the proposed algorithms. The solution of the ERao-3 algorithm has better compromise among the non-dimensional power output, thermal efficiency, and thermo-economic function. By the ERao-3 algorithm solution, the power output is increased by 6.4%, thermal efficiency is increased by 7.14%, and the thermo-economic function is increased by 70% compared to those of the solutions of the other algorithms compared. The SAP-Rao algorithm's performance in

Table 7.43 Pareto-optimal solutions obtained by the ERao-3 algorithm for the solar-assisted Stirling heat engine case study-2

Solution	ϕ	χ	A_r	T_H	T_h	P	η_{th}	F
1	1	0.47451	1.62478	1400	1000	0.3514687	0.3586402	0.2071931
2	1	0.4751	10	1400	850	0.6151175	0.3582596	0.1163736
3	1	0.45	10	1100	850	0.3427045	0.4292343	0.0648360
4	1	0.45291	1.77006	1100	861.682	0.2102446	0.4269176	0.1195524
5	1	0.53759	3.96958	1400	850	0.4729601	0.3155985	0.1750894
6	1	0.47884	7.9201	1400	850.581	0.5831853	0.3556895	0.1327132
7	1	0.51889	3.46544	1400	850	0.4537630	0.3283602	0.1825868
8	1	0.46253	5.47929	1400	894.458	0.5302809	0.3668325	0.1583746
9	1	0.45208	10	1244.99	850	0.4851544	0.4062189	0.0917860
10	1	0.53545	4.92092	1399.5	850	0.5050461	0.3171695	0.1624482
11	1	0.45919	2.09707	1400	993.046	0.3878196	0.3690938	0.2042506
12	1	0.49898	3.64363	1400	850	0.4601919	0.3419509	0.1796533
13	1	0.50407	6.35264	1400	870.121	0.5490857	0.3384836	0.1475022
14	1	0.50809	7.22694	1400	859.147	0.5659126	0.3357389	0.1381198
15	1	0.45	10	1345.76	850	0.5668061	0.3879677	0.1072336
16	1	0.45	2.90187	1400	972.377	0.4348128	0.3753697	0.1937964
17	1	0.45	7.3832	1207.84	850	0.4228139	0.4139878	0.1015348
18	1	0.45	1.88165	1335.52	912.826	0.3245572	0.3901581	0.1796687
19	1	0.46553	3.45372	1224.33	867.184	0.3555492	0.3996534	0.1433570
20	1	0.4591	2.16799	1347.44	860.828	0.3275047	0.3811583	0.1697673
21	1	0.4521	0.98559	1393.84	1000	0.2715280	0.3754085	0.1908946
22	1	0.45413	1.63117	1134.45	879.804	0.2209580	0.4215629	0.1300461
23	1	0.45	1.41582	1196.85	903.324	0.2329955	0.4156938	0.1450078
24	1	0.46382	1.45207	1340.21	1000	0.3075412	0.3793704	0.1895693
25	1	0.45	1.50705	1278.04	1000	0.2777241	0.4017240	0.1687390

terms of the hypervolume is much better than that achieved by other algorithms. Similarly, The AMTPG-Jaya algorithm's performance in terms of the spacing is much better than that achieved by other algorithms. The ERao-2 algorithm has achieved better performance in terms of coverage values. In addition, the performances of the modified versions are better or competitive to those of the basic algorithms as well as the NSGA-II algorithm.

Table 7.44 Best solutions obtained by various algorithms in MOO scenario of the solar-assisted Stirling heat engine case study-2

Algorithm		ϕ	χ	A_r	T_H	T_h	P	η_{th}	F
TOPSIS	NSGA-II	1.002271	0.45466	3.17578	1371.415	920.6022	0.436108	0.384027	0.184709
LINMAP		1.001516	0.45921	3.67126	1371.466	919.8346	0.454229	0.381995	0.176510
Fuzzy		1.000984	0.45146	4.34813	1329.38	913.8662	0.449328	0.396756	0.156916
Jaya		1	0.4867	2.5172	1400.0000	901.6943	0.410327	0.350354	0.197388
Rao-1		1	0.4500	2.4922	1399.4499	948.0930	0.408841	0.375507	0.197690
Rao-2		1	0.4521	10.0000	1391.6298	850.0000	0.602091	0.375995	0.113909
Rao-3		1	0.4500	4.4522	1366.5084	925.3221	0.476055	0.383319	0.163700
AMTPG-Jaya		1	0.4538	4.5484	1393.8580	923.8445	0.498183	0.374295	0.168915
SAP-Rao		1	0.4500	3.4480	1393.9301	981.6382	0.453162	0.376860	0.182896
ERao-1		1	0.4588	2.7567	1400.0000	969.2610	0.428509	0.369372	0.196436
ERao-2		1	0.4500	3.1075	1400.0000	1000.0000	0.439670	0.375379	0.188555
ERao-3		1.000023922	0.4500	2.9019	1400.0000	972.3766	0.434813	0.375370	0.193796

Source NSGA-II—Ahmadi et al. (2013b)
Result in boldface indicates better values

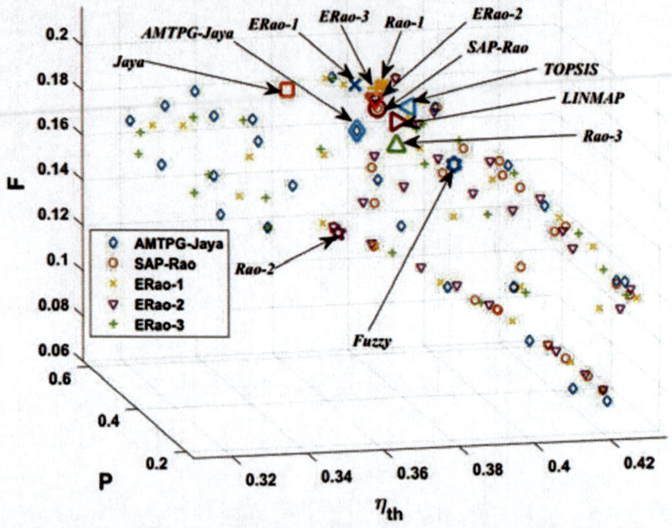

Fig. 7.3 Plot of Pareto-fronts of the proposed algorithms and solutions of the methods compared in solar-assisted Stirling heat engine system case study-2

7.2.3 Case Study-3

The description of this case study is presented in Sect. 2.2.2.3. The detailed description and thermodynamic analysis of the solar-assisted Stirling heat engine system considered in this case study were presented by Dai et al. (2018). The objective functions of this case study are power output (P), system efficiency (η), and ecological coefficient of performance ($ECOP$), which are given in Eqs. 2.35–2.37, respectively. These objective functions are maximization functions. Four decision variables are considered to assess the performance of the considered system. These decision variables and their ranges are as follows: $1000 \leq T_H \leq 1200$, $280 \leq T_L \leq 300$, $600 \leq T_h \leq 1000$, and $300 \leq T_c \leq 600$. To maintain the uniformity and to facilitate a fair comparison of the computational results of the proposed algorithms with previous works, characteristics of the solar-assisted Stirling system taken are the same as given by the Dai et al. (2018). The characteristic features of the system considered in this book are as follows: $C_v = 3.214\text{Jg}^{-1}\text{k}^{-1}$, $\mu_{2-3} = 0.3$, $\mu_{4-1} = 0.2$, $T_0 = 300\text{K}$, $m = 4g$, $\lambda = 2$, $k_1 = k_2 = 5000\frac{\text{K}}{\text{s}}$, $\alpha_{leak} = 12\frac{\text{W}}{\text{K.s}}$, $\alpha_c = \alpha_h = \frac{1000\ \text{W}}{\text{K.s}}$, $n = 1\text{mol}$ and $R = 8.314\text{Jmol}^{-1}\text{K}^{-1}$.

Dai et al. (2018) solved this MOO problem using the multi-objective PSO with crowding distance (MOPSOCD) algorithm and reported a Pareto-optimal solution using the TOPSIS decision-making method. The number of function evaluations taken by the MOPSOCD algorithm was not specified. However, in this book, the Jaya algorithm, Rao algorithms, and their modified versions are tested by taking 10,000 function evaluations as the termination criterion. The best Pareto-optimal solutions

Table 7.45 Ranks suggested by the MADM methods for different algorithm solutions presented in Table 7.44

Algorithm	SAW	WPM	TOPSIS	MTOPSIS	VIKOR	PROMETHEE	COPRAS	GRA	AR	CR	FR
TOPSIS	5	6	7	7	6	2	6	6	5.625	6.2	6
LINMAP	8	7	8	8	4	2	8	9	6.75	7.8	8
Bellman–Zadeh	11	11	12	12	3	5	12	1	8.375	11.6	12
Jaya	10	10	9	9	12	12	11	10	10.37	9.80	10
Rao-1	6	8	6	6	11	8	7	3	6.875	6.60	7
Rao-2	12	12	11	11	10	6	10	2	9.25	11.20	11
Rao-3	9	9	10	10	2	2	9	11	7.75	9.40	9
AMTPG-Jaya	7	2	1	1	1	9	2	12	4.375	2.60	2
SAP-Rao	4	3	4	4	5	4	4	8	4.5	3.80	4
ERao-1	2	5	3	3	9	11	3	5	5.125	3.20	3
ERao-2	3	4	5	5	7	7	5	7	5.375	4.40	5
ERao-3	1	1	2	2	8	10	1	4	3.625	1.40	1

AR average rank, *CR* corrected rank, *FR* final rank

Table 7.46 Spearman's rank correlation coefficients between different pairs of MADM method's rankings presented in Table 7.45

Method	SAW	WPM	TOPSIS	MTOPSIS	VIKOR	PROMETHEE	COPRAS	GRA
SAW	1	0.853	0.825	0.825	−0.098	−0.254	0.867	−0.014
WPM	0.853	1	0.937	0.937	0.238	−0.204	0.951	−0.315
TOPSIS	0.825	0.937	1	1	0.042	−0.437	0.965	−0.273
MTOPSIS	0.825	0.937	1	1	0.042	−0.437	0.965	−0.273
VIKOR	−0.098	0.238	0.042	0.042	1	0.570	0.112	−0.413
PROMETHEE	−0.254	−0.204	−0.437	−0.437	0.570	1	−0.289	−0.070
COPRAS	0.867	0.951	0.965	0.965	0.112	−0.289	1	−0.189
GRA	−0.014	−0.315	−0.273	−0.273	−0.413	−0.070	−0.189	1

Table 7.47 Hypervolume and spacing values of the Pareto-fronts obtained by the proposed algorithms in MOO for the solar-assisted Stirling engine case study-2

Algorithm	Hypervolume	Spacing
Jaya	0.00329951	0.08681974
Rao-1	0.00329721	0.08979312
Rao-2	0.00333040	0.09439404
Rao-3	0.00328939	0.08597348
AMTPG-Jaya	0.00333557	0.07115295
SAP-Rao	0.00348837	0.08190799
ERao-1	0.00333131	0.07647491
ERao-2	0.00343446	0.07422897
ERao-3	0.00333158	0.11455875

Results in the bold figure indicate better values

Table 7.48 Coverage (%) values of the Pareto-fronts obtained by the proposed algorithms in MOO for the solar-assisted Stirling engine case study-2

Algorithm	Jaya	Rao-1	Rao-2	Rao-3	AMTPG-Jaya	SAP-Rao	ERao-1	ERao-2	ERao-3
Jaya	–	20	16	12	12	4	4	0	8
Rao-1	16	–	28	4	12	4	4	4	16
Rao-2	8	8	–	4	8	0	12	0	4
Rao-3	8	24	24	–	16	12	8	4	16
AMTPG-Jaya	16	8	24	4	–	0	4	0	16
SAP-Rao	24	24	8	24	12	–	4	8	16
ERao-1	16	12	28	4	12	0	–	4	16
ERao-2	12	16	16	4	20	0	8	–	8
ERao-3	28	20	32	4	16	0	0	0	–

are reported using the average rank based on multiple decision-making methods. Similar to the previous case studies, in this case study also, first single-objective optimization is performed independently for each objective function, and then MOO is performed by the proposed algorithms. In all the computational experiments, the population size is maintained as 25 for all the algorithms. The elite size for the elitist Rao algorithms is taken as 20%.

The Jaya algorithm, Rao algorithms, and their modified versions have obtained identical solutions in all the single-objective optimization scenarios of this case study, and those solutions are presented in Table 7.49. The maximum power output, ECOP, and efficiency obtained by the proposed algorithms are 14.6026 kW, 1.0935, and 33.498%, respectively.

The Pareto-optimal solutions achieved by the Jaya algorithm, Rao algorithms, and their modified versions in MOO are presented in Tables 7.50, 7.51, 7.52, 7.53, 7.54, 7.55, 7.56, 7.57 and 7.58. Here, the power output is varied from 10.31 to 14.6 kW, the system efficiency is varied between 28.36% and 33.5%, and ECOP is varied from 0.63 to 1.0935. The best solutions from the Pareto-fronts obtained by

Table 7.49 Results obtained by the proposed algorithms in single-objective optimization scenarios of the solar-assisted Stirling heat engine case study-3

Objective	Algorithm	T_h (K)	T_c (K)	T_H (K)	T_L (K)	P (kW)	η (%)	ECOP
Maximize-P	MOPSOCD	874.3	455.4	1200	280	14.6	28.47	0.64
	Proposed algorithms	874.29534	455.35696	1200	280	14.6026	28.472	0.6351
Maximize-η	MOPSOCD	985.1	325.2	1119	280	12.23	33.49	0.92
	Proposed algorithms	999.9	325.92772	1135.9649	280	12.3397	33.498	0.9094
Maximize-ECOP	MOPSOCD	896.8	340.2	1000	300	10.86	32.74	1.09
	Proposed algorithms	896.8394	340.20277	1000	300	10.8632	32.7473	1.0935

Source MOPSOCD—Dai et al. (2018)

Table 7.50 Pareto-optimal solutions obtained by the Jaya algorithm for the solar-assisted Stirling heat engine case study-3

Solution	T_h (K)	T_c (K)	T_H (K)	T_L (K)	P (kW)	η (%)	ECOP
1	896.73	340.01	1000.00	300.00	10.8569	32.7473	1.0935
2	998.78	325.39	1136.32	280.00	12.3384	33.4976	0.9091
3	873.90	458.01	1200.00	280.00	14.6023	28.3561	0.6304
4	906.01	325.34	1000.00	291.04	10.5596	32.9467	1.0445
5	892.14	350.76	1005.14	290.95	11.7093	32.7967	1.0280
6	920.62	318.00	1014.57	280.00	10.9331	33.2978	0.9811
7	888.26	441.19	1200.00	280.00	14.5898	29.3014	0.6698
8	911.00	372.38	1200.00	285.74	14.0696	31.8430	0.8144
9	918.48	417.02	1200.00	280.00	14.4967	30.6522	0.7303
10	892.17	430.07	1200.00	283.90	14.5228	29.7895	0.7064
11	875.28	447.22	1200.00	280.00	14.5999	28.8169	0.6493
12	922.76	365.89	1121.67	298.60	13.0707	32.4468	0.9585
13	909.11	437.63	1200.00	291.84	14.4325	29.8281	0.7403
14	895.33	390.34	1093.86	299.43	13.2807	31.6452	0.9311
15	922.68	386.65	1117.96	293.58	13.4568	31.9927	0.9046
16	941.61	409.96	1200.00	280.00	14.4151	31.1895	0.7557
17	922.15	344.47	1074.99	281.98	12.6559	33.1404	0.9372
18	933.76	361.44	1175.55	280.00	13.8748	32.5168	0.8335
19	961.21	430.77	1200.00	289.45	14.3071	30.8157	0.7785
20	826.24	356.80	1000.00	300.00	12.0443	31.9825	1.0356
21	954.26	429.15	1181.42	285.47	14.2218	30.8405	0.7699
22	913.38	324.33	1000.00	284.94	10.7078	33.0993	1.0134
23	906.37	394.08	1155.77	300.00	13.7689	31.4039	0.8813
24	891.11	373.56	1077.55	300.00	12.9027	32.1111	0.9765
25	999.90	351.31	1200.00	280.00	13.6502	33.1160	0.8547

the proposed algorithms are identified based on the average rank method. Solution 12 of the Jaya algorithm, Solution 18 of the Rao-1 algorithm, Solution 7 of the Rao-2 algorithm, Solution 12 of the Rao-3 algorithm, Solution 16 of the AMTPG-Jaya algorithm, Solution 16 of the SAP-Rao algorithms, Solution 21 of the ERao-1 algorithm, Solution 13 of the ERao-2, and Solution 14 of the ERao-3 algorithm are identified as the best solutions from the respective algorithm Pareto-front. Now, these best solutions are compared with that of the MOPSOCD algorithm in Table 7.59.

In Table 7.59, the solutions with the best average rank from the Pareto-optimal solutions of the various algorithms are compared with the MOPSOCD algorithm solution. Figure 7.4 presents the Pareto-fronts achieved by the AMTPG-Jaya, SAP-Rao, and elitist Rao algorithms, including the optimal solutions reported for the MOPSOCD, Jaya, and Rao algorithms. The conflicting nature of these objectives

Table 7.51 Pareto-optimal solutions obtained by the Rao-1 algorithm for the solar-assisted Stirling heat engine case study-3

Solution	T_h (K)	T_c (K)	T_H (K)	T_L (K)	P (kW)	η (%)	ECOP
1	999.90	326.48	1141.40	280.00	12.4337	33.4962	0.9062
2	896.46	339.90	1000.00	300.00	10.8567	32.7472	1.0935
3	875.69	454.65	1200.00	280.00	14.6025	28.5298	0.6375
4	912.92	324.46	1000.00	289.29	10.4907	32.9644	1.0337
5	940.91	448.72	1200.00	298.11	14.3239	29.9717	0.7736
6	904.42	406.85	1200.00	280.43	14.4694	30.7733	0.7377
7	884.43	442.85	1200.00	280.00	14.5937	29.1664	0.6641
8	940.69	331.84	1080.84	280.00	12.3334	33.4059	0.9381
9	873.50	418.15	1200.00	281.32	14.5231	29.8806	0.7003
10	949.30	443.37	1199.90	287.30	14.3984	30.2463	0.7415
11	870.48	425.34	1200.00	280.00	14.5632	29.5606	0.6810
12	931.41	359.99	1200.00	280.00	14.0067	32.3895	0.8159
13	899.58	360.06	1000.00	297.70	11.4304	32.5316	1.0605
14	915.50	390.47	1125.50	300.00	13.4711	31.7630	0.9208
15	930.82	411.20	1172.30	300.00	13.9735	31.1555	0.8582
16	875.88	452.47	1199.92	280.00	14.6018	28.6213	0.6412
17	901.77	340.53	1000.00	287.15	11.3312	32.9934	1.0210
18	931.33	358.47	1125.00	300.00	12.8447	32.6183	0.9755
19	911.47	350.45	1011.73	299.94	11.2107	32.7245	1.0788
20	921.50	413.28	1136.75	298.78	13.7508	31.1166	0.8691
21	977.12	415.15	1200.00	294.86	14.1132	31.4702	0.8374
22	892.22	377.86	1131.16	280.70	13.8560	31.8428	0.8222
23	903.53	380.08	1040.42	294.73	12.5247	32.1561	0.9768
24	877.64	366.77	1067.56	297.27	12.8263	32.1765	0.9720
25	920.47	331.76	1191.76	280.00	13.3151	32.7384	0.8380

can be observed in Fig. 7.4. Furthermore, the power output of the system for the Jaya algorithm solution is highest, which is 1.1%, 2%, 10%, 15%, 0.6%, 4.7%, 5.4%, 0.5%, and 5.7% higher than that of the MOPSOCD, Rao-1, Rao-2, Rao-3, AMTPG-Jaya, SAP-Rao, ERao-1, ERao-2, and ERao-3 algorithms solutions, respectively. Similarly, the system efficiency for the Rao-2 algorithm solution is highest, which is 1.32%, 1.87%, 1.33%, 1.34%, 0.9%, 1.15%, 1.23%, 1.01%, and 1.12% higher than that of the MOPSOCD, Jaya, Rao-1, Rao-3, AMTPG-Jaya, SAP-Rao, ERao-1, ERao-2, and ERao-3 algorithms solutions, respectively. The ECOP of the Rao-3 algorithm solution is 6.6%, 13%, 11%, 7.2%, 12.4%, 7%, 6.2%, 11.6%, and 6.1% higher than that of the MOPSOCD, Jaya, Rao-1, Rao-2, AMTPG-Jaya, SAP-Rao, ERao-1, ERao-2, and ERao-3 algorithms solutions, respectively.

Table 7.52 Pareto-optimal solutions obtained by the Rao-2 algorithm for the solar-assisted Stirling heat engine case study-3

Solution	T_h (K)	T_c (K)	T_H (K)	T_L (K)	P (kW)	η (%)	ECOP
1	874.37	455.11	1200.00	280.00	14.6026	28.4835	0.6356
2	999.90	326.31	1135.66	280.00	12.3475	33.4979	0.9095
3	896.15	340.08	1000.00	300.00	10.8687	32.7472	1.0935
4	903.19	330.32	1000.00	296.64	10.5035	32.7928	1.0727
5	908.63	448.91	1200.00	280.00	14.5748	29.3839	0.6734
6	875.54	449.78	1200.00	280.00	14.6013	28.7210	0.6453
7	905.18	340.41	1036.45	289.90	11.8887	33.0520	1.0106
8	916.19	423.74	1200.00	280.00	14.5218	30.3955	0.7184
9	914.91	432.42	1200.00	288.58	14.4484	30.0983	0.7397
10	902.43	435.79	1200.00	280.00	14.5690	29.7538	0.6895
11	900.12	334.79	1000.00	290.71	11.0660	32.9967	1.0459
12	929.73	385.09	1200.00	297.87	13.9398	31.7167	0.8655
13	913.67	413.31	1200.00	287.32	14.3932	30.7105	0.7641
14	934.14	332.20	1083.46	288.86	12.0415	33.1894	0.9759
15	922.04	398.86	1142.21	300.00	13.6710	31.5369	0.8969
16	901.38	353.86	1000.00	298.57	11.2506	32.6403	1.0749
17	882.85	354.85	1141.51	291.22	13.2916	32.1649	0.8883
18	907.21	347.88	1111.45	292.44	12.8179	32.7016	0.9462
19	900.87	431.10	1200.00	300.00	14.3127	29.9265	0.7794
20	929.52	413.94	1200.00	300.00	14.1814	30.9245	0.8318
21	878.57	322.45	1006.92	281.00	11.5040	33.2636	0.9920
22	922.13	377.10	1067.81	300.00	12.6051	32.3407	0.9997
23	894.25	395.61	1200.00	291.47	14.2189	30.9657	0.7951
24	895.32	374.20	1000.00	300.00	11.6529	32.1732	1.0497
25	878.52	354.93	1118.39	291.93	13.1354	32.2356	0.9100

The decision-making methods ranking of the Pareto-optimal solutions obtained by different algorithms is shown in Table 7.60. Spearman's correlation coefficients for different pairs of rankings given by decision-making methods for different algorithm's solutions are shown in Table 7.61. The ranks given by the TOPSIS and MTOPSIS methods for each solution are identical. Furthermore, Spearman's correlation for all the pairs formed by all the MADM (except the pairs with VIKOR, PROMETHEE, and GRA) methods is greater than 0.5. Spearman's correlation values for all the pairs formed by the GRA methods are either negative or less than 0.5. Spearman's correlation values for the VIKOR-SAW, VIKOR-WPM, VIKOR-GRA, and VIKOR-COPRAS pairs are less than 0.5. Similarly, the PROMETHEE method has Spearman's correlation values less than 0.5 with the TOPSIS, MTOPSIS, and GRA methods. Hence, the ranks suggested by the VIKOR, PROMETHEE, and

Table 7.53 Pareto-optimal solutions obtained by the Rao-3 algorithm for the solar-assisted Stirling heat engine case study-3

Solution	T_h (K)	T_c (K)	T_H (K)	T_L (K)	P (kW)	η (%)	ECOP
1	897.74	340.47	1000.00	300.00	10.8600	32.7471	1.0935
2	996.99	327.09	1137.87	280.00	12.4336	33.4954	0.9081
3	874.02	454.09	1200.00	280.00	14.6025	28.5171	0.6369
4	907.51	312.27	1000.00	281.54	10.4627	33.1356	0.9929
5	906.57	391.25	1200.00	280.00	14.3785	31.2861	0.7604
6	911.57	414.80	1200.00	280.00	14.5003	30.6219	0.7288
7	887.85	428.94	1200.00	280.00	14.5685	29.7464	0.6892
8	958.43	388.14	1200.00	280.00	14.2409	31.9992	0.7958
9	899.96	395.18	1085.72	299.56	13.2166	31.5862	0.9338
10	837.93	341.55	1045.15	291.63	12.2937	32.3326	0.9656
11	897.97	453.01	1200.00	280.00	14.5901	29.0359	0.6585
12	891.97	354.23	1000.00	300.00	11.3788	32.6134	1.0831
13	837.89	328.64	1000.00	287.81	11.5739	32.7545	1.0085
14	999.90	327.61	1200.00	280.00	13.0735	33.3221	0.8661
15	897.59	420.11	1200.00	280.00	14.5362	30.2245	0.7106
16	844.15	361.69	1000.00	297.81	12.1487	32.1282	1.0314
17	901.84	340.84	1000.00	291.45	11.1833	32.9260	1.0459
18	868.04	368.09	1000.00	300.00	11.9835	32.1909	1.0510
19	998.52	382.99	1188.89	298.69	13.5721	32.4829	0.9202
20	996.01	355.50	1105.38	298.38	11.9584	32.9589	1.0026
21	996.00	359.99	1200.00	280.00	13.7998	32.9460	0.8454
22	947.83	396.11	1151.18	296.49	13.6732	31.8743	0.8934
23	983.90	377.16	1200.00	280.00	14.0477	32.5017	0.8218
24	999.90	382.26	1120.89	289.69	12.7322	32.5567	0.9156
25	999.90	350.84	1157.52	280.00	13.1917	33.2442	0.8827

GRA methods cannot be considered for calculating the average ranks. Now, the corrected average ranks are calculated, excluding the ranks suggested by the VIKOR, PROMETHEE, and GRA methods and presented in Table 7.60 as the corrected ranks.

The SAP-Rao algorithm solution has the least average rank of 1.8. Thus, it can be regarded as the best solution. The ERao-1, ERao-2, and ERao-3 algorithm's solutions have the next best average ranks, which are 2.2, 2.8, and 4, respectively. Here an observation can be made that the SAP-Rao algorithm solution has a better fitness value in none of the three objectives. However, it is ranked 1 because it has a better compromise among the objectives. Here, it can be noted that the MADM methods have ranked these solutions based on the relative importance of the solutions. If the MOPSOCD and SAP-Rao algorithms solutions are compared, the MOPSOCD solution has better ECOP, and the SAP-Rao algorithm solution has better power

Table 7.54 Pareto-optimal solutions obtained by the AMTPG-Jaya algorithm for the solar-assisted Stirling heat engine case study-3

Solution	T_h (K)	T_c (K)	T_H (K)	T_L (K)	P (kW)	η (%)	ECOP
1	873.65	455.35	1200.00	280.00	14.6026	28.4589	0.6345
2	896.94	340.25	1000.00	300.00	10.8633	32.7473	1.0935
3	999.90	326.28	1133.23	280.00	12.3110	33.4974	0.9109
4	896.16	314.26	1000.00	287.17	10.3098	32.9054	1.0149
5	889.45	446.63	1200.00	280.00	14.5943	29.1180	0.6620
6	890.50	433.74	1200.00	285.69	14.5120	29.6306	0.7063
7	927.36	373.50	1115.72	300.00	13.0939	32.3501	0.9644
8	892.14	375.51	1131.74	295.38	13.5013	31.8524	0.8976
9	958.36	413.90	1187.21	286.29	14.1962	31.3381	0.7963
10	901.97	423.86	1200.00	285.99	14.4698	30.1769	0.7328
11	975.49	340.10	1104.42	280.00	12.4927	33.3731	0.9208
12	911.14	384.66	1161.13	296.17	13.7677	31.7080	0.8766
13	914.02	413.91	1171.49	295.73	14.0958	30.8412	0.8209
14	966.79	411.06	1200.00	282.21	14.3020	31.4567	0.7786
15	923.98	428.87	1200.00	289.09	14.4157	30.3606	0.7545
16	964.68	359.77	1154.30	299.80	12.9979	32.7558	0.9638
17	895.24	324.42	1000.00	280.20	11.2797	33.2961	0.9950
18	892.16	402.93	1162.68	297.67	13.9670	30.9021	0.8377
19	898.96	411.84	1200.00	280.00	14.5039	30.5263	0.7244
20	870.21	371.56	1000.00	300.00	12.0120	32.1160	1.0455
21	917.99	322.01	1000.00	284.31	10.5438	33.0680	1.0069
22	857.33	342.53	1000.00	300.00	11.3787	32.5431	1.0777
23	881.56	361.70	1000.00	300.00	11.6965	32.4340	1.0694
24	874.12	350.96	1000.00	291.04	11.8643	32.7101	1.0273
25	940.99	385.29	1179.50	292.41	13.9064	31.9686	0.8629

output and system efficiency. In addition, combined improvement in the power output (1.1%) and system efficiency (0.17%) is relatively more significant than the descent in the ECOP (0.3%) by the SAP-Rao solution. Similarly, the SAP-Rao algorithm solution has better relative importance when compared to the solutions of the other algorithms.

The system's power output for the SAP-Rao algorithm solution is 1.1%, 5%, 9.7, 0.7, and 1% higher than that of the MOPSOCD, Rao-2, Rao-3, ERao-1, and ERao-3 solutions, respectively. Also, ECOP of the system for the SAP-Rao solution is 5.6%, 3.8%, 0.2%, 5.1%, and 4.3% higher than that of the Jaya, Rao-1, Rao-2, AMTPG-Jaya, and ERao-2 algorithms solutions, respectively. In addition, the thermal efficiency of the SAP-Rao algorithm solution is 0.17%, 0.71%, 0.18%, and 0.19% higher when compared to that of the MOPSOCD, Jaya, Rao-1, and Rao-3

Table 7.55 Pareto-optimal solutions obtained by the SAP-Rao algorithm for the solar-assisted Stirling heat engine case study-3

Solution	T_h (K)	T_c (K)	T_H (K)	T_L (K)	P (kW)	η (%)	ECOP
1	872.34	453.89	1200.00	280.00	14.6024	28.4900	0.6358
2	894.63	339.21	1000.00	300.00	10.8546	32.7457	1.0934
3	999.90	327.25	1141.72	280.00	12.4626	33.4953	0.9059
4	904.96	325.56	1000.00	294.76	10.3366	32.7880	1.0589
5	870.66	430.05	1200.00	280.00	14.5751	29.3879	0.6735
6	921.98	429.36	1200.00	280.00	14.5244	30.2944	0.7138
7	936.73	411.43	1200.00	293.59	14.2496	31.0967	0.8115
8	982.07	394.08	1176.26	292.45	13.7699	32.1577	0.8755
9	942.92	406.17	1200.00	280.00	14.3949	31.3194	0.7620
10	907.01	434.95	1200.00	280.00	14.5616	29.8613	0.6943
11	913.37	434.02	1200.00	297.75	14.3428	30.0340	0.7752
12	938.61	344.19	1037.21	294.24	11.3494	32.9614	1.0323
13	926.89	393.48	1115.41	298.82	13.3700	31.8488	0.9261
14	912.42	321.83	1000.00	284.08	10.6758	33.1246	1.0093
15	925.73	340.10	1042.67	284.95	11.9135	33.1872	0.9826
16	930.04	363.18	1080.46	300.00	12.4802	32.6758	1.0126
17	998.81	327.40	1135.20	284.59	12.1824	33.3915	0.9292
18	916.04	398.40	1200.00	293.41	14.1929	31.1971	0.8160
19	896.95	367.93	1070.23	300.00	12.6930	32.3321	0.9972
20	934.76	392.86	1193.86	295.91	14.0278	31.6220	0.8536
21	933.75	394.35	1171.12	294.27	13.9217	31.6842	0.8604
22	974.55	374.10	1167.07	292.72	13.5197	32.5908	0.9071
23	959.58	352.95	1120.68	290.17	12.8400	33.0333	0.9480
24	920.58	356.43	1066.89	287.84	12.6091	32.8680	0.9609
25	938.61	388.64	1185.93	289.89	14.0403	31.8295	0.8397

algorithms solutions, respectively. Hence, the SAP-Rao algorithm solution can be considered as the best solution.

Now, the performances of the Jaya algorithm, Rao algorithms, and their modified versions in the MOO scenario are evaluated based on the hypervolume, coverage, and spacing indicators. Table 7.62 presents the hypervolume and spacing values of the Pareto-fronts obtained by different algorithms in the MOO scenario. The Pareto-front obtained by the MOPSOCD algorithm was not reported. Thus, the performance metric values for the MOPSOCD algorithm were not presented here. Here, an observation can be made that the SAP-Rao algorithm Pareto-front has the highest hypervolume value, and the ERao-2 algorithm Pareto-front has the least spacing value. The SAP-Rao algorithm has 1.74%, 1.71%, 1.26%, 1.64%, 1%, 0.4%, and 0.39% higher hypervolume value when compared to that of the Jaya, Rao-1, Rao-2, Rao-3, ERao-1,

Table 7.56 Pareto-optimal solutions obtained by the ERao-1 algorithm for the solar-assisted Stirling heat engine case study-3

Solution	T_h (K)	T_c (K)	T_H (K)	T_L (K)	P (kW)	η (%)	ECOP
1	872.98	455.81	1200.00	280.00	14.6025	28.4259	0.6332
2	897.15	340.42	1000.00	300.00	10.8671	32.7473	1.0935
3	999.90	326.14	1134.35	280.00	12.3230	33.4978	0.9103
4	907.67	331.15	1000.00	296.60	10.4718	32.7900	1.0722
5	903.46	448.26	1200.00	280.00	14.5817	29.3161	0.6705
6	934.71	402.33	1189.11	295.62	14.0879	31.3869	0.8416
7	948.34	442.07	1200.00	295.86	14.3117	30.2980	0.7804
8	910.96	409.30	1200.00	280.01	14.4782	30.7915	0.7368
9	876.13	390.30	1165.64	280.00	14.2159	31.0540	0.7628
10	872.05	451.19	1200.00	280.00	14.6018	28.5929	0.6400
11	901.65	431.59	1200.00	281.34	14.5474	29.8935	0.7010
12	874.25	386.59	1016.89	300.00	12.4283	31.7038	0.9994
13	915.33	364.61	1200.00	295.81	13.6529	31.9894	0.8708
14	919.37	431.87	1200.00	280.00	14.5349	30.1688	0.7081
15	970.07	327.84	1063.36	286.68	11.1782	33.1839	0.9771
16	954.37	369.74	1183.52	285.98	13.8437	32.4707	0.8569
17	968.89	323.15	1083.72	280.00	11.7757	33.4592	0.9395
18	999.90	324.05	1200.00	286.45	12.5709	33.1041	0.8869
19	938.72	373.91	1053.93	300.00	12.0953	32.4755	1.0208
20	920.15	368.58	1152.13	295.45	13.4618	32.2461	0.9095
21	918.01	361.00	1070.01	300.00	12.3983	32.6516	1.0195
22	999.90	339.04	1113.62	289.37	11.9788	33.2449	0.9614
23	888.07	351.11	1093.53	294.56	12.7391	32.5124	0.9587
24	916.59	366.76	1127.50	300.00	13.1151	32.3266	0.9549
25	988.64	358.56	1179.83	294.46	13.2580	32.8981	0.9278

ERao-2, and ERao-3 algorithms, respectively. The AMTPG-Jaya, ERao-3, ERao-2, and ERao-1 algorithms have achieved the next better hypervolume values, respectively. Similarly, the ERao-2 algorithm has 20.5%, 30.8%, 16.7%, 25.1%, 0.5%, 34.8%, 27.3%, and 10.6% less spacing value when compared to that of the Jaya, Rao-1, Rao-2, Rao-3, AMTPG-Jaya, SAP-Rao, ERao-1, and ERao-3 algorithms, respectively. The performances of the proposed modified algorithms are better or competitive to that of the basic algorithms in terms of the hypervolume and spacing values.

The coverage values of the Pareto-fronts obtained by the Jaya algorithm, Rao algorithms, and their modified versions are presented in Table 7.63. The SAP-Rao algorithm has achieved better coverage values than the other algorithms compared. Only three solutions of the SAP-Rao algorithm are dominated by the solutions of

Table 7.57 Pareto-optimal solutions obtained by the ERao-2 algorithm for the solar-assisted Stirling heat engine case study-3

Solution	T_h (K)	T_c (K)	T_H (K)	T_L (K)	P (kW)	η (%)	ECOP
1	872.34	453.70	1200.00	280.00	14.6024	28.4977	0.6361
2	895.53	340.07	1000.00	300.00	10.8774	32.7470	1.0935
3	999.90	327.08	1136.86	280.00	12.3894	33.4973	0.9088
4	907.56	327.65	1003.43	296.46	10.3683	32.7603	1.0654
5	892.16	446.99	1200.00	280.00	14.5926	29.1553	0.6636
6	896.19	431.67	1200.00	280.00	14.5681	29.7943	0.6913
7	903.60	419.42	1200.00	280.00	14.5275	30.3450	0.7161
8	868.75	326.91	1000.00	289.98	11.1651	32.9411	1.0367
9	945.03	422.94	1200.00	280.00	14.4479	30.8312	0.7386
10	958.86	335.74	1100.22	293.76	11.9432	33.1306	0.9897
11	953.26	434.23	1200.00	297.68	14.2529	30.6226	0.8052
12	979.27	406.14	1194.13	280.00	14.1979	31.7411	0.7852
13	970.96	367.76	1142.30	300.00	13.0042	32.7205	0.9704
14	850.71	344.46	1000.00	289.78	11.9807	32.6685	1.0158
15	849.30	333.04	1000.00	300.00	10.9081	32.4208	1.0683
16	999.90	356.80	1167.84	290.08	13.1562	33.0611	0.9206
17	880.24	369.01	1030.42	300.00	12.3428	32.2456	1.0254
18	953.82	407.18	1200.00	280.13	14.3626	31.4198	0.7675
19	999.90	401.11	1200.00	300.00	13.8263	32.0488	0.8952
20	997.52	328.45	1098.89	286.45	11.5115	33.2549	0.9547
21	851.17	378.83	1109.87	300.00	13.3367	31.2502	0.8993
22	999.90	343.44	1175.34	280.00	13.2704	33.3106	0.8774
23	933.72	378.50	1185.67	290.01	13.9388	32.0303	0.8515
24	979.30	353.30	1078.49	300.00	11.5592	32.8751	1.0282
25	976.08	370.16	1200.00	280.00	14.0150	32.5957	0.8267

the other compared methods. From the coverage values, it can be observed that four solutions from the Pareto-fronts of the ERao-1, ERao-2, and Rao-2 algorithms are dominated solutions. Similarly, five solutions from the Pareto-fronts achieved by the Rao-1, Rao-3, AMTPG-Jaya, and ERao-3 algorithms are dominated, and six solutions of the Pareto-front achieved by the Jaya algorithm are dominated. The performance of the proposed modified algorithms is better or competitive to that of the basic algorithms in terms of the coverage values.

From the computational results of the solar-assisted Stirling heat engine case study-3, it can be observed that the performance of the considered system can be modified with the solutions achieved by the proposed algorithms. The solution of the SAP-Rao algorithm has better compromise among the power output, system efficiency, and ECOP. The SAP-Rao algorithm's solution has 9.7% higher power

Table 7.58 Pareto-optimal solutions obtained by the ERao-3 algorithm for the solar-assisted Stirling heat engine case study-3

Solution	T_h (K)	T_c (K)	T_H (K)	T_L (K)	P (kW)	η (%)	ECOP
1	897.69	341.15	1000.00	300.00	10.8873	32.7465	1.0934
2	999.90	328.74	1131.84	280.00	12.3646	33.4921	0.9114
3	873.55	456.99	1200.00	280.00	14.6025	28.3900	0.6317
4	898.56	325.36	1000.00	295.07	10.3949	32.7799	1.0605
5	917.21	389.35	1181.48	287.20	14.1168	31.5753	0.8154
6	949.41	328.16	1049.81	280.11	11.5437	33.3600	0.9580
7	954.39	395.17	1200.00	280.00	14.3008	31.7675	0.7841
8	849.67	363.52	1000.00	300.00	12.0732	32.1257	1.0462
9	970.06	382.34	1074.66	300.00	12.0694	32.3624	0.9957
10	931.80	415.02	1200.00	280.00	14.4609	30.9046	0.7421
11	855.32	403.64	1026.87	300.00	12.8718	30.8588	0.9339
12	914.18	314.74	1005.91	280.00	10.7075	33.2464	0.9850
13	999.90	369.97	1200.00	280.00	13.8938	32.7744	0.8362
14	922.96	361.01	1070.02	299.81	12.3618	32.6868	1.0207
15	882.08	407.89	1053.29	299.35	13.0467	31.0280	0.9207
16	885.93	375.54	1126.71	292.31	13.5579	31.8110	0.8819
17	987.55	338.70	1088.53	296.14	11.3769	33.0356	1.0068
18	864.82	441.46	1200.00	280.00	14.5932	28.8305	0.6499
19	908.28	420.45	1200.00	280.00	14.5247	30.3838	0.7178
20	999.90	332.15	1200.00	280.00	13.2206	33.3141	0.8656
21	910.89	442.44	1200.00	280.00	14.5660	29.6588	0.6853
22	894.08	420.81	1200.00	280.00	14.5416	30.1422	0.7068
23	999.90	350.87	1125.14	280.00	12.6923	33.2346	0.9000
24	999.90	378.06	1170.13	280.00	13.6431	32.6782	0.8448
25	968.24	342.54	1115.77	290.93	12.4269	33.1873	0.9655

output and 5.6% higher ECOP when compared to those of the solutions achieved by the other algorithms compared. The SAP-Rao algorithm's performance in terms of coverage and hypervolume is much better than those achieved by the other algorithms. Similarly, the ERao-2 algorithm's performance in terms of the spacing is much better than that achieved by other algorithms. In addition, the performances of the modified versions are better or competitive to those of the basic algorithms as well as the MOPSOCD algorithm. Here it can be observed that, in the three solar-assisted Stirling heat engine case studies, the performances of the proposed modified algorithms are superior or competitive to those of the basic Jaya and Rao algorithms. The next section presents the application of the Jaya and Rao algorithms along with their modified versions, to the solar-assisted Carnot-like heat engine case study.

Table 7.59 Best solutions obtained by various algorithms in MOO scenario of the solar-assisted Stirling heat engine case study-3

Algorithm	T_h (K)	T_c (K)	T_H (K)	T_L (K)	P (kW)	η (%)	ECOP
MOPSOCD	917	362.3	1070.4	300	12.44	32.62	1.016
Jaya	922.755	365.895	1121.67	298.6	**13.07068**	32.446789	0.95852
Rao-1	931.330	358.470	1125.00	300	12.84471	32.618340	0.97547
Rao-2	905.184	340.412	1036.45	289.9	11.88868	**33.051957**	1.01062
Rao-3	891.967	354.235	1000.00	300	11.37881	32.613382	**1.08310**
AMTPG-Jaya	964.678	359.767	1154.30	299.8	12.99792	32.755804	0.96379
SAP-Rao	930.039	363.177	1080.46	300	12.48016	32.675822	1.01262
ERao-1	918.009	361.004	1070.01	300	12.39834	32.651557	1.01949
ERao-2	970.964	367.758	1142.30	300	13.00417	32.720541	0.97044
ERao-3	922.959	361.006	1070.02	299.81	12.36179	32.686821	1.02074

Source MOPSOCD—Dai et al. (2018)
Bold figure indicates a better value

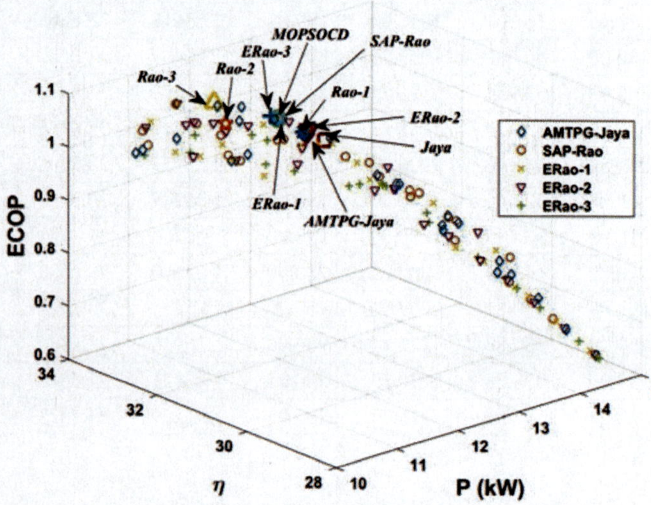

Fig. 7.4 Plot of Pareto-fronts of the proposed algorithms and solutions of the methods compared in solar-assisted Stirling heat engine system case study-3

7.3 Optimization of a Solar-Assisted Carnot-like Heat Engine System

The description of the selected solar-assisted Carnot-like heat engine system is presented in Sect. 2.2.3. The detailed description and thermodynamic analysis of the solar-assisted Carnot-like heat engine system considered in this book were presented

Table 7.60 Ranks suggested by the MADM methods for different algorithm solutions presented in Table 7.59

Algorithm	SAW	WPM	TOPSIS	MTOPSIS	VIKOR	PROMETHEE	COPRAS	GRA	AR	CR	FR
MOPSOCD	5	5	3	3	5	7	5	10	5.4	4.2	5
Jaya	8	8	8	8	10	10	8	5	8.1	8	8
Rao-1	7	7	7	7	8	8	7	6	7.1	7	7
Rao-2	10	10	10	10	1	5.5	10	1	7.2	10	10
Rao-3	9	9	9	9	9	9	9	4	8.4	9	9
AMTPG-Jaya	6	6	6	6	7	2.5	6	2	5.2	6	6
SAP-Rao	2	1	2	2	1	4	2	7	2.6	1.8	1
ERao-1	3	3	1	1	1	5.5	3	9	3.3	2.2	2
ERao-2	1	2	5	5	6	1	1	3	3.0	2.8	3
ERao-3	4	4	4	4	1	2.5	4	8	3.9	4	4

AR average rank, *CR* corrected rank, *FR* final rank

Table 7.61 Spearman's rank correlation coefficients between different pairs of MADM method's rankings presented in Table 7.60

Method	SAW	WPM	TOPSIS	MTOPSIS	VIKOR	PROMETHEE	COPRAS	GRA
SAW	1	0.988	0.855	0.855	0.438	0.671	1	−0.442
WPM	0.988	1	0.891	0.891	0.481	0.634	0.988	−0.491
TOPSIS	0.855	0.891	1	1	0.550	0.415	0.855	−0.758
MTOPSIS	0.855	0.891	1	1	0.550	0.415	0.855	−0.758
VIKOR	0.438	0.481	0.550	0.550	1	0.563	0.438	−0.300
PROMETHEE	0.671	0.634	0.415	0.415	0.563	1	0.671	0.152
COPRAS	1	0.988	0.855	0.855	0.438	0.671	1	−0.442
GRA	−0.442	−0.491	−0.758	−0.758	−0.300	0.152	−0.442	1

Table 7.62 Hypervolume and spacing values of the Pareto-fronts obtained by the proposed algorithms in MOO for the solar-assisted Stirling engine case study-3

Algorithm	Hypervolume	Spacing
Jaya	71.14207	0.058204
Rao-1	71.16683	0.066851
Rao-2	71.47948	0.055525
Rao-3	71.21845	0.061742
AMTPG-Jaya	72.32720	0.046464
SAP-Rao	**72.38306**	0.070969
ERao-1	71.66467	0.063585
ERao-2	72.09578	**0.046245**
ERao-3	72.09954	0.051722

Results in the bold figure indicate better values

Table 7.63 Coverage (%) values of the Pareto-fronts obtained by the proposed algorithms in MOO for the solar-assisted Stirling engine case study-3

Algorithm	Jaya	Rao-1	Rao-2	Rao-3	AMTPG-Jaya	SAP-Rao	ERao-1	ERao-2	ERao-3
Jaya	–	20	0	4	12	4	4	4	12
Rao-1	4	–	0	12	12	12	4	16	12
Rao-2	12	16	–	20	16	12	16	8	20
Rao-3	16	16	16	–	16	8	12	16	20
AMTPG-Jaya	12	16	8	16	–	4	16	4	20
SAP-Rao	16	16	16	12	20	–	16	4	8
ERao-1	12	8	8	12	12	8	–	8	20
ERao-2	12	20	8	16	12	8	12	–	12
ERao-3	24	16	4	4	8	8	8	12	–

by Sayyaadi et al. (2015). The three objective functions considered in this case study are non-dimensional power output (W), non-dimensional ecological function (E), and thermal efficiency (η). These three objective functions are maximization functions and are presented in Eqs. 2.40–2.42. The decision variables are the temperature ratio (γ), the convective coefficient ratio (ϕ), the operating temperature ratio (τ), and the heat source to the heat sink allocation parameter (β), and their ranges are as follows:$0.1 \le \phi \le 0.6$,$0.1 \le \beta \le 0.6$, $0.2 \le \tau \le 0.4$, and $0.7 \le \gamma \le 0.8$.

Sayyaadi et al. (2015) employed NSGA-II to this MOO case study for finding the optimal decision variables and documented three non-dominated solutions using Bellman–Zadeh, TOPSIS, and LINMAP decision-making methods based on the similarity to the ideal solutions. The number of function evaluations taken by NSGA-II to achieve convergence was not specified. However, in this work, the proposed algorithms are executed using only 10,000 function evaluations as the termination criterion and reported the Pareto-optimal solutions using the average rank based on the multiple decision-making methods. Firstly, single-objective optimization is performed for each objective function independently, and then multi-objective optimization is performed in this work. In all the computational experiments, the population size is maintained as 25 for all the algorithms. The elite size for the elitist Rao algorithms is taken as 20%.

Table 7.64 presents the single-objective optimization results of this case study. The Jaya algorithm, Rao algorithms, and their modified versions have obtained identical solutions in maximizing thermal efficiency. However, the proposed algorithms have obtained different combinations of design variables, which have resulted in the same objective values in the remaining two single-objective optimization scenarios. The maximum non-dimensional power output, ecological function, and efficiency obtained by the proposed algorithms are 0.2, 0.120645329, and 0.7225921, respectively.

Table 7.64 Results obtained by the proposed algorithms in single-objective optimization scenarios of the solar-assisted Carnot-like heat engine case study

Objective	Algorithm	ϕ	β	τ	γ	η	W	E
Maximize-η	NSGA-II	0.1023	0.103	0.2	0.8	0.7216	0.0587	0.0523
	Proposed algorithms	0.1	0.1	0.2	0.8	0.722592	0.05711	0.050995
Maximize-W	NSGA-II	0.3638	0.5552	0.2	0.8	0.4993	0.2	0.0795
	Jaya	0.5338884	0.496649	0.2	0.8	0.500009	0.2	0.0800006
	Rao-1	0.392703	0.544478	0.2	0.8	0.499998	0.2	0.0799992
	Rao-2	0.6	0.474255	0.2	0.8	0.499999	0.2	0.0799995
	Rao-3	0.2459854	0.594178	0.2	0.8	0.499999	0.2	0.0799999
	AMTPG-Jaya	0.2288	0.6	0.2	0.8	0.499999	0.2	0.0799999
	SAP-Rao	0.2403719	0.596079	0.2	0.8	0.500002	0.2	0.0800001
	ERao-1	0.2287994	0.6	0.2	0.8	0.500001	0.2	0.0800001
	ERao-2	0.4991504	0.508417	0.2	0.8	0.500001	0.2	0.0800001
	ERao-3	0.2862074	0.580554	0.2	0.8	0.499999	0.2	0.0799994
Maximize-E	NSGA-II	0.2328	0.4008	0.2	0.8	0.6157	0.1721	0.1206
	Jaya	0.4574605	0.325392	0.2	0.8	0.612700	0.1737	0.1206453
	Rao-1	0.6	0.277104	0.2	0.8	0.612701	0.1737	0.1206453
	Rao-2	0.54593	0.2954205	0.2	0.8	0.612701	0.1737	0.1206453
	Rao-3	0.6	0.2771036	0.2	0.8	0.612701	0.1737	0.1206453
	AMTPG-Jaya	0.3340455	0.3671949	0.2	0.8	0.612702	0.1737	0.1206453
	SAP-Rao	0.3754643	0.3531646	0.2	0.8	0.612702	0.1737	0.1206453
	ERao-1	0.1400834	0.4329009	0.2	0.8	0.612702	0.1737	0.1206453
	ERao-2	0.5715787	0.2867318	0.2	0.8	0.612701	0.1737	0.1206453
	ERao-3	0.4772283	0.3186931	0.2	0.8	0.612701	0.1737	0.1206453

Source NSGA-II—Sayyaadi et al. (2015)

The Pareto-optimal solutions achieved by the Jaya algorithm, Rao algorithms, and their modified versions in MOO are presented in Tables 7.65, 7.66, 7.67, 7.68, 7.69, 7.70, 7.71, 7.72 and 7.73.

Here, the non-dimensional power output is varied from 0.057 to 0.2, the thermal efficiency is varied between 0.49996 and 0.7226, and ecological function is varied from 0.05099 to 0.12065. The best solutions from the Pareto-fronts obtained by the proposed algorithms are identified based on the average rank method. Solution 7 of the Jaya algorithm, Solution 25 of the Rao-1 algorithm, Solution 1 of the Rao-2 algorithm, Solution 17 of the Rao-3 algorithm, Solution 21 of the AMTPG-Jaya algorithm, Solution 10 of the SAP-Rao algorithms, Solution 5 of the ERao-1 algorithm,

Table 7.65 Pareto-optimal solutions obtained by the Jaya algorithm for the solar-assisted Carnot-like heat engine case study

Solution	ϕ	β	τ	γ	η	W	E
1	0.3344	0.3671	0.2	0.8	0.6127	0.1738	0.1206
2	0.3800	0.5488	0.2	0.8	0.5000	0.2000	0.0800
3	0.1000	0.1016	0.2	0.8	0.7222	0.0578	0.0515
4	0.3166	0.1000	0.2	0.8	0.7049	0.0862	0.0746
5	0.1475	0.1051	0.2	0.8	0.7177	0.0657	0.0582
6	0.1000	0.3144	0.2	0.8	0.6635	0.1364	0.1084
7	0.2483	0.4250	0.2	0.8	0.5995	0.1802	0.1200
8	0.4796	0.1000	0.2	0.8	0.6899	0.1069	0.0899
9	0.4308	0.1417	0.2	0.8	0.6826	0.1159	0.0960
10	0.1182	0.4289	0.2	0.8	0.6177	0.1710	0.1205
11	0.1000	0.3543	0.2	0.8	0.6496	0.1489	0.1144
12	0.1206	0.1000	0.2	0.8	0.7210	0.0600	0.0534
13	0.3207	0.3161	0.2	0.8	0.6358	0.1595	0.1183
14	0.2718	0.4955	0.2	0.8	0.5586	0.1938	0.1100
15	0.1781	0.3734	0.2	0.8	0.6323	0.1619	0.1190
16	0.2103	0.1000	0.2	0.8	0.7138	0.0722	0.0635
17	0.1955	0.4792	0.2	0.8	0.5816	0.1873	0.1169
18	0.1000	0.1543	0.2	0.8	0.7097	0.0788	0.0688
19	0.2714	0.2914	0.2	0.8	0.6514	0.1474	0.1138
20	0.4170	0.1000	0.2	0.8	0.6958	0.0991	0.0843
21	0.3563	0.1000	0.2	0.8	0.7014	0.0914	0.0785
22	0.3762	0.4403	0.2	0.8	0.5697	0.1910	0.1138
23	0.2705	0.2262	0.2	0.8	0.6734	0.1264	0.1026
24	0.2181	0.1000	0.2	0.8	0.7132	0.0732	0.0643
25	0.2210	0.2560	0.2	0.8	0.6692	0.1307	0.1052

Table 7.66 Pareto-optimal solutions obtained by the Rao-1 algorithm for the solar-assisted Carnot-like heat engine case study

Solution	ϕ	β	τ	γ	η	W	E
1	0.5466	0.4923	0.2	0.8	0.5000	0.2000	0.0800
2	0.5261	0.3021	0.2	0.8	0.6127	0.1738	0.1206
3	0.1004	0.1000	0.2	0.8	0.7226	0.0572	0.0510
4	0.2143	0.2977	0.2	0.8	0.6560	0.1434	0.1119
5	0.1000	0.1236	0.2	0.8	0.7171	0.0667	0.0590
6	0.2363	0.1181	0.2	0.8	0.7072	0.0827	0.0719
7	0.5494	0.1198	0.2	0.8	0.6771	0.1223	0.1001
8	0.3214	0.1136	0.2	0.8	0.7009	0.0921	0.0790
9	0.2318	0.1034	0.2	0.8	0.7112	0.0764	0.0669
10	0.2140	0.1000	0.2	0.8	0.7135	0.0727	0.0639
11	0.3385	0.3338	0.2	0.8	0.6263	0.1658	0.1199
12	0.1000	0.3946	0.2	0.8	0.6344	0.1605	0.1186
13	0.2764	0.1389	0.2	0.8	0.6982	0.0959	0.0819
14	0.6000	0.4443	0.2	0.8	0.5212	0.1993	0.0927
15	0.3442	0.2901	0.2	0.8	0.6428	0.1543	0.1166
16	0.4894	0.4387	0.2	0.8	0.5486	0.1958	0.1061
17	0.3999	0.4557	0.2	0.8	0.5565	0.1942	0.1093
18	0.2168	0.1730	0.2	0.8	0.6944	0.1010	0.0857
19	0.4068	0.1174	0.2	0.8	0.6920	0.1042	0.0880
20	0.5813	0.4422	0.2	0.8	0.5268	0.1988	0.0957
21	0.6000	0.3572	0.2	0.8	0.5736	0.1898	0.1149
22	0.1000	0.2380	0.2	0.8	0.6873	0.1103	0.0922
23	0.4614	0.3902	0.2	0.8	0.5810	0.1875	0.1168
24	0.6000	0.4189	0.2	0.8	0.5378	0.1975	0.1012
25	0.2039	0.4481	0.2	0.8	0.5957	0.1819	0.1195

Solution 11 of the ERao-2, and Solution 9 of the ERao-3 algorithm are identified as the best solutions from the respective algorithm Pareto-front.

In Table 7.74, the solutions with the best average rank from the Pareto-optimal solutions of the various algorithms are compared with the NSGA-II (TOPSIS, LINMAP, and Fuzzy) algorithm solutions. Figure 7.5 presents the Pareto-fronts achieved by the AMTPG-Jaya, SAP-Rao, and elitist Rao algorithms, including the optimal solutions reported for the NSGA-II, Jaya, and Rao algorithms. The conflicting nature of these objectives can be observed in Fig. 7.5. Furthermore, the ecological function of the system for the Rao-2 algorithm solution is highest, which is 2.7%, 2.6%, 3.9%, 0.6%, 1%, 1.3%, 0.5%, 0.4%, 0.4%, 0.3%, and 0.3% higher than that of the NSGA-II (TOPSIS, LINMAP, and Fuzzy), Jaya, Rao-1, Rao-3, AMTPG-Jaya, SAP-Rao, ERao-1, ERao-2, and ERao-3 algorithms solutions, respectively.

Table 7.67 Pareto-optimal solutions obtained by the Rao-2 algorithm for the solar-assisted Carnot-like heat engine case study

Solution	ϕ	β	τ	γ	η	W	E
1	0.1777	0.4201	0.2	0.8	0.6127	0.1738	0.1206
2	0.2288	0.6000	0.2	0.8	0.5000	0.2000	0.0800
3	0.1030	0.1000	0.2	0.8	0.7224	0.0575	0.0513
4	0.1000	0.3238	0.2	0.8	0.6604	0.1394	0.1099
5	0.1000	0.2395	0.2	0.8	0.6868	0.1108	0.0926
6	0.1079	0.3854	0.2	0.8	0.6369	0.1587	0.1181
7	0.1269	0.5599	0.2	0.8	0.5496	0.1956	0.1065
8	0.2780	0.1000	0.2	0.8	0.7082	0.0812	0.0707
9	0.1000	0.2209	0.2	0.8	0.6921	0.1041	0.0879
10	0.2523	0.4385	0.2	0.8	0.5924	0.1833	0.1190
11	0.1130	0.1698	0.2	0.8	0.7046	0.0866	0.0749
12	0.1642	0.4969	0.2	0.8	0.5779	0.1885	0.1161
13	0.1000	0.2910	0.2	0.8	0.6712	0.1287	0.1040
14	0.1000	0.1411	0.2	0.8	0.7129	0.0736	0.0646
15	0.1574	0.5209	0.2	0.8	0.5662	0.1919	0.1127
16	0.1929	0.1175	0.2	0.8	0.7110	0.0767	0.0671
17	0.1000	0.2565	0.2	0.8	0.6818	0.1169	0.0966
18	0.3980	0.1000	0.2	0.8	0.6976	0.0967	0.0825
19	0.2412	0.5087	0.2	0.8	0.5569	0.1941	0.1094
20	0.1987	0.6000	0.2	0.8	0.5074	0.1999	0.0846
21	0.1000	0.4163	0.2	0.8	0.6256	0.1663	0.1199
22	0.1990	0.5616	0.2	0.8	0.5334	0.1981	0.0991
23	0.1643	0.6000	0.2	0.8	0.5156	0.1996	0.0895
24	0.1983	0.3087	0.2	0.8	0.6540	0.1451	0.1127
25	0.1815	0.1000	0.2	0.8	0.7162	0.0683	0.0603

However, these solutions are non-dominated. Hence, to identify the best solution among the solutions of different algorithms, which has the best compromise among the objectives, the MADM methods based average ranks are calculated and presented in Table 7.75.

The decision-making methods ranking of the Pareto-optimal solutions obtained by different algorithms is shown in Table 7.75. Spearman's correlation coefficients for different pairs of rankings given by decision-making methods for different algorithm's solutions are shown in Table 7.76. The ranks given by the SAW and COPRAS methods for each solution are identical. Similarly, the ranks given by the VIKOR, PROMETHEE, and GRA methods are equal. The ranks given by the TOPSIS and MTOPSIS methods for each solution are identical. Furthermore, Spearman's correlation for all the pairs formed by all the MADM (except the pairs with TOPSIS and

Table 7.68 Pareto-optimal solutions obtained by the Rao-3 algorithm for the solar-assisted Carnot-like heat engine case study

Solution	ϕ	β	τ	γ	η	W	E
1	0.2290	0.6000	0.2	0.8	0.5000	0.2000	0.0800
2	0.1000	0.4464	0.2	0.8	0.6127	0.1738	0.1206
3	0.1000	0.1000	0.2	0.8	0.7226	0.0571	0.0510
4	0.1886	0.2961	0.2	0.8	0.6595	0.1402	0.1103
5	0.1866	0.2409	0.2	0.8	0.6777	0.1217	0.0997
6	0.2764	0.3532	0.2	0.8	0.6270	0.1654	0.1198
7	0.1094	0.3804	0.2	0.8	0.6387	0.1574	0.1177
8	0.3035	0.1000	0.2	0.8	0.7060	0.0845	0.0733
9	0.1270	0.2234	0.2	0.8	0.6888	0.1083	0.0908
10	0.1165	0.1000	0.2	0.8	0.7213	0.0594	0.0529
11	0.2725	0.1270	0.2	0.8	0.7017	0.0909	0.0781
12	0.1000	0.3043	0.2	0.8	0.6669	0.1331	0.1066
13	0.2274	0.1630	0.2	0.8	0.6962	0.0986	0.0839
14	0.1000	0.5318	0.2	0.8	0.5708	0.1907	0.1141
15	0.1669	0.1000	0.2	0.8	0.7173	0.0663	0.0587
16	0.1269	0.2114	0.2	0.8	0.6922	0.1040	0.0878
17	0.4685	0.3646	0.2	0.8	0.5927	0.1831	0.1191
18	0.1699	0.4763	0.2	0.8	0.5875	0.1852	0.1182
19	0.1000	0.5928	0.2	0.8	0.5349	0.1979	0.0998
20	0.1412	0.5475	0.2	0.8	0.5541	0.1948	0.1083
21	0.1182	0.6000	0.2	0.8	0.5262	0.1988	0.0954
22	0.1000	0.2879	0.2	0.8	0.6722	0.1277	0.1034
23	0.1729	0.6000	0.2	0.8	0.5136	0.1997	0.0883
24	0.1307	0.5093	0.2	0.8	0.5773	0.1887	0.1159
25	0.1000	0.1416	0.2	0.8	0.7128	0.0739	0.0648

MTOPSIS) methods is greater than 0.5. The TOPSIS and MTOPSIS methods have Spearman's correlation values less than 0.5 with the VIKOR, PROMETHEE, and GRA methods. Hence, the ranks suggested by the TOPSIS and MTOPSIS methods cannot be considered for calculating the average ranks. Now, the corrected average ranks are calculated, excluding the ranks suggested by the TOPSIS and MTOPSIS methods and presented in Table 7.75 as the corrected ranks.

The AMTPG-Jaya algorithm solution has the least average rank, which is 3.5. Thus, it can be regarded as the best solution. The SAP-Rao, ERao-1, and ERao-3 algorithm's solutions have the next best and the same average rank, which is 4. The SAP-Rao and ERao-3 algorithms solutions are ranked three (least among the ranks suggested to these three solutions) by three methods each, whereas the ERao-1 algorithm solution ranked four by all the methods. Hence, the ERao-1 solution is

Table 7.69 Pareto-optimal solutions obtained by the AMTPG-Jaya algorithm for the solar-assisted Carnot-like heat engine case study

Solution	ϕ	β	τ	γ	η	W	E
1	0.2288	0.6000	0.2	0.8	0.5000	0.2000	0.0800
2	0.2731	0.3878	0.2	0.8	0.6127	0.1738	0.1206
3	0.1017	0.1000	0.2	0.8	0.7225	0.0573	0.0512
4	0.1000	0.2429	0.2	0.8	0.6858	0.1121	0.0934
5	0.2844	0.2553	0.2	0.8	0.6624	0.1375	0.1090
6	0.1836	0.5552	0.2	0.8	0.5407	0.1971	0.1026
7	0.2301	0.2368	0.2	0.8	0.6743	0.1253	0.1020
8	0.1274	0.1698	0.2	0.8	0.7034	0.0884	0.0763
9	0.1080	0.5362	0.2	0.8	0.5670	0.1917	0.1129
10	0.1624	0.1850	0.2	0.8	0.6962	0.0986	0.0839
11	0.1005	0.2241	0.2	0.8	0.6912	0.1053	0.0888
12	0.2294	0.3279	0.2	0.8	0.6432	0.1540	0.1165
13	0.2123	0.1000	0.2	0.8	0.7137	0.0724	0.0637
14	0.2458	0.5074	0.2	0.8	0.5568	0.1942	0.1094
15	0.1578	0.6000	0.2	0.8	0.5171	0.1995	0.0904
16	0.2379	0.1000	0.2	0.8	0.7116	0.0759	0.0664
17	0.1000	0.1578	0.2	0.8	0.7088	0.0802	0.0699
18	0.2245	0.2392	0.2	0.8	0.6742	0.1255	0.1021
19	0.2073	0.4684	0.2	0.8	0.5851	0.1861	0.1177
20	0.2121	0.4724	0.2	0.8	0.5822	0.1871	0.1171
21	0.1013	0.4741	0.2	0.8	0.5999	0.1801	0.1200
22	0.2931	0.1000	0.2	0.8	0.7069	0.0832	0.0722
23	0.4032	0.4980	0.2	0.8	0.5298	0.1985	0.0973
24	0.3183	0.1374	0.2	0.8	0.6948	0.1006	0.0854
25	0.1196	0.3905	0.2	0.8	0.6334	0.1612	0.1188

assigned with a final rank of four in Table 7.75. The SAP-Rao and ERao-3 solutions are ranked 3 and 5 by three different methods each. Thus, the values of the objectives are compared for these two solutions. The ERao-3 solution has better values in two objectives (η and E), and the SAP-Rao solution has better value in one objective (W). Hence, the ERao-3 and SAP-Rao solutions are assigned with final ranks of 2 and 3, respectively.

Here an observation can be made that the AMTPG-Jaya algorithm solution has a better compromise among the three objectives. The non-dimensional power output for the AMTPG-Jaya algorithm solution is 17.6%, 3.62%, 0.47%, 0.49%, 1.09%, and 0.73% higher than that of the NSGA-II (Fuzzy), Rao-2, SAP-Rao, ERao-1, ERao-2, and ERao-3 solutions, respectively. Also, thermal efficiency for the AMTPG-Jaya solution is 2.7%, 2.6%, 0.1%, 0.7%, and 1.2% higher than that of the NSGA-II

Table 7.70 Pareto-optimal solutions obtained by the SAP-Rao algorithm for the solar-assisted Carnot-like heat engine case study

Solution	ϕ	β	τ	γ	η	W	E
1	0.6000	0.2770	0.2	0.8	0.6127	0.1737	0.1206
2	0.6000	0.4743	0.2	0.8	0.5000	0.2000	0.0800
3	0.1000	0.1000	0.2	0.8	0.7226	0.0571	0.0510
4	0.1000	0.2131	0.2	0.8	0.6943	0.1012	0.0858
5	0.1000	0.3861	0.2	0.8	0.6377	0.1581	0.1179
6	0.1555	0.2192	0.2	0.8	0.6873	0.1103	0.0922
7	0.1782	0.1212	0.2	0.8	0.7113	0.0762	0.0667
8	0.4453	0.4081	0.2	0.8	0.5745	0.1896	0.1151
9	0.6000	0.4314	0.2	0.8	0.5298	0.1985	0.0972
10	0.4829	0.3409	0.2	0.8	0.6017	0.1792	0.1202
11	0.5922	0.4537	0.2	0.8	0.5166	0.1995	0.0901
12	0.1941	0.1371	0.2	0.8	0.7060	0.0846	0.0733
13	0.1661	0.3253	0.2	0.8	0.6520	0.1469	0.1135
14	0.1192	0.1831	0.2	0.8	0.7006	0.0924	0.0793
15	0.4147	0.4699	0.2	0.8	0.5450	0.1964	0.1045
16	0.3740	0.1590	0.2	0.8	0.6832	0.1153	0.0955
17	0.1000	0.2944	0.2	0.8	0.6701	0.1299	0.1047
18	0.1346	0.3916	0.2	0.8	0.6309	0.1629	0.1192
19	0.1000	0.2769	0.2	0.8	0.6756	0.1239	0.1011
20	0.5982	0.1618	0.2	0.8	0.6580	0.1416	0.1110
21	0.1000	0.1342	0.2	0.8	0.7146	0.0709	0.0624
22	0.5381	0.3933	0.2	0.8	0.5654	0.1921	0.1124
23	0.3376	0.4163	0.2	0.8	0.5891	0.1846	0.1185
24	0.4408	0.4473	0.2	0.8	0.5533	0.1949	0.1080
25	0.1938	0.2272	0.2	0.8	0.6811	0.1177	0.0972

(TOPSIS and LINMAP), Jaya, Rao-1, and Rao-3 algorithms solutions, respectively. In addition, the ecological function of the AMTPG-Jaya algorithm solution is 2.2%, 2%, 3.3%, 0.4%, and 0.8% higher when compared to that of the NSGA-II (TOPSIS, LINMAP, and Fuzzy), Rao-1, and Rao-3 algorithms solutions, respectively. Hence, the AMTPG-Jaya algorithm solution can be considered as the best solution.

Now, the performances of the Jaya algorithm, Rao algorithms, and their modified versions in the MOO scenario are evaluated based on the hypervolume, coverage, and spacing indicators. Table 7.77 presents the hypervolume and spacing values of the Pareto-fronts obtained by different algorithms in the MOO scenario. Here, an observation can be made that the ERao-1 algorithm's Pareto-front has the highest hypervolume value, and the Rao-2 algorithm's Pareto-front has the least spacing value. The ERao-1 algorithm has 3.79%, 0.12%, 0.13%, 0.1%, and 0.13% higher

Table 7.71 Pareto-optimal solutions obtained by the ERao-1 algorithm for the solar-assisted Carnot-like heat engine case study

Solution	ϕ	β	τ	γ	η	W	E
1	0.1000	0.4463	0.2	0.8	0.6128	0.1737	0.1206
2	0.2288	0.6000	0.2	0.8	0.5000	0.2000	0.0800
3	0.1000	0.1000	0.2	0.8	0.7226	0.0571	0.0510
4	0.1000	0.1747	0.2	0.8	0.7045	0.0868	0.0750
5	0.3586	0.3828	0.2	0.8	0.6018	0.1792	0.1202
6	0.1014	0.1523	0.2	0.8	0.7101	0.0783	0.0683
7	0.1465	0.1208	0.2	0.8	0.7140	0.0718	0.0632
8	0.1905	0.3015	0.2	0.8	0.6575	0.1421	0.1113
9	0.1003	0.2850	0.2	0.8	0.6730	0.1268	0.1028
10	0.1000	0.6000	0.2	0.8	0.5303	0.1984	0.0975
11	0.1130	0.1000	0.2	0.8	0.7216	0.0589	0.0525
12	0.1257	0.5306	0.2	0.8	0.5667	0.1918	0.1128
13	0.1000	0.2406	0.2	0.8	0.6865	0.1113	0.0929
14	0.1614	0.1000	0.2	0.8	0.7178	0.0655	0.0580
15	0.2520	0.3577	0.2	0.8	0.6286	0.1644	0.1196
16	0.1000	0.1996	0.2	0.8	0.6979	0.0962	0.0822
17	0.3128	0.2365	0.2	0.8	0.6655	0.1345	0.1073
18	0.2119	0.3173	0.2	0.8	0.6493	0.1492	0.1145
19	0.2863	0.5131	0.2	0.8	0.5452	0.1964	0.1046
20	0.1000	0.1480	0.2	0.8	0.7113	0.0764	0.0668
21	0.5550	0.3558	0.2	0.8	0.5824	0.1870	0.1171
22	0.3733	0.5284	0.2	0.8	0.5162	0.1996	0.0898
23	0.2037	0.2812	0.2	0.8	0.6629	0.1370	0.1087
24	0.4408	0.4449	0.2	0.8	0.5548	0.1946	0.1086
25	0.3082	0.4503	0.2	0.8	0.5767	0.1889	0.1158

hypervolume value when compared to that of the Jaya, Rao-1, Rao-2, Rao-3, and AMTPG-Jaya algorithms, respectively. The ERao-3, ERao-2, SAP-Rao, and Rao-3 algorithms have achieved the next better hypervolume values, respectively. Similarly, the Rao-2 algorithm has 71%, 30.5%, 2.5%, 46.5%, 26.4%, 41%, 21%, and 29% less spacing value when compared to that of the Jaya, Rao-1, Rao-3, AMTPG-Jaya, SAP-Rao, ERao-1, ERao-2, and ERao-3 algorithms, respectively. The performances of the proposed modified algorithms are better or competitive to that of the basic algorithms in terms of the hypervolume and spacing values.

The coverage values of the Pareto-fronts obtained by the Jaya algorithm, Rao algorithms, and their modified versions are presented in Table 7.78. The AMTPG-Jaya, Rao-1, and ERao-2 algorithms have achieved better coverage values than the other algorithms compared. Not a single solution of these algorithms is dominated by the

Table 7.72 Pareto-optimal solutions obtained by the ERao-2 algorithm for the solar-assisted Carnot-like heat engine case study

Solution	ϕ	β	τ	γ	η	W	E
1	0.2287	0.6000	0.2	0.8	0.5000	0.2000	0.0800
2	0.3869	0.3493	0.2	0.8	0.6127	0.1738	0.1206
3	0.1027	0.1000	0.2	0.8	0.7224	0.0575	0.0513
4	0.5153	0.1000	0.2	0.8	0.6865	0.1113	0.0929
5	0.1456	0.1000	0.2	0.8	0.7190	0.0634	0.0563
6	0.4480	0.1000	0.2	0.8	0.6930	0.1030	0.0871
7	0.2696	0.2556	0.2	0.8	0.6640	0.1360	0.1081
8	0.6000	0.1819	0.2	0.8	0.6507	0.1480	0.1140
9	0.6000	0.2338	0.2	0.8	0.6310	0.1628	0.1192
10	0.2019	0.1000	0.2	0.8	0.7145	0.0710	0.0625
11	0.3902	0.3673	0.2	0.8	0.6041	0.1781	0.1203
12	0.2940	0.1000	0.2	0.8	0.7068	0.0833	0.0723
13	0.1794	0.5347	0.2	0.8	0.5540	0.1948	0.1083
14	0.6000	0.4520	0.2	0.8	0.5159	0.1996	0.0897
15	0.6000	0.4421	0.2	0.8	0.5226	0.1991	0.0935
16	0.4196	0.4027	0.2	0.8	0.5819	0.1872	0.1170
17	0.3379	0.1000	0.2	0.8	0.7030	0.0890	0.0767
18	0.2696	0.5276	0.2	0.8	0.5398	0.1972	0.1022
19	0.6000	0.4245	0.2	0.8	0.5342	0.1980	0.0995
20	0.2216	0.3414	0.2	0.8	0.6391	0.1571	0.1176
21	0.6000	0.1000	0.2	0.8	0.6779	0.1214	0.0995
22	0.6000	0.1038	0.2	0.8	0.6767	0.1227	0.1004
23	0.3539	0.4550	0.2	0.8	0.5658	0.1920	0.1125
24	0.3881	0.1068	0.2	0.8	0.6967	0.0980	0.0835
25	0.2353	0.4524	0.2	0.8	0.5883	0.1848	0.1183

Pareto-fronts of the other algorithms. In addition, one solution from the Pareto-fronts of the Jaya, Rao-2, SAP-Rao, ERao-1, and ERao-3 algorithms are dominated. Similarly, two solutions from the Pareto-front achieved by the Rao-3 algorithm are dominated. The performance of the proposed modified algorithms is better or competitive to that of the basic algorithms in terms of the coverage values.

From the computational results of the solar-assisted Carnot-like heat engine case study, it can be observed that the performance of the considered system can be improved with the solutions achieved by the proposed algorithms. The solution of the AMTPG-Jaya algorithm has better compromise among the non-dimensional power output, efficiency, and ecological function. The AMTPG-Jaya algorithm's solution has 17% higher power output, 3.3% higher ecological function, and 2.7% higher

Table 7.73 Pareto-optimal solutions obtained by the ERao-3 algorithm for the solar-assisted Carnot-like heat engine case study

Solution	ϕ	β	τ	γ	η	W	E
1	0.2288	0.6000	0.2	0.8	0.5000	0.2000	0.0800
2	0.2245	0.4042	0.2	0.8	0.6127	0.1737	0.1206
3	0.1000	0.1000	0.2	0.8	0.7226	0.0571	0.0510
4	0.1000	0.1607	0.2	0.8	0.7081	0.0814	0.0708
5	0.1065	0.1000	0.2	0.8	0.7221	0.0580	0.0517
6	0.1915	0.1000	0.2	0.8	0.7154	0.0696	0.0614
7	0.1000	0.3289	0.2	0.8	0.6586	0.1410	0.1108
8	0.2270	0.1000	0.2	0.8	0.7125	0.0744	0.0653
9	0.2398	0.4211	0.2	0.8	0.6027	0.1787	0.1202
10	0.4158	0.1000	0.2	0.8	0.6960	0.0990	0.0842
11	0.2133	0.2251	0.2	0.8	0.6797	0.1193	0.0982
12	0.2541	0.5299	0.2	0.8	0.5416	0.1970	0.1030
13	0.1838	0.5257	0.2	0.8	0.5584	0.1938	0.1100
14	0.2167	0.5709	0.2	0.8	0.5233	0.1991	0.0938
15	0.1634	0.1559	0.2	0.8	0.7038	0.0878	0.0758
16	0.1973	0.5607	0.2	0.8	0.5344	0.1980	0.0996
17	0.3347	0.3528	0.2	0.8	0.6189	0.1703	0.1205
18	0.1000	0.2919	0.2	0.8	0.6709	0.1290	0.1042
19	0.2045	0.5966	0.2	0.8	0.5084	0.1999	0.0852
20	0.2121	0.3277	0.2	0.8	0.6454	0.1523	0.1158
21	0.2354	0.3606	0.2	0.8	0.6297	0.1637	0.1194
22	0.3118	0.1605	0.2	0.8	0.6889	0.1082	0.0908
23	0.1494	0.4904	0.2	0.8	0.5839	0.1865	0.1174
24	0.3490	0.1000	0.2	0.8	0.7020	0.0904	0.0778
25	0.3667	0.4445	0.2	0.8	0.5692	0.1911	0.1136

thermal efficiency when compared to those of the solutions achieved by other algorithms compared. In addition, the performances of the modified versions are better or competitive to those of the basic algorithms as well as the NSGA-II algorithm. The next section presents the Jaya and Rao algorithms' application to the selected bio-energy systems, along with their modified versions.

Table 7.74 Best solutions obtained by various algorithms in MOO scenario of the solar-assisted Carnot-like heat engine case study

Algorithm		ϕ	β	τ	γ	η	W	E
TOPSIS	NSGA-II	0.4447	0.3903	0.2	0.8	0.583877	0.186475	0.117451
LINMAP		0.3684	0.4149	0.2	0.8	0.584513	0.186248	0.117585
Fuzzy		0.4401	0.2533	0.2	0.8	0.644401	0.153089	0.116124
Jaya		0.248	0.425	0.20	0.80	0.599529	0.180211	0.119952
Rao-1		0.204	0.448	0.20	0.8	0.595680	0.181886	0.119499
Rao-2		0.178	0.420	0.20	0.80	0.612702	0.173763	0.120645
Rao-3		0.468	0.365	0.20	0.8	0.592698	0.183122	0.119073
AMTPG-Jaya		0.101	0.474	0.20	0.80	0.599867	0.180060	0.119987
SAP-Rao		0.483	0.341	0.20	0.799999937	0.601716	0.179219	0.120160
ERao-1		0.359	0.383	0.20	0.8	0.601801	0.179179	0.120168
ERao-2		0.390	0.367	0.20	0.8	0.604056	0.178123	0.120343
ERao-3		0.240	0.421	0.20	0.80	0.602729	0.178749	0.120245

Source NSGA-II—Sayyaadi et al. (2015)
Bold figure indicates a better value

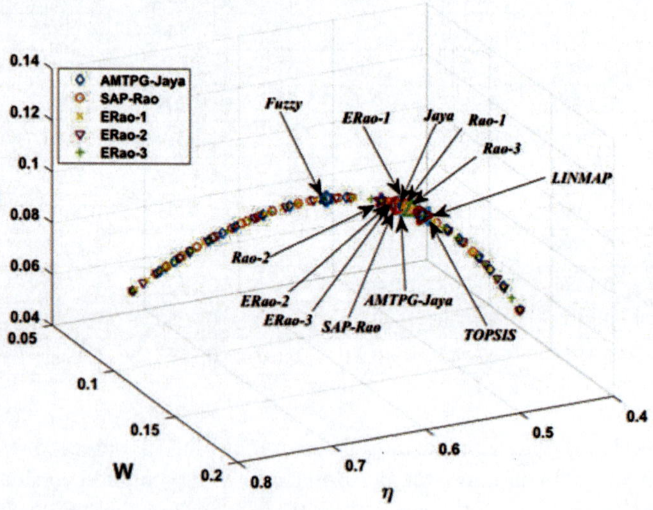

Fig. 7.5 Plot of Pareto-fronts of the proposed algorithms and solutions of the methods compared in solar-assisted Carnot-like heat engine case study

Table 7.75 Ranks suggested by the MADM methods for different algorithm solutions presented in Table 7.74

Algorithm	SAW	WPM	TOPSIS	MTOPSIS	VIKOR	PROMETHEE	COPRAS	GRA	AR	CR	FR
TOPSIS	11	11	10	10	11	11	11	11	10.8	11	11
LINMAP	10	10	9	9	10	10	10	10	9.8	10	10
Fuzzy	12	12	12	12	12	12	12	12	12.0	12	12
Jaya	2	2	3	3	7	7	2	7	4.1	4.5	6
Rao-1	6	7	1	1	8	8	6	8	5.6	7.2	8
Rao-2	9	9	11	11	1	1	9	1	6.5	5	7
Rao-3	8	8	2	2	9	9	8	9	6.9	8.5	9
AMTPG-Jaya	1	1	4	4	6	6	1	6	3.6	3.5	1
SAP-Rao	3	3	5	5	5	5	3	5	4.3	4	3
ERao-1	4	4	6	6	4	4	4	4	4.5	4	4
ERao-2	7	6	8	8	2	2	7	2	5.3	4.3	5
ERao-3	5	5	7	7	3	3	5	3	4.8	4	2

AR average rank, *CR* corrected rank, *FR* final rank

Table 7.76 Spearman's rank correlation coefficients between different pairs of MADM method's rankings presented in Table 7.75

Method	SAW	WPM	TOPSIS	MTOPSIS	VIKOR	PROMETHEE	COPRAS	GRA
SAW	1	0.9	0.7	0.7	0.5	0.5	1	0.5
WPM	0.9	1	0.6	0.6	0.5	0.5	0.9	0.5
TOPSIS	0.7	0.6	1	1	0.0	0.0	0.7	0.0
MTOPSIS	0.7	0.6	1	1	0.0	0.0	0.7	0.0
VIKOR	0.5	0.5	0.0	0.0	1	1	0.5	1
PROMETHEE	0.5	0.5	0.0	0.0	1	1	0.5	1
COPRAS	1	0.9	0.7	0.7	0.5	0.5	1	0.5
GRA	0.5	0.5	0.0	0.0	1	1	0.5	1

Table 7.77 Hypervolume and spacing values of the Pareto-fronts obtained by the proposed algorithms in MOO for the solar-assisted Carnot-like engine case study

Algorithm	Hypervolume	Spacing
Jaya	4.39053E-03	0.139464
Rao-1	4.55133E-03	0.057768
Rao-2	4.55104E-03	**0.040155**
Rao-3	4.55232E-03	0.041179
AMTPG-Jaya	4.55097E-03	0.074990
SAP-Rao	4.55484E-03	0.054526
ERao-1	**4.55675E-03**	0.068134
ERao-2	4.55517E-03	0.050868
ERao-3	4.55632E-03	0.056517

Results in the bold figure indicate better values

Table 7.78 Coverage (%) values of the Pareto-fronts obtained by the proposed algorithms in MOO for the solar-assisted Carnot-like heat engine case study

Algorithm	Jaya	Rao-1	Rao-2	Rao-3	AMTPG-Jaya	SAP-Rao	ERao-1	ERao-2	ERao-3
Jaya	–	0	0	4	0	4	0	0	0
Rao-1	0	–	0	4	0	4	0	0	0
Rao-2	4	0	–	4	0	4	0	0	0
Rao-3	0	0	0	–	0	0	0	0	0
AMTPG-Jaya	4	0	4	4	–	4	0	0	0
SAP-Rao	0	0	0	4	0	–	0	0	0
ERao-1	0	0	0	8	0	4	–	0	4
ERao-2	0	0	0	4	0	0	0	–	0
ERao-3	0	0	0	8	0	4	4	0	–

References

Ahmadi, M. H., Mohammadi, A. H., Dehghani, S., & Barranco-Jiménez, M. A. (2013a). Multi-objective thermodynamic-based optimization of output power of solar dish-Stirling engine by implementing an evolutionary algorithm. *Energy Conversion and Management*, *75*, 438–445. https://doi.org/10.1016/j.enconman.2013.06.030

Ahmadi, M. H., Sayyaadi, H., Mohammadi, A. H., & Barranco-Jimenez, M. A. (2013b). Thermo-economic multi-objective optimization of solar dish-Stirling engine by implementing evolutionary algorithm. *Energy Conversion and Management*, *73*, 370–380. https://doi.org/10.1016/j.enconman.2013.05.031

Dai, D., Yuan, F., Long, R., Liu, Z., & Liu, W. (2018). Performance analysis and multi-objective optimization of a Stirling engine based on MOPSOCD. *International Journal of Thermal Sciences, 124*, 399–406. https://doi.org/10.1016/j.ijthermalsci.2017.10.030

Li, Y., Liao, S., & Liu, G. (2015). Thermo-economic multi-objective optimization for a solar-dish Brayton system using NSGA-II and decision making. *International Journal of Electrical Power & Energy Systems, 64*, 167–175. https://doi.org/10.1016/j.ijepes.2014.07.027

Sayyaadi, H., Ahmadi, M. H., & Dehghani, S. (2015). Optimal design of a solar-driven heat engine based on thermal and ecological criteria. *Journal of Energy Engineering, 141*(3), 4014012. https://doi.org/10.1061/(asce)ey.1943-7897.0000191

Chapter 8
Optimization of the Selected Bio-Energy Systems

Abstract This chapter presents the applications of different versions of Jaya and Rao algorithms to the optimization problems of the biodiesel engine system and microalgae-based biomass cultivation process. Three multiobjective optimization case studies of biodiesel engine design and a multiobjective optimization case study of microalgae cultivation process optimization are considered for optimization to see if there can be any improvement in the performances of the selected systems. The optimization is carried out using the Jaya algorithm, Rao algorithms, and their modified versions. Furthermore, to identify the best solutions from the Pareto-fronts, the average rank method described in Chap. 5 is employed. Computational results revealed that the performances of the modified Jaya and Rao algorithms are superior to those of the other algorithms. Also, the performances of the selected systems are improved by the solutions of the proposed algorithms.

8.1 Design Optimization of the Single-Cylinder Direct-Injection Diesel Engine

The description of the selected single-cylinder direct-injection diesel engine system is presented in Sect. 2.3.1. The detailed description and specifications of this system were presented by Dhingra et al. (2014). This is a case study of a single-cylinder direct-injection compression ignition engine that runs using Jatropha biodiesel blends. The seven objectives of this case study are: minimizing the combustion parameters such as brake specific fuel consumption (BSFC—kg/kWh) and peak cylinder pressure ($P_{\max}$—*bar*); maximizing the performance in terms of brake-thermal efficiency (BTE—*N-m*); and minimizing the emissions such as carbon mono oxide emission (CO—%), nitrogen oxides emission (NO_x—ppm), hydrocarbon emission (HC—ppm), and smoke emission opacity (S_m). The design variables are biodiesel blending ratio (X_1), load torque (X_2), and compression ratio (X_3), and their ranges are $11.25 \le X_1 \le 33.75$ (% *V/V*), $7.5 \le X_2 \le 12.5$ (*N-m*), and $13.5 \le X_3 \le 16.5$ (*V/V*). The regression models of this case study's objectives proposed by Dhingra et al. (2014) are as given below:

V. R. Ravipudi et al., *Design Optimization of Renewable Energy Systems Using Advanced Optimization Algorithms*, Green Energy and Technology,
https://doi.org/10.1007/978-3-030-95589-2_8

$$\begin{aligned}\mathrm{BSFC} = &-46.68493 + 0.13685X_1 + 1.32378X_2 + 5.31712X_3\\&- 3.21186\times10^{-3}X_1^2 - 0.056480X_2^2 - 0.16911X_3^2\\&+ 2.33817\times10^{-3}X_1X_2 - 1.13137\times10^{-3}X_1X_3\\&- 0.022062X_2X_3\end{aligned} \quad (8.1)$$

$$\begin{aligned}\mathrm{BTE} = &-2400.88522 + 10.28829X_1 + 71.43483X_2\\&+ 259.72937X_3 - 0.18203X_1^2 - 4.07476X_2^2\\&- 9.07056X_3^2 - 0.19143X_1X_2 - 4.19026\times10^{-3}X_1X_3\\&+ 1.26091X_2X_3\end{aligned} \quad (8.2)$$

$$\begin{aligned}P_{\max} = &-3669.50268 + 10.41183X_1 + 129.7155X_2\\&+ 396.37136X_3 - 0.28479X_1^2 - 5.68691X_2^2\\&- 12.90809X_3^2 - 0.037712X_1X_2 + 0.18856X_1X_3\\&- 1.03709X_2X_3\end{aligned} \quad (8.3)$$

$$\begin{aligned}\mathrm{Sqrt(CO)} = &-102.47076 + 0.41566X_1 + 1.86746X_2\\&+ 11.94069X_3 - 7.55818\times10^{-3}X_1^2 - 0.11396X_2^2\\&- 0.40724X_3^2 - 3.75222\times10^{-3}X_1X_2\\&- 2.57927\times10^{-3}X_1X_3 + 0.035673X_2X_3\end{aligned} \quad (8.4)$$

$$\begin{aligned}\mathrm{Sqrt}(\mathrm{NO}_x) = &-765.06345 + 3.33115X_1 + 12.59961X_2\\&+ 91.17741X_3 - 0.062945X_1^2 - 0.70774X_2^2\\&- 3.08084X_3^2 - 0.026616X_1X_2 - 0.01625X_1X_3\\&+ 0.1666X_2X_3\end{aligned} \quad (8.5)$$

$$\begin{aligned}\log_{10}(\mathrm{HC}) = &-182.12527 + 0.57186X_1 + 3.99162X_2\\&+ 21.03397X_3 - 0.012787X_1^2 - 0.19348X_2^2\\&- 0.70153X_3^2 + 8.11844\times10^{-5}X_1X_2\\&+ 2.71724\times10^{-4}X_1X_3 + 1.69324\times10^{-3}X_2X_3\end{aligned} \quad (8.6)$$

$$\begin{aligned}\mathrm{Sqrt}(S_m) = &133.78384 - 0.22401X_1 - 3.48682X_2\\&- 15.04873X_3 + 0.010848X_1^2 + 0.18082X_2^2\\&+ 0.49577X_3^2 + 3.59009\times10^{-3}X_1X_2\\&- 5.06832\times10^{-3}X_1X_3 + 0.041469X_2X_3\end{aligned} \quad (8.7)$$

Dhingra et al. (2014) presented the optimum design parameters using the NSGA-II algorithm and reported 30 Pareto-optimal solutions after 50,000 function evaluations in the simultaneous optimization of these seven objectives. Now, the proposed algorithms, along with basic Jaya and Rao algorithms, are used to find the optimal parameters. For a fair comparison of the proposed algorithms with the NSGA-II, the proposed and other considered algorithms are executed for 50,000 function evaluations. In multi-objective optimization using the Jaya algorithm, Rao algorithms, and their modified versions, a *priori* approach is followed. In multi-objective optimization through a *priori* approach, the combined objective function of this case study is given by the following equation:

$$Z_{combined} = w_1 \frac{BSFC}{BSFC_{min}} - w_2 \frac{BTE}{BTE_{max}} + w_3 \frac{P_{max}}{P_{maxmin}} + w_4 \frac{CO}{CO_{min}} + w_5 \frac{NO_x}{NO_{xmin}} + w_6 \frac{HC}{HC_{min}} + w_7 \frac{S_m}{S_{mmin}} \quad (8.8)$$

Here, the terms $BSFC_{min}$, $P_{max\ min}$, CO_{min}, $NO_{x\ min}$, HC_{min}, and $S_{m\ min}$ represents the objective function values achieved in single-objective optimization (minimization) of BSFC, P_{max}, CO, NO_x, HC, and S_m, respectively. Similarly, the term BTE_{max} represents the maximum BTE value achieved in single-objective optimization (maximization) of BTE. The weights of the objectives are varied between zero and one randomly such that $w_1 + w_2 + w_3 + w_4 + w_5 + w_6 + w_7 = 1$.

Dhingra et al. (2014) presented 30 Pareto-optimal solutions in this case study. Hence, for a fair comparison of spacing values, in this work also 30 Pareto-optimal solutions are found. By using different combinations of the weights, the algorithms are executed 30 times. For each combination of the weights, a single Pareto-optimal solution is found by executing the algorithm. The population size for all the algorithms is taken as 50 in each execution. The elite population size is taken as 20% of the population for the elitist Rao algorithms. Furthermore, the solutions obtained by various algorithms are non-dominated. Hence, the average rank method was used to identify the best solution from the Pareto-front of the respective algorithm. The best solution from the Pareto-front of the NSGA-II algorithm was not identified by Dhingra et al. (2014). Hence, the best solution from the NSGA-II Pareto-front is also selected using the average rank method and reported in this work.

The Pareto-optimal solutions achieved by the Jaya algorithm, Rao algorithms, and their modified versions in this case study are presented in Tables 8.1, 8.2, 8.3, 8.4, 8.5, 8.6, 8.7, 8.8 and 8.9. Similar to the previous case studies, the best solutions from these Pareto-fronts are identified. Solution 29 of the NSGA-II algorithm, Solution 15 of the Jaya algorithm, Solution 11 of the AMTPG-Jaya algorithm, Solution 30 of the Rao-1 algorithm, Solution 18 of the Rao-2 algorithm, Solution 22 of the Rao-3 algorithm, Solution 25 of the ERao-1 algorithm, Solution 2 of the ERao-2 algorithm, Solution 23 of the ERao-3 algorithm, and Solution 12 of the SAP-Rao algorithm are identified as the best solutions from the respective algorithm's Pareto-front. Now, these best solutions are compared with those of the NSGA-II algorithm in Table 8.10.

Table 8.1 Pareto-optimal solutions obtained by the Jaya algorithm for the single-cylinder direct-injection diesel engine case study

Solution	X_1	X_2	X_3	BSFC	BTE	$P_{\max}$	CO (%)	NO_x	HC	S_m
1	16.3402	11.8641	16.2418	0.1	23.0257	38.6459	0.15229	174.967	8.14667	116.8354
2	15.6022	11.7793	16.2331	0.1	22.2531	37.9039	0.12906	166.392	7.05306	112.5002
3	15.647	11.782	16.2357	0.1	22.2842	37.9481	0.12981	166.699	7.08652	112.7368
4	17.3386	11.8013	16.2317	0.16337	25.537	43.8038	0.24389	197.422	12.8797	116.6346
5	16.2086	11.8476	16.2419	0.1	22.886	38.5201	0.14778	173.338	7.92446	116.0407
6	16.7768	11.8776	16.2369	0.11508	23.8532	40.0643	0.18045	182.992	9.52364	118.0359
7	16.2128	11.8471	16.2427	0.1	22.8829	38.5229	0.14763	173.294	7.91757	116.0551
8	15.7734	11.7953	16.2387	0.1	22.4136	38.0801	0.13344	168.072	7.25007	113.4601
9	15.761	11.795	16.2376	0.1	22.4088	38.0688	0.13336	168.03	7.24631	113.3987
10	15.7364	11.7925	16.237	0.1	22.3846	38.0435	0.13268	167.773	7.21554	113.259
11	17.208	11.8194	16.2336	0.15196	25.1413	42.9257	0.22805	193.969	11.9926	116.9767
12	15.372	11.7557	16.2265	0.1	22.0135	37.6556	0.1226	163.917	6.77064	111.198
13	17.1168	11.8318	16.2347	0.14407	24.8671	42.3166	0.21741	191.598	11.4156	117.2086
14	16.8395	11.8693	16.2367	0.12039	24.0395	40.478	0.18696	184.555	9.84479	117.887
15	15.372	11.755	16.227	0.1	22.006	37.654	0.1224	163.83	6.7597	111.187
16	15.8525	11.8039	16.2402	0.1	22.4957	38.1617	0.1358	168.956	7.3581	113.9172
17	16.8714	11.865	16.2366	0.12309	24.1343	40.6886	0.19033	185.353	10.0126	117.8107
18	15.3736	11.7551	16.2271	0.1	22.0092	37.656	0.12245	163.865	6.76381	111.1998
19	16.5586	11.8952	16.2382	0.1	23.2753	38.8494	0.16088	177.997	8.58057	118.2026
20	16.1849	11.8441	16.2423	0.1	22.8566	38.4964	0.14681	172.994	7.87793	115.8926
21	16.5661	11.8965	16.2379	0.1	23.2847	38.8562	0.16122	178.115	8.59798	118.2512

(continued)

Table 8.1 (continued)

Solution	X_1	X_2	X_3	BSFC	BTE	P_{max}	CO (%)	NO_x	HC	S_m
22	16.9288	11.8573	16.2363	0.12797	24.3052	41.0682	0.19647	186.798	10.3227	117.6724
23	15.6679	11.7846	16.2359	0.1	22.3089	37.9708	0.13052	166.964	7.11823	112.8598
24	17.0952	11.8348	16.2349	0.14221	24.8023	42.1726	0.21494	191.041	11.2836	117.2628
25	16.7624	11.8795	16.237	0.11386	23.8103	39.969	0.17897	182.633	9.45124	118.0701
26	15.5027	11.7686	16.2307	0.1	22.1468	37.7969	0.12614	165.282	6.92413	111.9316
27	16.5914	11.9003	16.2373	0.1	23.314	38.8791	0.16226	178.48	8.65208	118.4122
28	15.9692	11.8171	16.2418	0.1	22.6186	38.2807	0.13943	170.303	7.5264	114.5996
29	16.4252	11.8757	16.2409	0.1	23.1207	38.7261	0.15549	176.103	8.30633	117.3608
30	16.7865	11.8763	16.2369	0.1159	23.882	40.1282	0.18145	183.233	9.57255	118.013

Table 8.2 Pareto-optimal solutions obtained by the AMTPG-Jaya algorithm for the single-cylinder direct-injection diesel engine case study

Solution	X_1	X_2	X_3	BSFC	BTE	P_{max}	CO	NO_x	HC	S_m
1	16.589	11.8999	16.2374	0.1	23.3112	38.877	0.16216	178.445	8.64685	118.397
2	15.7072	11.7897	16.2361	0.1	22.3569	38.0136	0.13191	167.481	7.18097	113.0941
3	16.2279	11.8496	16.2423	0.1	22.9036	38.5383	0.14832	173.537	7.95092	116.1524
4	16.2905	11.8573	16.2424	0.1	22.9691	38.5982	0.15042	174.298	8.05413	116.5288
5	15.441	11.7631	16.2284	0.10001	22.0884	37.7318	0.12461	164.686	6.85748	111.5893
6	16.5741	11.8976	16.2378	0.1	23.2936	38.8634	0.16153	178.225	8.61424	118.3015
7	15.7837	11.7984	16.2374	0.1	22.4392	38.0938	0.13428	168.366	7.28817	113.5386
8	15.9675	11.8171	16.2416	0.1	22.6182	38.2792	0.13943	170.3	7.52632	114.5911
9	16.6281	11.8973	16.2371	0.10253	23.411	39.083	0.16548	179.315	8.80476	118.3853
10	16.4734	11.8826	16.2401	0.1	23.1759	38.771	0.15738	176.772	8.40211	117.6627
11	15.263	11.746	16.222	0.1	21.908	37.538	0.1199	162.85	6.6552	110.6
12	16.4961	11.8858	16.2397	0.1	23.2012	38.7919	0.15826	177.081	8.44667	117.8042
13	16.0203	11.8245	16.2413	0.1	22.6824	38.334	0.14142	171.024	7.61991	114.9138
14	15.7273	11.7903	16.2378	0.1	22.3655	38.032	0.13207	167.557	7.18806	113.1943
15	16.1045	11.8348	16.2417	0.1	22.774	38.4181	0.14425	172.057	7.75425	115.4162
16	16.5164	11.8887	16.2393	0.1	23.2248	38.8106	0.15908	177.37	8.48864	117.9325
17	16.5006	11.8862	16.2398	0.1	23.2049	38.7959	0.15838	177.124	8.45274	117.8302
18	16.1819	11.8433	16.2426	0.1	22.8501	38.4929	0.14659	172.917	7.86711	115.87
19	17.1317	11.8298	16.2345	0.14536	24.9117	42.4157	0.21912	191.983	11.5075	117.1711
20	15.9913	11.8215	16.2407	0.1	22.6546	38.3053	0.1406	170.717	7.581	114.7465
21	17.2407	11.8149	16.2332	0.1548	25.24	43.1447	0.23194	194.826	12.2076	116.8922

(continued)

Table 8.2 (continued)

Solution	X_1	X_2	X_3	BSFC	BTE	P_{max}	CO	NO_x	HC	S_m
22	15.9711	11.8215	16.2387	0.1	22.6505	38.2881	0.14059	170.694	7.58052	114.6506
23	17.2735	11.8103	16.2327	0.15767	25.3395	43.3655	0.23591	195.693	12.4285	116.8065
24	16.6809	11.8903	16.2371	0.10698	23.568	39.4312	0.17072	180.615	9.05316	118.2619
25	15.6502	11.7842	16.2344	0.1	22.3017	37.9546	0.1304	166.902	7.11284	112.7725
26	17.0948	11.8348	16.2349	0.14218	24.8013	42.1704	0.2149	191.032	11.2816	117.2637
27	17.2638	11.8117	16.2328	0.15682	25.3099	43.3	0.23473	195.436	12.3625	116.832
28	15.8955	11.8122	16.2382	0.1	22.5664	38.2107	0.13805	169.76	7.46188	114.2011
29	16.4863	11.8844	16.2399	0.1	23.1903	38.7829	0.15788	176.947	8.42728	117.7433
30	15.8122	11.799	16.2398	0.1	22.4504	38.1196	0.13447	168.463	7.29703	113.6792

Table 8.3 Pareto-optimal solutions obtained by the Rao-1 algorithm for the single-cylinder direct-injection diesel engine case study

Solution	X_1	X_2	X_3	BSFC	BTE	P_{max}	CO	NO_x	HC	S_m
1	15.5953	11.7824	16.2302	0.1	22.2741	37.9027	0.12984	166.649	7.0881	112.496
2	16.5377	11.8918	16.2389	0.1	23.249	38.8301	0.15993	177.669	8.53216	118.0669
3	15.8479	11.8042	16.2395	0.1	22.4972	38.1582	0.13589	168.98	7.362	113.8986
4	17.3107	11.8052	16.2321	0.16092	25.4522	43.6157	0.24045	196.679	12.684	116.7087
5	16.7671	11.8789	16.237	0.11426	23.8242	40	0.17945	182.75	9.47478	118.059
6	17.1826	11.8228	16.2339	0.14976	25.0648	42.7558	0.22505	193.306	11.8286	117.0418
7	17.3206	11.8038	16.232	0.16179	25.4822	43.6824	0.24166	196.942	12.753	116.6824
8	16.195	11.8442	16.2432	0.1	22.859	38.5051	0.14683	173.011	7.87913	115.9408
9	17.1256	11.8306	16.2346	0.14483	24.8935	42.3753	0.21842	191.826	11.4699	117.1864
10	16.4931	11.8861	16.2392	0.1	23.2028	38.7898	0.15834	177.105	8.45066	117.7928
11	17.0519	11.8407	16.2353	0.13849	24.6726	41.8845	0.21003	189.928	11.0242	117.3708
12	15.4602	11.7625	16.2308	0.1	22.0894	37.7483	0.1245	164.672	6.85244	111.6749
13	16.9412	11.8556	16.2362	0.12903	24.3423	41.1508	0.19782	187.113	10.3915	117.6421
14	15.7474	11.7975	16.2345	0.1	22.4236	38.0607	0.13399	168.227	7.27489	113.3572
15	16.0083	11.8201	16.2433	0.1	22.6484	38.318	0.14024	170.618	7.56486	114.814
16	16.9958	11.8483	16.2358	0.1337	24.5052	41.5127	0.20379	188.498	10.6986	117.5089
17	16.2472	11.8522	16.2421	0.1	22.9254	38.5572	0.14902	173.791	7.98546	116.2701
18	16.5337	11.8917	16.2386	0.1	23.2476	38.8268	0.15991	177.656	8.53059	118.0461
19	16.6524	11.8941	16.2371	0.10457	23.4833	39.2432	0.16788	179.913	8.9182	118.3286
20	15.9283	11.814	16.2401	0.10001	22.5868	38.2417	0.13856	169.966	7.4856	114.3738
21	16.1931	11.8474	16.2406	0.1	22.8814	38.507	0.14771	173.299	7.92086	115.964

(continued)

Table 8.3 (continued)

Solution	X_1	X_2	X_3	BSFC	BTE	P_{max}	CO	NO_x	HC	S_m
22	16.8582	11.8667	16.2367	0.12198	24.0952	40.6018	0.18894	185.024	9.94312	117.8422
23	16.7039	11.8873	16.2371	0.10892	23.6364	39.5831	0.17302	181.184	9.16379	118.2079
24	15.6407	11.7842	16.2335	0.1	22.2989	37.9461	0.13037	166.882	7.11163	112.7269
25	16.6823	11.8902	16.2371	0.1071	23.5722	39.4405	0.17086	180.65	9.0599	118.2586
26	16.7538	11.8807	16.237	0.11314	23.7848	39.9124	0.17809	182.42	9.40854	118.0903
27	16.5473	11.8931	16.2388	0.1	23.2594	38.8388	0.1603	177.797	8.55083	118.1265
28	15.8361	11.8006	16.2409	0.1	22.4681	38.1428	0.13493	168.644	7.31807	113.8086
29	17.0702	11.8382	16.2352	0.14007	24.7275	42.0064	0.2121	190.398	11.1332	117.3252
30	15.357	11.75	16.229	0.1	21.963	37.631	0.121	163.36	6.702	111.071

Table 8.4 Pareto-optimal solutions obtained by the Rao-2 algorithm for the single-cylinder direct-injection diesel engine case study

Solution	X_1	X_2	X_3	BSFC	BTE	P_{max}	CO	NO_x	HC	S_m
1	17.2824	11.8091	16.2326	0.15844	25.3663	43.4251	0.23699	195.928	12.4889	116.7832
2	17.2802	11.8094	16.2326	0.15825	25.3596	43.4103	0.23672	195.869	12.4739	116.789
3	16.5939	11.9007	16.2372	0.1	23.3171	38.8814	0.16237	178.519	8.65789	118.4283
4	16.5666	11.8964	16.238	0.1	23.2846	38.8567	0.16121	178.113	8.59767	118.2537
5	17.3282	11.8027	16.2318	0.16246	25.5053	43.7335	0.2426	197.144	12.8062	116.6623
6	17.1426	11.8283	16.2344	0.1463	24.9446	42.4887	0.22039	192.266	11.5757	117.1435
7	15.3801	11.7569	16.2265	0.10001	22.025	37.6655	0.12293	164.037	6.7847	111.2464
8	16.997	11.8481	16.2358	0.1338	24.5088	41.5206	0.20393	188.529	10.7055	117.506
9	16.5034	11.8869	16.2395	0.1	23.2104	38.7987	0.15858	177.195	8.46316	117.8515
10	16.4495	11.8782	16.2412	0.1	23.1422	38.748	0.15618	176.356	8.34172	117.5038
11	17.2064	11.8196	16.2336	0.15182	25.1366	42.9151	0.22786	193.928	11.9824	116.9807
12	16.6793	11.8905	16.2371	0.10685	23.5634	39.421	0.17056	180.577	9.04583	118.2655
13	17.0555	11.8402	16.2353	0.1388	24.6834	41.9085	0.21044	190.021	11.0456	117.3618
14	16.7719	11.8783	16.237	0.11467	23.8386	40.032	0.17995	182.87	9.49908	118.0475
15	17.1421	11.8284	16.2344	0.14626	24.943	42.4852	0.22033	192.253	11.5724	117.1448
16	15.499	11.7653	16.2328	0.1	22.1206	37.7879	0.12528	164.981	6.88591	111.8833
17	15.908	11.8079	16.2426	0.1	22.5382	38.2156	0.13693	169.399	7.41049	114.22
18	15.283	11.751	16.221	0.1	21.949	37.564	0.1211	163.3	6.708	110.732
19	17.0718	11.838	16.2352	0.14021	24.7324	42.0174	0.21229	190.441	11.143	117.3211
20	15.8931	11.8135	16.2371	0.1	22.5748	38.2104	0.13838	169.867	7.47695	114.2018
21	16.4232	11.8757	16.2407	0.1	23.1204	38.7245	0.15549	176.102	8.30639	117.3513

(continued)

Table 8.4 (continued)

Solution	X_1	X_2	X_3	BSFC	BTE	P_{max}	CO	NO_x	HC	S_m
22	17.225	11.817	16.2334	0.15344	25.1926	43.0395	0.23007	194.414	12.1039	116.9328
23	16.4392	11.8782	16.2403	0.1	23.1401	38.7397	0.15617	176.34	8.34054	117.4534
24	16.9269	11.8575	16.2363	0.12782	24.2997	41.0561	0.19627	186.752	10.3126	117.6768
25	16.2259	11.8492	16.2424	0.1	22.9004	38.5362	0.14821	173.499	7.94565	116.1391
26	15.4719	11.763	16.2316	0.1	22.0963	37.7596	0.12465	164.737	6.85885	111.7344
27	16.9159	11.859	16.2364	0.12688	24.267	40.9833	0.19509	186.474	10.2525	117.7034
28	16.6361	11.8963	16.2371	0.1032	23.4349	39.1359	0.16627	179.513	8.84205	118.3666
29	16.6239	11.8979	16.2371	0.10218	23.3988	39.0558	0.16508	179.214	8.78565	118.3949
30	16.9652	11.8524	16.236	0.13108	24.4137	41.3094	0.20042	187.719	10.5249	117.5839

Table 8.5 Pareto-optimal solutions obtained by the Rao-3 algorithm for the single-cylinder direct-injection diesel engine case study

Solution	X_1	X_2	X_3	BSFC	BTE	P_{max}	CO	NO_x	HC	S_m
1	17.053	11.8405	16.2353	0.13859	24.6761	41.8922	0.21016	189.958	11.031	117.3679
2	15.8465	11.7995	16.2428	0.1	22.4624	38.1502	0.13465	168.561	7.30527	113.8472
3	16.4707	11.8819	16.2403	0.1	23.171	38.7682	0.1572	176.71	8.39302	117.6429
4	16.8741	11.8646	16.2366	0.12333	24.1425	40.7069	0.19062	185.423	10.0274	117.804
5	15.9876	11.8202	16.2413	0.1	22.6445	38.3004	0.14025	170.598	7.56492	114.7163
6	15.5124	11.7719	16.2294	0.10001	22.1739	37.8115	0.12699	165.583	6.96179	112.0065
7	15.98	11.8155	16.244	0.1	22.6094	38.2878	0.13901	170.175	7.50722	114.6355
8	15.5056	11.7666	16.2325	0.10001	22.1322	37.7965	0.12562	165.106	6.90103	111.9254
9	16.3735	11.8685	16.2416	0.1	23.0622	38.6774	0.15351	175.4	8.20716	117.0397
10	16.2724	11.854	16.2431	0.1	22.9428	38.5798	0.14952	173.982	8.01015	116.4091
11	17.1303	11.83	16.2346	0.14524	24.9078	42.4069	0.21897	191.949	11.4994	117.1744
12	17.2495	11.8136	16.2331	0.15557	25.2667	43.2041	0.23301	195.059	12.2666	116.8692
13	16.5825	11.899	16.2375	0.1	23.304	38.8711	0.16191	178.356	8.63365	118.356
14	16.6335	11.8966	16.2371	0.10298	23.427	39.1185	0.16601	179.448	8.82976	118.3727
15	15.4233	11.7616	16.2276	0.1	22.0723	37.7129	0.1242	164.525	6.83972	111.4929
16	15.4359	11.7625	16.2283	0.10001	22.0824	37.7263	0.12445	164.624	6.85038	111.5598
17	16.1982	11.845	16.2429	0.1	22.8652	38.5084	0.14705	173.086	7.88948	115.9642
18	16.3509	11.8659	16.2415	0.1	23.0402	38.6565	0.15279	175.142	8.17136	116.905
19	17.2983	11.8069	16.2323	0.15984	25.4146	43.5322	0.23893	196.35	12.5982	116.7414
20	17.1689	11.8247	16.2341	0.14858	25.0238	42.6646	0.22345	192.95	11.7416	117.0766
21	16.0688	11.8282	16.2432	0.1	22.7195	38.3797	0.14245	171.421	7.66906	115.1814

(continued)

Table 8.5 (continued)

Solution	X_1	X_2	X_3	BSFC	BTE	P_{max}	CO	NO_x	HC	S_m
22	15.356	11.75	16.229	0.1	21.968	37.632	0.1212	163.42	6.7102	111.075
23	16.1202	11.8356	16.2425	0.1	22.7834	38.4322	0.14449	172.154	7.76604	115.4995
24	16.9902	11.849	16.2359	0.13321	24.4883	41.475	0.20317	188.354	10.6662	117.5228
25	17.1271	11.8304	16.2346	0.14496	24.898	42.3851	0.21859	191.864	11.479	117.1827
26	16.6385	11.8959	16.2371	0.10341	23.4422	39.1521	0.16651	179.573	8.85351	118.3608
27	16.715	11.8858	16.2371	0.10986	23.6694	39.6562	0.17414	181.458	9.21758	118.1818
28	15.7669	11.7967	16.237	0.1	22.4226	38.0767	0.13381	168.188	7.26672	113.4425
29	17.0368	11.8427	16.2355	0.1372	24.6276	41.7845	0.20834	189.543	10.9356	117.4081
30	15.5978	11.7809	16.2315	0.1	22.2639	37.9028	0.12946	166.525	7.07136	112.4941

Table 8.6 Pareto-optimal solutions obtained by the ERao-1 algorithm for the single-cylinder direct-injection diesel engine case study

Solution	X_1	X_2	X_3	BSFC	BTE	P_{max}	CO	NO_x	HC	S_m
1	17.2796	11.8095	16.2326	0.1582	25.3578	43.4063	0.23664	195.854	12.4698	116.7906
2	15.8354	11.8013	16.2403	0.1	22.473	38.1432	0.13511	168.705	7.32639	113.812
3	16.2465	11.8517	16.2424	0.1	22.922	38.556	0.1489	173.75	7.97946	116.2627
4	16.3932	11.8719	16.2408	0.1	23.0893	38.6967	0.15445	175.73	8.25406	117.1691
5	16.1699	11.8433	16.2415	0.1	22.8479	38.483	0.14659	172.904	7.86679	115.8126
6	16.5794	11.8984	16.2377	0.1	23.2996	38.8683	0.16174	178.3	8.62535	118.3352
7	16.0447	11.8271	16.2417	0.1	22.7066	38.358	0.14215	171.293	7.65426	115.0557
8	16.4941	11.8857	16.2396	0.1	23.2007	38.7903	0.15825	177.077	8.44619	117.7944
9	15.7818	11.7961	16.2389	0.1	22.4221	38.0887	0.13368	168.162	7.26093	113.5076
10	16.0572	11.827	16.2429	0.1	22.7089	38.3685	0.14214	171.305	7.65431	115.1148
11	17.0543	11.8403	16.2353	0.13871	24.68	41.901	0.21031	189.992	11.0389	117.3646
12	16.1414	11.838	16.2428	0.1	22.8043	38.4528	0.14513	172.389	7.79679	115.624
13	16.4357	11.8773	16.2406	0.1	23.1337	38.7361	0.15594	176.261	8.32896	117.4279
14	16.7449	11.8818	16.237	0.11238	23.7583	39.8537	0.17718	182.199	9.36438	118.1114
15	15.827	11.8024	16.2388	0.1	22.4788	38.1376	0.13538	168.785	7.33854	113.7815
16	17.3075	11.8056	16.2322	0.16064	25.4425	43.5943	0.24005	196.594	12.6619	116.7171
17	17.0476	11.8412	16.2354	0.13813	24.66	41.8565	0.20956	189.82	10.9993	117.3812
18	16.7757	11.8778	16.237	0.11499	23.8499	40.0569	0.18033	182.964	9.51803	118.0386
19	16.0517	11.827	16.2424	0.1	22.707	38.3637	0.14211	171.289	7.65274	115.0879
20	15.8426	11.803	16.2398	0.1	22.4872	38.1519	0.13557	168.866	7.34722	113.8622
21	16.828	11.8708	16.2368	0.11942	24.0054	40.4023	0.18576	184.268	9.7852	117.9143

(continued)

Table 8.6 (continued)

Solution	X_1	X_2	X_3	BSFC	BTE	$P_{\max}$	CO	NO_x	HC	S_m
22	15.7222	11.7891	16.2381	0.1	22.3557	38.0257	0.13177	167.447	7.17419	113.1589
23	16.6102	11.8997	16.2371	0.10102	23.3578	38.9649	0.16373	178.876	8.72216	118.427
24	16.6613	11.8929	16.2371	0.10533	23.5099	39.3022	0.16877	180.133	8.96035	118.3077
25	15.287	11.748	16.223	0.1	21.934	37.564	0.1206	163.12	6.6848	110.732
26	15.5268	11.7704	16.2319	0.1	22.1665	37.8216	0.12663	165.478	6.94567	112.0614
27	16.5992	11.9011	16.2371	0.10009	23.3251	38.8924	0.16265	178.606	8.67181	118.4526
28	15.697	11.7878	16.2365	0.1	22.34	38.0015	0.1314	167.295	7.15767	113.0278
29	17.282	11.8091	16.2326	0.15841	25.365	43.4223	0.23693	195.917	12.486	116.7844
30	17.2285	11.8165	16.2333	0.15374	25.2032	43.063	0.23049	194.506	12.127	116.9238

Table 8.7 Pareto-optimal solutions obtained by the ERao-2 algorithm for the single-cylinder direct-injection diesel engine case study

Solution	X_1	X_2	X_3	BSFC	BTE	P_{max}	CO	NO_x	HC	S_m
1	16.5122	11.8883	16.2393	0.1	23.221	38.8069	0.15896	177.325	8.48212	117.9075
2	16.014	11.634	16.363	0.1035	21.34	38.243	0.0977	155.61	5.7295	112.813
3	15.5787	11.7759	16.2333	0.1	22.2209	37.8772	0.12812	166.045	7.01131	112.3569
4	15.5859	11.7776	16.2327	0.1	22.2362	37.8866	0.12859	166.215	7.0324	112.4077
5	16.7848	11.8765	16.2369	0.11576	23.8769	40.1169	0.18127	183.19	9.5639	118.017
6	16.9125	11.8595	16.2364	0.12658	24.2566	40.9603	0.19471	186.387	10.2335	117.7118
7	15.4728	11.7652	16.2301	0.1	22.1128	37.764	0.1252	164.927	6.88316	111.7589
8	16.3177	11.8607	16.2423	0.1	22.9979	38.624	0.15136	174.635	8.10047	116.6935
9	16.0103	11.8226	16.2417	0.1	22.6666	38.3229	0.14091	170.842	7.59576	114.8476
10	15.3888	11.756	16.228	0.1	22.0205	37.6716	0.12271	163.974	6.77508	111.2798
11	16.5203	11.8896	16.239	0.1	23.2315	38.8145	0.15933	177.456	8.5013	117.9607
12	16.9924	11.8487	16.2358	0.1334	24.4948	41.4895	0.20341	188.41	10.6787	117.5175
13	16.3439	11.865	16.2415	0.1	23.0327	38.6499	0.15255	175.053	8.15906	116.8624
14	15.9647	11.8173	16.2412	0.1	22.6185	38.277	0.13946	170.308	7.52773	114.5793
15	17.1252	11.8307	16.2346	0.1448	24.8922	42.3724	0.21837	191.815	11.4673	117.1875
16	17.323	11.8034	16.2319	0.162	25.4895	43.6985	0.24196	197.006	12.7698	116.6761
17	16.2843	11.8567	16.2422	0.1	22.9642	38.5925	0.15027	174.242	8.04672	116.4936
18	16.8876	11.8628	16.2365	0.12447	24.1826	40.7959	0.19205	185.761	10.0992	117.7717
19	16.8552	11.8672	16.2367	0.12172	24.0861	40.5815	0.18861	184.947	9.92691	117.8495
20	17.3186	11.8041	16.232	0.16161	25.4762	43.6689	0.24142	196.889	12.739	116.6878
21	15.9223	11.8116	16.2413	0.1	22.5681	38.2329	0.13792	169.746	7.45611	114.3231

(continued)

Table 8.7 (continued)

Solution	X_1	X_2	X_3	BSFC	BTE	$P_{\max}$	CO	NO_x	HC	S_m
22	16.086	11.8323	16.2417	0.1	22.7527	38.3995	0.14358	171.814	7.72225	115.3038
23	15.6742	11.7852	16.2361	0.1	22.3149	37.9773	0.13068	167.026	7.12548	112.8954
24	15.9376	11.8138	16.2412	0.1	22.5873	38.249	0.13851	169.96	7.4834	114.4168
25	15.6208	11.7786	16.2356	0.1	22.252	37.9192	0.12888	166.354	7.04509	112.5807
26	15.6233	11.7797	16.235	0.1	22.2609	37.9232	0.12917	166.455	7.05805	112.6029
27	16.9596	11.8531	16.2361	0.1306	24.3972	41.2727	0.19982	187.579	10.4939	117.5974
28	17.1294	11.8301	16.2346	0.14516	24.905	42.4008	0.21886	191.925	11.4936	117.1768
29	16.1184	11.8352	16.2427	0.1	22.7797	38.4302	0.14436	172.11	7.76002	115.4867
30	16.915	11.8591	16.2364	0.1268	24.2642	40.9771	0.19499	186.451	10.2474	117.7057

Table 8.8 Pareto-optimal solutions obtained by the ERao-3 algorithm for the single-cylinder direct-injection diesel engine case study

Solution	X_1	X_2	X_3	BSFC	BTE	P_{max}	CO	NO_x	HC	S_m
1	16.914	11.8593	16.2364	0.12672	24.2613	40.9708	0.19488	186.427	10.2422	117.708
2	16.6787	11.8906	16.2371	0.10679	23.5615	39.4167	0.1705	180.561	9.04272	118.267
3	16.9641	11.8525	16.236	0.13098	24.4104	41.302	0.2003	187.691	10.5187	117.5866
4	16.5952	11.9008	16.2372	0.1	23.3185	38.8826	0.16242	178.537	8.66048	118.4367
5	17.0863	11.836	16.235	0.14145	24.7758	42.1137	0.21393	190.813	11.2301	117.285
6	16.3435	11.8645	16.2418	0.1	23.0292	38.649	0.15241	175.008	8.1524	116.8553
7	17.3256	11.8031	16.2319	0.16223	25.4974	43.7161	0.24228	197.076	12.7881	116.6691
8	17.175	11.8239	16.234	0.1491	25.0419	42.7049	0.22416	193.108	11.78	117.0612
9	15.5565	11.7743	16.2321	0.1	22.2036	37.8548	0.12769	165.872	6.99224	112.2379
10	16.5397	11.8924	16.2387	0.1	23.253	38.8321	0.16009	177.721	8.53997	118.0819
11	15.3693	11.7551	16.2267	0.1	22.0075	37.6519	0.12243	163.852	6.7629	111.1792
12	16.0728	11.8303	16.242	0.1	22.7352	38.3857	0.14301	171.612	7.6955	115.2205
13	17.3102	11.8052	16.2321	0.16087	25.4505	43.6119	0.24038	196.664	12.6801	116.7101
14	16.9818	11.8502	16.2359	0.13249	24.4632	41.4192	0.20224	188.14	10.6184	117.5434
15	16.9207	11.8584	16.2363	0.12729	24.2813	41.0151	0.1956	186.596	10.2787	117.6918
16	16.5182	11.8895	16.2389	0.1	23.2299	38.8126	0.15928	177.436	8.49857	117.9483
17	16.5341	11.8916	16.2387	0.1	23.2469	38.8271	0.15987	177.645	8.52892	118.0472
18	15.7344	11.7906	16.2382	0.1	22.37	38.0388	0.13218	167.601	7.19284	113.2313
19	17.2308	11.8162	16.2333	0.15394	25.2102	43.0785	0.23076	194.567	12.1422	116.9178
20	16.6908	11.889	16.2371	0.10781	23.5974	39.4966	0.17171	180.86	9.10064	118.2386
21	16.297	11.8583	16.2422	0.1	22.9777	38.6047	0.15071	174.4	8.06837	116.5705

(continued)

Table 8.8 (continued)

Solution	X_1	X_2	X_3	BSFC	BTE	P_{max}	CO	NO_x	HC	S_m
22	17.3485	11.7999	16.2315	0.16424	25.5671	43.8707	0.24512	197.687	12.9501	116.6081
23	15.727	11.719	16.284	0.102	21.907	38.052	0.1166	162.08	6.5024	112.385
24	17.0656	11.8388	16.2352	0.13967	24.7136	41.9755	0.21157	190.279	11.1054	117.3368
25	16.6427	11.8954	16.2371	0.10376	23.4545	39.1794	0.16692	179.675	8.87281	118.3512
26	15.7438	11.7924	16.2379	0.1	22.3851	38.0497	0.13265	167.77	7.214	113.292
27	15.7236	11.79	16.2376	0.1	22.3624	38.0282	0.13199	167.526	7.18438	113.1738
28	16.1841	11.8425	16.2434	0.1	22.845	38.4938	0.14637	172.848	7.85704	115.873
29	17.3512	11.7995	16.2315	0.16447	25.5752	43.8886	0.24545	197.758	12.9691	116.601
30	15.9353	11.8134	16.2413	0.1	22.5838	38.2465	0.1384	169.92	7.47821	114.4019

Table 8.9 Pareto-optimal solutions obtained by the SAP-Rao algorithm for the single-cylinder direct-injection diesel engine case study

Solution	X_1	X_2	X_3	BSFC	BTE	P_{max}	CO	NO_x	HC	S_m
1	16.6691	11.8919	16.2371	0.10599	23.5331	39.3538	0.16954	180.326	8.99731	118.289
2	15.9034	11.8114	16.2397	0.1	22.5618	38.2162	0.13783	169.693	7.45157	114.23
3	17.3444	11.8005	16.2316	0.16388	25.5546	43.8429	0.24461	197.577	12.9208	116.619
4	17.3289	11.8026	16.2318	0.16251	25.5073	43.7379	0.24268	197.162	12.8109	116.661
5	15.8558	11.804	16.2404	0.1	22.4975	38.1649	0.13584	168.974	7.35994	113.934
6	16.9089	11.86	16.2364	0.12628	24.246	40.9367	0.19433	186.297	10.2141	117.72
7	16.3171	11.8614	16.2417	0.1	23.0027	38.6243	0.15155	174.698	8.10981	116.697
8	15.3124	11.7547	16.2209	0.1	21.9889	37.5992	0.12222	163.716	6.75548	110.909
9	16.5316	11.8917	16.2385	0.1	23.2471	38.8252	0.1599	177.651	8.53	118.036
10	17.1936	11.8213	16.2338	0.15072	25.0981	42.8297	0.22635	193.594	11.8997	117.013
11	16.9405	11.8557	16.2362	0.12898	24.3403	41.1462	0.19774	187.096	10.3876	117.644
12	15.247	11.747	16.22	0.1	21.909	37.525	0.1201	162.9	6.6633	110.53
13	17.0192	11.8451	16.2356	0.13569	24.5749	41.6675	0.20638	189.093	10.8329	117.452
14	16.173	11.8426	16.2423	0.1	22.8436	38.4848	0.1464	172.846	7.85808	115.821
15	16.0149	11.8243	16.2409	0.1	22.6804	38.3293	0.14138	171.006	7.6181	114.887
16	17.1224	11.8311	16.2346	0.14455	24.8838	42.3538	0.21805	191.742	11.45	117.195
17	16.548	11.8936	16.2385	0.1	23.2628	38.8397	0.16043	177.842	8.55778	118.135
18	16.2295	11.849	16.2428	0.1	22.8997	38.539	0.14816	173.486	7.94326	116.154
19	15.4404	11.7654	16.2266	0.10001	22.1056	37.7353	0.12519	164.887	6.88345	111.608
20	17.05	11.8409	16.2354	0.13833	24.6671	41.8723	0.20983	189.881	11.0133	117.375
21	15.5717	11.777	16.2317	0.1	22.2277	37.873	0.1284	166.134	7.02411	112.334

(continued)

Table 8.9 (continued)

Solution	X_1	X_2	X_3	BSFC	BTE	P_{max}	CO	NO_x	HC	S_m
22	16.8775	11.8642	16.2366	0.12361	24.1524	40.7288	0.19097	185.506	10.045	117.796
23	16.4984	11.8862	16.2396	0.1	23.2046	38.7941	0.15838	177.124	8.45288	117.82
24	16.0253	11.8267	16.2401	0.1	22.6996	38.3412	0.14203	171.231	7.64823	114.959
25	17.2144	11.8185	16.2335	0.15252	25.1608	42.9689	0.22881	194.138	12.0347	116.96
26	15.589	11.7807	16.2308	0.1	22.2596	37.8944	0.12938	166.487	7.0677	112.45
27	16.3666	11.8678	16.2415	0.1	23.0558	38.671	0.1533	175.325	8.19681	116.699
28	16.9537	11.8539	16.2361	0.1301	24.3796	41.2335	0.19918	187.429	10.4609	117.612
29	15.9446	11.8146	16.2412	0.1	22.5951	38.2565	0.13874	170.046	7.49442	114.458
30	15.4099	11.7598	16.2275	0.1	22.0548	37.6976	0.12371	164.341	6.81843	111.413

Table 8.10 Best solutions obtained by various algorithms in MOO scenario of the single-cylinder direct-injection diesel engine case study

Algorithm	X_1	X_2	X_3	BSFC	BTE	P_{max}	CO	NO_x	HC	S_m	Spacing
NSGA-II	33.75	7.5	13.6	0.112	21.37	**14.73**	0.75	**11.56**	48.97	154.5	0.4003
Jaya	15.3716	11.7547	16.2272	0.1000002	**22.01**	37.65	0.1224	163.8	6.7597	111.2	0.1857
Rao1	15.3568	11.7496	16.2293	0.1000019	21.96	37.63	0.1210	163.4	6.7020	111.1	0.1264
Rao2	15.2827	11.7505	16.2207	0.1000030	21.95	37.56	0.1211	163.3	6.7080	110.7	0.1277
Rao3	15.3563	11.7503	16.2287	0.1000022	21.97	37.63	0.1212	163.4	6.7102	111.1	0.1581
AMTPG-Jaya	15.2634	11.7459	16.2221	0.1000009	21.91	37.54	0.1199	162.9	6.6552	110.6	0.0911
SAP-Rao	15.2472	11.7467	16.2198	0.1000040	21.91	37.53	0.1201	162.9	6.6633	**110.5**	0.1057
ERao-1	15.2868	11.7484	16.2227	**0.1000002**	21.93	37.56	0.1206	163.1	6.6848	110.7	0.1174
ERao-2	16.0140	11.6339	16.3633	0.1035358	21.34	38.24	**0.0977**	155.6	**5.7295**	112.8	**0.0905**
ERao-3	15.7274	11.7189	16.2840	0.1020057	21.91	38.05	0.1166	162.1	6.5024	112.4	0.1079

Source NSGA-II:—Dhingra et al. (2014)
Result in boldface indicates better values

Now, to identify the best solution among the solutions of different algorithms given in Table 8.10, which has the best compromise among the objectives, the MADM methods based average ranks are calculated and presented in Table 8.11. The Spearman's correlation coefficients for different pairs of rankings given by decision-making methods for different algorithm's solutions are shown in Table 8.12. The ranks given by the SAW, WPM, TOPSIS, and MTOPSIS methods for each solution are identical. Furthermore, the Spearman's correlation for all the pairs formed by all the MADM (except the pairs with GRA) methods is greater than 0.5. The Spearman's correlation values for the pairs formed by the GRA method are less than 0.5. Hence, the ranks suggested by the GRA method cannot be considered for calculating the average ranks. Therefore, the corrected average ranks are calculated, excluding the ranks suggested by the GRA method, and presented in Table 8.11 as the corrected ranks.

The AMTPG-Jaya algorithm's solution has achieved the least average rank of 2.29. Thus, it can be regarded as the best solution. The AMTPG-Jaya algorithm's solution ranked 1 by two methods (PROMETHEE and VIKOR), ranked two by COPRAS, and ranked three by remaining methods. Hence, it has achieved the least average rank. The ERao-3, SAP-Rao, ERao-2, and ERao-1 algorithm's solutions have achieved the next best average rank values, respectively.

Here, it can be noted that the MADM methods have ranked these solutions based on the relative importance of the solutions and the AMTPG-Jaya solution has the best compromise among the compared solutions. The brake-thermal efficiency for the AMTPG-Jaya algorithm solution is 2.5 and 2.7% higher than that of the NSGA-II and ERao-2 solutions, respectively. Also, the BSFC value for the AMTPG-Jaya algorithm's solution is 10.7, 3.4 and 2% lesser than that of the NSGA-II, ERao-2 and ERao-3 solutions, respectively. By the AMTPG-Jaya solution, CO emissions are 84, 2, 0.9, 1, 1.1, 0.1 and 0.6% lesser when compared to that of the NSGA-II, Jaya, Rao-1, Rao-2, Rao-3, SAP-Rao, and ERao-1 solutions, respectively. Similarly, for the AMTPG-Jaya solution, NO_x emissions are lesser compared to those of the Jaya, Rao-1, Rao-2, Rao-3, SAP-Rao, and ERao-1 solutions. By the AMTPG-Jaya solution, HC emissions are reduced about 86.4, 1.5, 0.7, 0.8, 0.8, 0.1 and 0.4% when compared to that of the NSGA-II, Jaya, Rao-1, Rao-2, Rao-3, SAP-Rao, and ERao-1 solutions, respectively. Similarly, by the AMTPG-Jaya solution, peak cylinder pressure is reduced by 1.8, 1.4, 0.3, 0.2, 0.1, 0.2 and 0.1%, when compared to that of the ERao-2, ERao-3, Jaya, Rao-1, Rao-2, Rao-3, and ERao-1 solutions, respectively. Hence, the AMTPG-Jaya algorithm solution achieved the least average rank.

Furthermore, the spacing values of the Pareto-fronts obtained by the algorithms compared are also presented in Table 8.10. The ERao-2 algorithm has the least spacing value, which is 77.4, 51.3, 28.4, 29.2, 42.8, 0.6, 14.4, 22.9 and 16.1% lesser value when compared to that of the NSGA-II, Jaya, Rao-1, Rao-2, Rao-3, AMTPG-Jaya, SAP-Rao, ERao-1, and ERao-3 algorithms, respectively. The performances of the proposed modified algorithms are better or competitive to that of the basic algorithms in terms of the spacing metric. The coverage metric values of the Pareto-fronts obtained by the Jaya algorithm, Rao algorithms, and their modified versions

Table 8.11 Ranks suggested by the MADM methods for different algorithm solutions presented in Table 8.10

Algorithm	SAW	WPM	TOPSIS	MTOPSIS	VIKOR	PROMETHEE	COPRAS	GRA	AR	CR	FR
NSGA-II	10	10	10	10	10	10	10	10	10.0	10.00	10
Jaya	9	9	9	9	8	9	9	1	7.88	8.86	9
Rao-1	7	7	7	7	6	4	7	3	6.00	6.43	7
Rao-2	6	6	6	6	5	6	6	4	5.63	5.86	6
Rao-3	8	8	8	8	7	8	8	2	7.13	7.86	8
AMTPG-Jaya	3	3	3	3	1	1	2	7	2.88	2.29	1
SAP-Rao	4	4	4	4	1	2	3	6	3.50	3.14	3
ERao-1	5	5	5	5	1	3	4	5	4.13	4.00	5
ERao-2	1	1	1	1	9	7	1	9	3.75	3.00	4
ERao-3	2	2	2	2	1	5	5	8	3.38	2.71	2

AR-Average Rank; CR-Corrected Rank; FR-Final Rank

Table 8.12 Spearman's rank correlation coefficients between different pairs of MADM method's rankings presented in Table 8.11

Method	SAW	WPM	TOPSIS	MTOPSIS	VIKOR	PROMETHEE	COPRAS	GRA
SAW	1	1	1	1	0.52	0.60	0.93	−0.45
WPM	1	1	1	1	0.52	0.60	0.93	−0.45
TOPSIS	1	1	1	1	0.52	0.60	0.93	−0.45
MTOPSIS	1	1	1	1	0.52	0.60	0.93	−0.45
VIKOR	0.52	0.52	0.52	0.52	1	0.88	0.52	0.04
PROMETHEE	0.60	0.60	0.60	0.60	0.88	1	0.71	−0.05
COPRAS	0.93	0.93	0.93	0.93	0.52	0.71	1	−0.38
GRA	−0.45	−0.45	−0.45	−0.45	0.04	−0.05	−0.38	1

are presented in Table 8.13. All the algorithms have performed well in terms of coverage.

From the computational results of the single-cylinder direct-injection diesel engine case study, it can be observed that the performance of the considered system can be improved with the solutions achieved by the proposed algorithms. The solution of the AMTPG-Jaya algorithm has better compromise among the seven objectives. The performance objectives BTE, P_{max}, CO, NO_x, HC, BSFC, and smoke values obtained by the ERao-3 solution are 21.908 N-*m*, 37.538 bar, 0.1199%, 162.85 ppm, 6.6552 ppm, 0.10000012 kg/kWh, and 110.6, respectively. The ERao-2 algorithm has achieved the least spacing value (0.09049), which is much better than that achieved by other algorithms compared, including the NSGA-II. All the algorithms have achieved better performance in terms of coverage values. In addition, the performances of the modified versions in this case study are better or competitive to those of the basic algorithms as well as the NSGA-II algorithm.

8.2 Design Optimization of a Turbocharged DI Diesel Engine

The description of the selected turbocharged DI diesel engine system is presented in Sect. 2.3.2. In this case study, a turbocharged DI diesel engine using biodiesel and diesel blends was considered. The waste vegetable cooking oil is considered as the source of biodiesel. The specific fuel properties and engine specifications were presented in Shirneshan et al. (2016).

The six objectives of this case study are: maximizing the brake power (P) and brake torque (T); minimizing the brake-specific fuel consumption (BSFC) and emissions such as carbon monooxide emission (CO), nitrogen oxides emission (NO_x), and hydrocarbon emission (HC). The design variables were the percentage of biodiesel in fuel (X_1), engine speed (X_2), and engine load (X_3), and their ranges were $0 \leq$

Table 8.13 Coverage (%) values of the Pareto-fronts obtained by the proposed algorithms in MOO for the single-cylinder direct-injection diesel engine case study

Algorithm	Jaya	Rao-1	Rao-2	Rao-3	AMTPG-Jaya	SAP-Rao	ERao-1	ERao-2	ERao-3	NSGA-II
Jaya	–	0	0	0	0	0	0	0	0	0
AMTPG-Jaya	0	–	0	0	0	0	0	0	0	0
Rao1	0	0	–	0	0	0	0	0	0	0
Rao2	0	0	0	–	0	0	0	0	0	0
Rao3	0	0	0	0	–	0	0	0	0	0
SAP-Rao	0	0	0	0	0	–	0	0	0	3.33
ERao1	0	0	0	0	0	0	–	0	0	0
ERao2	0	0	0	0	0	0	0	–	0	0
ERao3	0	0	0	0	0	0	0	0	–	0
NSGA-II	0	0	0	0	0	0	0	0	0	–

$X_1 \leq 100$ (%), $1000 \leq X_2 \leq 2800$(rpm), and $25 \leq X_3 \leq 100$ (%). The regression models of this case study's objectives are as follows:

$$\begin{aligned} P =& -47.32 - 0.08X_1 + 0.056X_2 + 0.205X_3 + 0.0002X_1^2 \\ & - 1.4 \times 10^{-5}X_2^2 - 5.499 \times 10^{-4}X_3^2 + 0.0004X_1X_3 \\ & + 0.0002X_2X_3 \end{aligned} \tag{8.9}$$

$$\begin{aligned} T =& -299.277 - 0.524X_1 + 0.302X_2 + 4.654X_3 \\ & + 0.00362X_1^2 - 7.401 \times 10^{-5}X_2^2 - 0.00637X_3^2 \\ & - 3.771 \times 10^{-4}X_2X_3 \end{aligned} \tag{8.10}$$

$$\begin{aligned} \text{BSFC} =& 298.74 + 0.5X_1 - 0.088X_2 - 0.236X_3 \\ & + 0.0014X_1^2 + 2.67 \times 10^{-5}X_2^2 - 0.00018X_1X_2 \\ & - 0.00336X_1X_3 \end{aligned} \tag{8.11}$$

$$\begin{aligned} \text{CO} =& 0.109 - 3.43 \times 10^{-4}X_1 - 3.96 \times 10^{-5}X_2 \\ & - 7.53 \times 10^{-4}X_3 + 6.48 \times 10^{-7}X_1^2 + 6.32 \times 10^{-9}X_2^2 \\ & + 2.93 \times 10^{-6}X_3^2 + 1.69 \times 10^{-6}X_1X_3 + 7.33 \times 10^{-8}X_2X_3 \end{aligned} \tag{8.12}$$

$$\begin{aligned} \text{NO}_x =& 216.71 - 0.264X_1 + 0.158X_2 + 0.755X_3 \\ & + 0.0114X_1^2 - 5.37 \times 10^{-5}X_2^2 + 0.0558X_3^2 \\ & - 0.00188X_2X_3 \end{aligned} \tag{8.13}$$

$$\begin{aligned} \text{HC} =& 56.38 - 0.028X_1 + 0.019X_2 - 0.554X_3 \\ & - 5.19 \times 10^{-4}X_1^2 - 2.1 \times 10^{-6}X_2^2 + 0.0026X_3^2 \\ & + 3.14 \times 10^{-5}X_1X_2 - 0.0011X_1X_3 + 8.38 \times 10^{-5}X_2X_3 \end{aligned} \tag{8.14}$$

Shirneshan et al. (2016) presented the optimum design parameters using the ABC algorithm and reported a Pareto-optimal solution through a *priori* articulation of preferences method. Now, the proposed algorithms, along with basic Jaya and Rao algorithms, are used to find the optimal parameters. The function evaluations taken by the ABC algorithm were not specified. However, the proposed and other considered algorithms are executed for 50,000 function evaluations. In multi-objective optimization using the Jaya algorithm, Rao algorithms, and their modified versions, a *priori* approach is followed. In multi-objective optimization through a *priori* approach, the combined objective function of this case study is given by the following equation:

$$Z_{\text{combined}} = -w_1\frac{P}{P_{\max}} - w_2\frac{T}{T_{\max}} + w_3\frac{\text{BSFC}}{\text{BSFC}_{\min}} + w_4\frac{\text{CO}}{\text{CO}_{\min}} + w_5\frac{\text{NO}_x}{\text{NO}_{x\,\min}} + w_6\frac{\text{HC}}{\text{HC}_{\min}} \quad (8.15)$$

Here, the terms $\text{BSFC}_{\min}$, $\text{CO}_{\min}$, $\text{NO}_{x\,\min}$, and $\text{HC}_{\min}$ represent the objective function values achieved in single-objective optimization (minimization) of BSFC, CO, NO_x, and HC, respectively. Similarly, the terms $P_{\max}$ and $T_{\max}$ represent the maximum power and torque value achieved in single-objective optimization (maximization) of *P* and *T*, respectively. The weights of the objectives are varied between zero and one randomly such that $w_1 + w_2 + w_3 + w_4 + w_5 + w_6 = 1$.

Shirneshan et al. (2016) presented a single Pareto-optimal solution in this case study using a *priori* approach. Similar to the previous case study, for 30 combinations of the weights, 30 Pareto-optimal solutions are identified. By using different combinations of the weights, the algorithms are executed 30 times. For each combination of the weights, a single Pareto-optimal solution is found by executing the algorithm. The population size for all the algorithms is taken as 50 in each execution. The elite population size is taken as 20% of the population for the elitist Rao algorithms. Furthermore, the solutions obtained by various algorithms are non-dominated. Hence, the average rank method was used to identify the best solution from the Pareto-front of the respective algorithm. The Pareto-optimal solutions achieved by the Jaya algorithm, Rao algorithms, and their modified versions in this case study are presented in Tables 8.14, 8.15, 8.16, 8.17, 8.18, 8.19, 8.20, 8.21 and 8.22

Similar to the previous case studies, the best solutions from these Pareto-fronts are identified. Solution 26 of the Jaya algorithm, Solution 24 of the AMTPG-Jaya algorithm, Solution 18 of the Rao-1 algorithm, Solution 10 of the Rao-2 algorithm, Solution 7 of the Rao-3 algorithm, Solution 8 of the ERao-1 algorithm, Solution 16 of the ERao-2 algorithm, Solution 28 of the ERao-3 algorithm, and Solution 16 of the SAP-Rao algorithm are identified as the best solutions from the respective algorithm Pareto-front. Now, these best solutions are compared with those of the NSGA-II algorithm in Table 8.23.

In Table 8.23, the solutions with the best average rank from the Pareto-optimal solutions of the various algorithms are compared with that reported by the ABC algorithm. The ERao-3 algorithm has achieved higher brake power and least CO and NO_x emission values. The brake power for the ERao-3 algorithm solution is 3.4, 3.6, 1.9, 3.4, 3.8, 2.7, 2.0, 1.9 and 2.7% higher than that of the ABC, Jaya, Rao-1, Rao-2, Rao-3, AMTPG-Jaya, SAP-Rao, ERao-1, and ERao-2 algorithms solutions, respectively. The CO emissions are 13.8, 2.9, 1.2, 2.7, 3.1, 1.9, 1.3, 1.2 and 1.9% lesser than that of the ABC, Jaya, Rao-1, Rao-2, Rao-3, AMTPG-Jaya, SAP-Rao, ERao-1, and ERao-2 algorithms solutions, respectively. Similarly, the NO_x emissions are 6.6, 6.2, 3.7, 6, 6.5, 4.9, 3.8, 3.6 and 4.9% lesser than that of the ABC, Jaya, Rao-1, Rao-2, Rao-3, AMTPG-Jaya, SAP-Rao, ERao-1, and ERao-2 algorithms solutions, respectively. The Rao-3 algorithm has the least HC emissions value, which is 0.9, 0.3, 3.3, 0.6, 1.9, 3.2, 3.4, 1.9 and 7.1% lesser than that of the ABC, Jaya, Rao-1, Rao-2, AMTPG-Jaya, SAP-Rao, ERao-1, ERao-2 and ERao-3

Table 8.14 Pareto-optimal solutions obtained by the Jaya algorithm for the turbocharged DI diesel engine case study

SN	X_1 (%)	X_2(rpm)	X_3 (%)	P (kW)	T (N.m)	BSFC (g/kWh)	CO(%)	NO_x (ppm)	HC (ppm)
1	63.211	2800.000	100.000	68.992	243.538	222.170	0.01378	374.064	114.869
2	100.000	1902.421	93.284	56.109	302.987	204.358	0.01274	632.901	73.666
3	66.829	2800.000	100.000	68.941	243.346	221.599	0.01346	378.473	114.443
4	48.408	2800.000	100.000	69.253	245.314	224.890	0.01529	359.136	116.468
5	100.000	1000.000	97.606	26.876	269.277	227.610	0.02597	830.404	40.769
6	66.169	2800.000	100.000	68.950	243.374	221.700	0.01352	377.646	114.522
7	52.973	2800.000	100.000	69.163	244.597	223.986	0.01479	363.207	115.999
8	100.000	1094.200	94.685	30.182	270.848	224.562	0.02398	789.873	44.251
9	97.438	2555.298	97.538	66.880	271.835	210.454	0.01057	488.258	99.936
10	50.579	2800.000	100.000	69.209	244.954	224.453	0.01505	361.013	116.247
11	100.000	1000.000	100.000	27.681	276.503	226.240	0.02613	854.110	40.610
12	89.241	2665.030	100.000	68.813	263.184	213.227	0.01152	456.090	105.883
13	100.000	1000.000	96.500	26.502	265.915	228.242	0.02590	819.671	40.853
14	90.845	2644.588	99.058	68.150	263.768	212.870	0.01123	458.908	104.719
15	51.683	2800.000	100.000	69.188	244.784	224.235	0.01493	362.009	116.133
16	60.828	2800.000	100.000	69.028	243.717	222.567	0.01401	371.324	115.141
17	100.000	2014.258	93.599	58.498	301.261	204.018	0.01188	609.767	77.933
18	80.039	2800.000	100.000	68.801	243.447	219.823	0.01242	397.103	112.775
19	53.744	2800.000	100.000	69.149	244.491	223.839	0.01471	363.941	115.917
20	100.000	1000.000	100.000	27.681	276.503	226.240	0.02613	854.110	40.610
21	100.000	1000.000	99.450	27.497	274.850	226.555	0.02609	848.609	40.644
22	100.000	1000.000	95.463	26.150	262.748	228.835	0.02585	809.730	40.937
23	69.737	2800.000	100.000	68.904	243.259	221.166	0.01321	382.232	114.092
24	100.000	1776.095	93.095	53.090	303.144	205.449	0.01393	658.575	68.911
25	59.571	2800.000	100.000	69.048	243.828	222.782	0.01413	369.930	115.283
26	100.000	2257.301	94.806	62.839	292.388	205.284	0.01065	558.130	87.419
27	56.640	2800.000	100.000	69.097	244.130	223.302	0.01442	366.822	115.606
28	100.000	1000.000	96.626	26.544	266.298	228.170	0.02591	820.884	40.843
29	54.189	2800.000	100.000	69.141	244.431	223.755	0.01467	364.372	115.870
30	100.000	2487.519	100.000	68.104	285.694	207.076	0.01091	530.902	96.903

algorithms solutions, respectively. The ABC algorithm's solution has higher brake torque and least BSFC values. Hence, to identify the best solution among the solutions of different algorithms, which has the best compromise among the objectives, the MADM methods based average ranks are calculated and presented in Table 8.24.

Table 8.24 presents the decision making methods ranking of the Pareto-optimal solutions obtained by different algorithms. The Spearman's correlation coefficients

Table 8.15 Pareto-optimal solutions obtained by the AMTPG-Jaya algorithm for the turbocharged DI diesel engine case study

SN	X_1 (%)	X_2(rpm)	X_3 (%)	P (kW)	T (N.m)	BSFC (g/kWh)	CO (%)	NO_x(ppm)	HC (ppm)
1	100.000	1000.000	96.027	26.341	264.473	228.512	0.02588	815.124	40.890
2	100.000	1117.024	94.570	31.144	272.843	223.556	0.02355	785.648	45.052
3	99.568	2528.323	97.115	66.487	274.191	209.866	0.01040	497.620	98.473
4	100.000	1000.000	100.000	27.681	276.503	226.240	0.02613	854.110	40.610
5	50.495	2800.000	100.000	69.211	244.967	224.469	0.01506	360.939	116.256
6	49.828	2800.000	100.000	69.224	245.075	224.603	0.01513	360.351	116.324
7	100.000	1000.000	100.000	27.681	276.503	226.240	0.02613	854.110	40.610
8	100.000	1000.000	100.000	27.681	276.503	226.240	0.02613	854.110	40.610
9	100.000	1210.407	94.140	34.932	280.341	219.706	0.02185	768.256	48.347
10	99.905	2524.135	97.051	66.426	274.554	209.781	0.01037	499.096	98.245
11	100.000	2533.346	97.140	66.526	273.791	209.997	0.01038	497.175	98.611
12	85.095	2734.527	100.000	68.863	253.334	216.363	0.01190	426.709	109.360
13	58.162	2800.000	100.000	69.071	243.966	223.029	0.01426	368.411	115.440
14	100.000	1000.000	99.252	27.430	274.253	226.668	0.02608	846.634	40.656
15	100.000	1000.000	98.492	27.175	271.960	227.103	0.02602	839.103	40.707
16	100.000	1000.000	96.667	26.558	266.425	228.146	0.02591	821.287	40.839
17	100.000	1481.879	93.282	44.637	296.105	210.936	0.01753	716.620	58.079
18	100.000	2294.013	95.055	63.401	290.438	205.712	0.01054	550.164	88.880
19	57.246	2800.000	100.000	69.087	244.063	223.192	0.01436	367.448	115.540
20	100.000	1429.958	93.399	42.928	293.782	212.336	0.01828	726.629	56.199
21	100.000	1000.000	95.626	26.205	263.247	228.742	0.02586	811.285	40.923
22	48.795	2800.000	100.000	69.245	245.247	224.811	0.01525	359.463	116.429
23	100.000	1000.000	99.608	27.550	275.326	226.464	0.02610	850.187	40.634
24	100.000	2446.946	96.292	65.500	280.589	208.152	0.01033	516.514	95.054
25	100.000	1000.000	100.000	27.681	276.503	226.240	0.02613	854.110	40.610
26	53.242	2800.000	100.000	69.158	244.559	223.934	0.01477	363.461	115.970
27	100.000	1000.000	100.000	27.681	276.503	226.240	0.02613	854.110	40.610
28	100.000	1000.000	99.771	27.604	275.815	226.371	0.02611	851.816	40.624
29	86.791	2705.215	100.000	68.858	257.555	214.994	0.01173	439.026	107.903
30	100.000	1949.309	93.399	57.142	302.444	204.144	0.01236	623.249	75.448

for different pairs of rankings for different algorithm's solutions are shown in Table 8.25. The ranks given by the pairs SAW-WPM and TOPSIS-MTOPSIS methods for each solution are identical. Furthermore, the Spearman's correlation for all the pairs of decision-making methods (except with the PROMETHEE and GRA methods) is greater than 0.5. Spearman's correlation values for the pairs consisting of PROMETHEE or GRA methods are either negative or nearer to zero. Hence, the ranks

Table 8.16 Pareto-optimal solutions obtained by the Rao-1 algorithm for the turbocharged DI diesel engine case study

SN	X_1 (%)	X_2(rpm)	X_3 (%)	P (kW)	T (N.m)	BSFC (g/kWh)	CO (%)	NO_x (ppm)	HC(ppm)
1	68.315	2800.000	100.000	68.922	243.294	221.375	0.01333	380.370	114.265
2	100.000	1000.000	100.000	27.681	276.503	226.240	0.02613	854.110	40.610
3	49.884	2800.000	100.000	69.223	245.065	224.591	0.01513	360.400	116.318
4	62.218	2800.000	100.000	69.007	243.608	222.333	0.01388	372.906	114.983
5	95.832	2576.231	97.878	67.182	269.984	210.957	0.01071	481.160	101.065
6	100.000	1000.000	96.363	26.455	265.497	228.320	0.02589	818.351	40.863
7	57.669	2800.000	100.000	69.079	244.017	223.117	0.01431	367.890	115.494
8	100.000	1000.000	100.000	27.681	276.503	226.240	0.02613	854.110	40.610
9	100.000	1000.000	100.000	27.681	276.503	226.240	0.02613	854.110	40.610
10	100.000	2200.541	94.457	61.924	295.088	204.745	0.01086	570.359	85.175
11	55.833	2800.000	100.000	69.111	244.225	223.449	0.01450	365.999	115.694
12	49.093	2800.000	100.000	69.239	245.196	224.750	0.01521	359.717	116.399
13	54.832	2800.000	100.000	69.129	244.348	223.634	0.01460	365.001	115.802
14	100.000	2250.038	94.759	62.725	292.755	205.207	0.01067	559.700	87.131
15	100.000	1000.000	100.000	27.681	276.503	226.240	0.02613	854.110	40.610
16	73.391	2800.000	100.000	68.863	243.238	220.656	0.01291	387.229	113.637
17	79.438	2800.000	100.000	68.806	243.415	219.894	0.01246	396.169	112.855
18	100.000	2325.908	95.286	63.870	288.614	206.133	0.01047	543.208	90.155
19	100.000	1000.000	98.851	27.295	273.044	226.897	0.02605	842.654	40.683
20	100.000	2318.004	95.227	63.756	289.077	206.024	0.01048	544.935	89.839
21	99.596	2527.976	97.109	66.482	274.221	209.859	0.01039	497.742	98.454
22	59.817	2800.000	100.000	69.044	243.805	222.740	0.01410	370.200	115.255
23	100.000	1000.000	100.000	27.681	276.503	226.240	0.02613	854.110	40.610
24	64.026	2800.000	100.000	68.980	243.487	222.038	0.01371	375.031	114.774
25	59.643	2800.000	100.000	69.047	243.821	222.770	0.01412	370.009	115.275
26	100.000	2181.335	94.348	61.601	295.915	204.596	0.01094	574.474	84.420
27	68.206	2800.000	100.000	68.923	243.297	221.391	0.01334	380.229	114.278
28	100.000	1000.000	97.883	26.969	270.116	227.451	0.02598	833.112	40.749
29	56.199	2800.000	100.000	69.105	244.181	223.382	0.01446	366.370	115.654
30	100.000	1460.115	93.328	43.929	295.171	211.507	0.01784	720.824	57.290

suggested by the PROMETHEE and GRA methods cannot be considered for calculating the average ranks. Now, the corrected average ranks are calculated, excluding the ranks suggested by the PROMETHEE and GRA methods, and presented in Table 8.24 as the corrected ranks.

The ERao-1 algorithm's solution has the best compromise among the objectives and has achieved the least average rank of 1.33. Hence, the ERao-1 algorithm's

Table 8.17 Pareto-optimal solutions obtained by the Rao-2 algorithm for the turbocharged DI diesel engine case study

SN	X_1 (%)	X_2(rpm)	X_3 (%)	P (kW)	T (N.m)	BSFC (g/kWh)	CO (%)	NO_x(ppm)	HC (ppm)
1	100.000	1000.000	100.000	27.681	276.503	226.240	0.02613	854.110	40.610
2	100.000	1000.000	98.123	27.050	270.845	227.313	0.02600	835.475	40.732
3	56.458	2800.000	100.000	69.100	244.151	223.335	0.01443	366.635	115.626
4	100.000	1682.108	93.063	50.612	302.025	206.752	0.01495	677.374	65.416
5	100.000	1000.000	96.612	26.540	266.257	228.178	0.02591	820.756	40.844
6	100.000	1000.000	100.000	27.681	276.503	226.240	0.02613	854.110	40.610
7	100.000	1131.410	94.500	31.743	274.068	222.935	0.02328	782.980	45.558
8	100.000	1000.000	100.000	27.681	276.503	226.240	0.02613	854.110	40.610
9	100.000	2476.862	96.573	65.868	278.336	208.753	0.01033	509.845	96.280
10	100.000	2264.319	94.852	62.948	292.028	205.361	0.01063	556.610	87.697
11	48.682	2800.000	100.000	69.248	245.266	224.834	0.01526	359.367	116.440
12	100.000	1000.000	100.000	27.681	276.503	226.240	0.02613	854.110	40.610
13	79.390	2800.000	100.000	68.806	243.412	219.899	0.01246	396.095	112.861
14	100.000	1000.000	100.000	27.681	276.503	226.240	0.02613	854.110	40.610
15	48.160	2800.000	100.000	69.258	245.357	224.941	0.01532	358.929	116.492
16	100.000	1491.586	93.263	44.950	296.504	210.688	0.01739	714.742	58.431
17	100.000	2407.655	95.942	64.997	283.385	207.424	0.01034	525.231	93.453
18	51.066	2800.000	100.000	69.200	244.878	224.356	0.01500	361.449	116.197
19	100.000	1000.000	95.560	26.183	263.043	228.780	0.02585	810.650	40.929
20	100.000	1000.000	100.000	27.681	276.503	226.240	0.02613	854.110	40.610
21	100.000	1549.446	93.166	46.762	298.642	209.308	0.01660	703.494	60.539
22	77.682	2800.000	100.000	68.821	243.336	220.104	0.01259	393.487	113.086
23	100.000	2302.161	95.113	63.523	289.983	205.815	0.01052	548.390	89.205
24	100.000	1000.000	100.000	27.681	276.503	226.240	0.02613	854.110	40.610
25	100.000	1000.000	100.000	27.681	276.503	226.240	0.02613	854.110	40.610
26	63.187	2800.000	100.000	68.992	243.540	222.174	0.01379	374.036	114.871
27	100.000	1000.000	100.000	27.681	276.503	226.240	0.02613	854.110	40.610
28	100.000	1000.000	100.000	27.681	276.503	226.240	0.02613	854.110	40.610
29	73.569	2800.000	100.000	68.861	243.239	220.632	0.01290	387.482	113.615
30	100.000	1000.000	98.668	27.234	272.491	227.002	0.02603	840.838	40.695

solution can be considered as the best solution. The brake power for the ERao-1 algorithm's solution is 1.4, 1.7, 1.5, 1.8, 0.8, 0.1 and 0.8% higher than that of the ABC, Jaya, Rao-2, Rao-3, AMTPG-Jaya, SAP-Rao, and ERao-2 solutions, respectively. The CO emissions for the ERao-1 algorithm solution is 12.8, 1.7, 1.5, 1.9, 0.7, 0.1 and 0.7% lesser than that of the ABC, Jaya, Rao-2, Rao-3, AMTPG-Jaya, SAP-Rao and ERao-2 solutions, respectively. The ERao-1 algorithm solution's NO_x emissions

Table 8.18 Pareto-optimal solutions obtained by the Rao-3 algorithm for the turbocharged DI diesel engine case study

SN	X_1 (%)	X_2(rpm)	X_3 (%)	P (kW)	T (N.m)	BSFC (g/kWh)	CO(%)	NO_x(ppm)	HC (ppm)
1	100.000	1000.000	100.000	27.681	276.503	226.240	0.02613	854.110	40.610
2	78.555	2800.000	100.000	68.813	243.372	219.998	0.01252	394.812	112.971
3	56.931	2800.000	100.000	69.092	244.098	223.249	0.01439	367.121	115.575
4	73.003	2800.000	100.000	68.867	243.236	220.708	0.01294	386.685	113.686
5	100.000	1000.000	100.000	27.681	276.503	226.240	0.02613	854.110	40.610
6	72.965	2800.000	100.000	68.867	243.235	220.713	0.01295	386.631	113.691
7	100.000	2250.989	94.765	62.740	292.707	205.217	0.01067	559.495	87.168
8	55.058	2800.000	100.000	69.125	244.320	223.592	0.01458	365.225	115.777
9	48.748	2800.000	100.000	69.246	245.255	224.821	0.01525	359.423	116.433
10	100.000	1311.368	93.751	38.765	287.250	216.025	0.02013	749.248	51.939
11	100.000	1000.000	97.654	26.892	269.423	227.582	0.02597	830.874	40.766
12	50.323	2800.000	100.000	69.215	244.995	224.503	0.01508	360.786	116.273
13	100.000	1000.000	96.407	26.470	265.632	228.295	0.02590	818.777	40.860
14	100.000	1000.000	100.000	27.681	276.503	226.240	0.02613	854.110	40.610
15	100.000	1000.000	100.000	27.681	276.503	226.240	0.02613	854.110	40.610
16	100.000	1000.000	96.939	26.650	267.251	227.991	0.02593	823.914	40.819
17	100.000	2228.404	94.623	62.380	293.811	204.991	0.01075	564.369	86.274
18	100.000	1000.000	100.000	27.681	276.503	226.240	0.02613	854.110	40.610
19	57.056	2800.000	100.000	69.090	244.084	223.226	0.01437	367.251	115.561
20	66.034	2800.000	100.000	68.952	243.380	221.721	0.01353	377.478	114.538
21	50.655	2800.000	100.000	69.208	244.942	224.438	0.01504	361.081	116.239
22	100.000	1000.000	98.552	27.195	272.141	227.068	0.02603	839.696	40.703
23	100.000	1000.000	98.105	27.044	270.789	227.324	0.02600	835.291	40.733
24	68.638	2800.000	100.000	68.918	243.285	221.327	0.01330	380.790	114.225
25	100.000	1216.670	94.114	35.178	280.806	219.463	0.02174	767.083	48.569
26	100.000	1000.000	100.000	27.681	276.503	226.240	0.02613	854.110	40.610
27	100.000	2344.910	95.430	64.142	287.469	206.406	0.01043	539.049	90.918
28	60.715	2800.000	100.000	69.030	243.726	222.586	0.01402	371.197	115.154
29	100.000	1000.000	100.000	27.681	276.503	226.240	0.02613	854.110	40.610
30	100.000	1000.000	100.000	27.681	276.503	226.240	0.02613	854.110	40.610

are 3, 2.7, 2.4, 3, 1.3, 0.2 and 1.3% lesser than those of the ABC, Jaya, Rao-2, Rao-3, AMTPG-Jaya, SAP-Rao and ERao-2 solutions, respectively.

The spacing values of the Pareto-fronts obtained by different algorithms in the MOO scenario are also presented in Table 8.23. Here, an observation can be made that the SAP-Rao algorithm's Pareto-front has the least spacing value. The SAP-Rao algorithm has a better spacing, which is 34.2, 67.4, 29.3, 23.3, 49.6, 64.3, 52.2 and

Table 8.19 Pareto-optimal solutions obtained by the ERao-1 algorithm for the turbocharged DI diesel engine case study

SN	X_1 (%)	X_2(rpm)	X_3 (%)	P (kW)	T (N.m)	BSFC (g/kWh)	CO(%)	NO_x(ppm)	HC (ppm)
1	84.151	2751.443	100.000	68.855	250.853	217.182	0.01200	419.645	110.195
2	78.260	2800.000	100.000	68.816	243.360	220.034	0.01254	394.362	113.010
3	71.624	2800.000	100.000	68.882	243.236	220.898	0.01305	384.775	113.859
4	100.000	1000.000	95.448	26.145	262.702	228.844	0.02585	809.588	40.938
5	67.683	2800.000	100.000	68.930	243.314	221.469	0.01338	379.556	114.341
6	71.908	2800.000	100.000	68.879	243.235	220.858	0.01303	385.165	113.823
7	51.511	2800.000	100.000	69.191	244.810	224.269	0.01495	361.852	116.151
8	100.000	2326.932	95.293	63.885	288.553	206.147	0.01046	542.984	90.196
9	100.000	2100.578	93.943	60.168	298.913	204.155	0.01134	591.644	81.267
10	50.185	2800.000	100.000	69.217	245.017	224.531	0.01509	360.664	116.288
11	100.000	1000.000	96.405	26.470	265.626	228.296	0.02590	818.758	40.860
12	70.205	2800.000	100.000	68.899	243.251	221.099	0.01317	382.855	114.034
13	68.016	2800.000	100.000	68.926	243.303	221.419	0.01336	379.984	114.301
14	100.000	1966.759	93.448	57.515	302.176	204.091	0.01223	619.640	76.114
15	100.000	1000.000	100.000	27.681	276.503	226.240	0.02613	854.110	40.610
16	100.000	1314.952	93.738	38.896	287.472	215.904	0.02007	748.569	52.067
17	100.000	1000.000	95.531	26.173	262.956	228.796	0.02585	810.378	40.931
18	60.018	2800.000	100.000	69.041	243.787	222.705	0.01408	370.422	115.233
19	100.000	2110.106	93.987	60.344	298.600	204.191	0.01129	589.629	81.637
20	84.684	2741.845	100.000	68.860	252.265	216.714	0.01194	423.650	109.721
21	100.000	1000.000	97.470	26.830	268.864	227.687	0.02596	829.075	40 .779
22	50.062	2800.000	100.000	69.220	245.037	224.556	0.01511	360.556	116.300
23	51.924	2800.000	100.000	69.183	244.748	224.188	0.01491	362.230	116.108
24	52.637	2800.000	100.000	69.170	244.645	224.050	0.01483	362.891	116.034
25	100.000	1000.000	100.000	27.681	276.503	226.240	0.02613	854.110	40.610
26	100.000	1000.000	100.000	27.681	276.503	226.240	0.02613	854.110	40.610
27	55.694	2800.000	100.000	69.114	244.242	223.475	0.01451	365.859	115.709
28	100.000	2231.141	94.640	62.425	293.680	205.017	0.01074	563.779	86.383
29	100.000	2446.645	96.289	65.497	280.611	208.146	0.01033	516.581	95.042
30	48.251	2800.000	100.000	69.257	245.341	224.922	0.01531	359.004	116.483

16.4% lesser when compared to that of the Jaya, Rao-1, Rao-2, Rao-3, AMTPG-Jaya, ERao-1, ERao-2 and ERao-3 algorithms, respectively.

The coverage metric values of the Pareto-fronts obtained by the Jaya algorithm, Rao algorithms, and their modified versions are presented in Table 8.26. The Jaya algorithm has achieved better coverage values than the other algorithms compared.

Table 8.20 Pareto-optimal solutions obtained by the ERao-2 algorithm for the turbocharged DI diesel engine case study

SN	X_1 (%)	X_2 (rpm)	X_3 (%)	*P* (kW)	*T* (N.m)	BSFC (g/kWh)	CO (%)	NO_x (ppm)	HC (ppm)
1	100.000	1000.000	100.000	27.681	276.503	226.240	0.02613	854.110	40.610
2	100.000	1000.000	100.000	27.681	276.503	226.240	0.02613	854.110	40.610
3	100.000	1000.000	100.000	27.681	276.503	226.240	0.02613	854.110	40.610
4	100.000	1917.874	93.319	56.455	302.837	204.276	0.01261	629.727	74.252
5	100.000	1650.765	93.072	49.740	301.416	207.280	0.01532	683.588	64.258
6	100.000	1000.000	95.559	26.182	263.040	228.781	0.02585	810.641	40.929
7	100.000	1000.000	97.324	26.781	268.422	227.771	0.02595	827.658	40.790
8	50.235	2800.000	100.000	69.216	245.009	224.521	0.01509	360.709	116.282
9	70.206	2800.000	100.000	68.899	243.251	221.098	0.01317	382.857	114.034
10	50.491	2800.000	100.000	69.211	244.968	224.470	0.01506	360.935	116.256
11	100.000	1690.678	93.063	50.846	302.171	206.615	0.01485	675.670	65.733
12	100.000	1735.952	93.071	52.056	302.795	205.954	0.01435	666.635	67.414
13	100.000	1000.000	100.000	27.681	276.503	226.240	0.02613	854.110	40.610
14	96.310	2569.952	97.775	67.092	270.541	210.802	0.01067	483.274	100.727
15	100.000	2307.103	95.148	63.596	289.704	205.879	0.01051	547.313	89.402
16	100.000	2293.782	95.053	63.398	290.451	205.709	0.01054	550.214	88.870
17	100.000	2518.855	96.990	66.362	274.993	209.665	0.01036	500.435	98.011
18	60.213	2800.000	100.000	69.038	243.770	222.671	0.01407	370.637	115.211
19	52.679	2800.000	100.000	69.169	244.639	224.042	0.01482	362.931	116.030
20	100.000	1000.000	100.000	27.681	276.503	226.240	0.02613	854.110	40.610
21	100.000	1000.000	100.000	27.681	276.503	226.240	0.02613	854.110	40.610
22	59.462	2800.000	100.000	69.050	243.838	222.801	0.01414	369.812	115.295
23	100.000	1000.000	100.000	27.681	276.503	226.240	0.02613	854.110	40.610
24	100.000	1000.000	95.470	26.152	262.768	228.831	0.02585	809.792	40.936
25	53.670	2800.000	100.000	69.150	244.501	223.853	0.01472	363.871	115.925
26	51.685	2800.000	100.000	69.188	244.784	224.235	0.01493	362.010	116.133
27	88.363	2680.645	99.722	68.650	260.370	214.040	0.01155	447.690	106.632
28	72.044	2800.000	100.000	68.877	243.235	220.840	0.01302	385.352	113.806
29	100.000	1214.641	94.123	35.098	280.656	219.542	0.02177	767.463	48.497
30	100.000	1000.000	95.219	26.067	261.999	228.975	0.02584	807.405	40.957

From the coverage values, it can be observed that two solutions from the Pareto-front of the Jaya, and three solutions from the Pareto-fronts of the SAP-Rao and ERao-1 algorithm are dominated. Similarly, five solutions are dominated from the Pareto-fronts achieved by the AMTPG-Jaya, Rao-1, and ERao-3 algorithms; six solutions are dominated from the Pareto-front achieved by the ERao-2 algorithms; eight solutions are dominated from the Pareto-front achieved by the Rao-3 algorithm;

Table 8.21 Pareto-optimal solutions obtained by the ERao-3 algorithm for the turbocharged DI diesel engine case study

SN	X_1 (%)	X_2 (rpm)	X_3 (%)	P (kW)	T (N.m)	BSFC (g/kWh)	CO (%)	NO_x (ppm)	HC (ppm)
1	100.000	1990.415	93.520	58.010	301.754	204.041	0.01205	614.732	77.019
2	100.000	1000.000	97.588	26.870	269.224	227.619	0.02597	830.234	40.770
3	100.000	1000.000	97.497	26.839	268.946	227.672	0.02596	829.340	40.777
4	100.000	1000.000	100.000	27.681	276.503	226.240	0.02613	854.110	40.610
5	100.000	1046.716	94.936	28.135	266.492	226.738	0.02490	798.631	42.588
6	100.000	1668.053	93.066	50.224	301.766	206.983	0.01512	680.164	64.896
7	100.000	1936.894	93.366	56.873	302.614	204.191	0.01246	625.811	74.975
8	54.062	2800.000	100.000	69.143	244.448	223.779	0.01468	364.248	115.884
9	100.000	1000.000	100.000	27.681	276.503	226.240	0.02613	854.110	40.610
10	100.000	1000.000	100.000	27.681	276.503	226.240	0.02613	854.110	40.610
11	100.000	2508.413	96.884	66.241	275.844	209.431	0.01035	502.780	97.579
12	100.000	1000.000	100.000	27.681	276.503	226.240	0.02613	854.110	40.610
13	85.361	2729.840	100.000	68.863	254.016	216.140	0.01187	428.672	109.128
14	100.000	1344.416	93.641	39.962	289.241	214.928	0.01960	742.977	53.122
15	100.000	1277.442	93.873	37.507	285.067	217.207	0.02069	755.660	50.728
16	100.000	2493.833	96.739	66.071	277.011	209.112	0.01034	506.048	96.978
17	48.919	2800.000	100.000	69.243	245.226	224.786	0.01523	359.568	116.416
18	53.264	2800.000	100.000	69.158	244.556	223.930	0.01476	363.483	115.968
19	82.958	2773.446	100.000	68.833	247.575	218.279	0.01214	410.500	111.275
20	65.597	2800.000	100.000	68.958	243.401	221.789	0.01357	376.938	114.590
21	100.000	2507.748	96.877	66.234	275.898	209.416	0.01035	502.929	97.552
22	100.000	1000.000	100.000	27.681	276.503	226.240	0.02613	854.110	40.610
23	100.000	1000.000	95.664	26.218	263.363	228.720	0.02586	811.649	40.920
24	63.347	2800.000	100.000	68.990	243.529	222.148	0.01377	374.225	114.853
25	100.000	2129.819	94.080	60.701	297.917	204.280	0.01119	585.451	82.405
26	100.000	1579.868	93.128	47.682	299.605	208.648	0.01620	697.546	61.652
27	58.890	2800.000	100.000	69.059	243.893	222.901	0.01419	369.190	115.359
28	100.000	2416.454	96.019	65.112	282.775	207.581	0.01034	523.283	93.811
29	100.000	1000.000	100.000	27.681	276.503	226.240	0.02613	854.110	40.610
30	80.621	2800.000	100.000	68.796	243.480	219.757	0.01238	398.015	112.697

10 solutions are dominated from the Pareto-front achieved by the Rao-2 algorithm. The performances of the proposed modified algorithms are better or competitive to that of the basic algorithms in terms of coverage and spacing values.

From the computational results of the turbocharged DI diesel engine case study, it can be observed that the solutions obtained by the proposed algorithms are better than those reported in the literature. The solution of the ERao-1 algorithm has better

Table 8.22 Pareto-optimal solutions obtained by the SAP-Rao algorithm for the turbocharged DI diesel engine case study

SN	X_1 (%)	X_2(rpm)	X_3 (%)	P (kW)	T (N.m)	BSFC (g/kWh)	CO (%)	NO_x (ppm)	HC (ppm)
1	100.000	1570.341	93.135	47.395	299.305	208.852	0.01632	699.381	61.303
2	53.929	2800.000	100.000	69.146	244.466	223.804	0.01469	364.120	115.898
3	97.232	2557.269	97.574	66.912	271.663	210.498	0.01058	487.510	100.052
4	50.815	2800.000	100.000	69.205	244.917	224.406	0.01502	361.224	116.223
5	100.000	1986.985	93.508	57.939	301.816	204.048	0.01208	615.436	76.887
6	100.000	1000.000	100.000	27.681	276.503	226.240	0.02613	854.110	40.610
7	79.240	2800.000	100.000	68.807	243.405	219.917	0.01247	395.863	112.881
8	100.000	2511.125	96.917	66.276	275.637	209.488	0.01035	502.207	97.692
9	69.134	2800.000	100.000	68.912	243.272	221.254	0.01326	381.437	114.165
10	100.000	1661.957	93.070	50.055	301.653	207.085	0.01519	681.389	64.671
11	100.000	2110.318	93.978	60.342	298.566	204.198	0.01129	589.511	81.646
12	100.000	1000.000	95.332	26.105	262.347	228.910	0.02584	808.483	40.948
13	100.000	1025.188	95.055	27.186	264.426	227.761	0.02533	802.587	41.837
14	100.000	1000.000	98.383	27.138	271.631	227.165	0.02602	838.030	40.714
15	49.783	2800.000	100.000	69.225	245.082	224.612	0.01514	360.312	116.329
16	100.000	2323.141	95.267	63.831	288.780	206.094	0.01047	543.823	90.044
17	49.242	2800.000	100.000	69.236	245.171	224.720	0.01520	359.845	116.383
18	100.000	2440.745	96.235	65.422	281.042	208.033	0.01033	517.892	94.801
19	48.294	2800.000	100.000	69.256	245.334	224.913	0.01530	359.041	116.479
20	100.000	1000.000	96.583	26.530	266.168	228.195	0.02591	820.471	40.846
21	80.379	2800.000	100.000	68.798	243.466	219.784	0.01239	397.636	112.729
22	100.000	1000.000	100.000	27.681	276.503	226.240	0.02613	854.110	40.610
23	100.000	1043.465	94.956	27.993	266.191	226.889	0.02497	799.251	42.475
24	100.000	1000.000	100.000	27.681	276.503	226.240	0.02613	854.110	40.610
25	48.297	2800.000	100.000	69.256	245.333	224.913	0.01530	359.044	116.479
26	100.000	2193.723	94.420	61.811	295.391	204.689	0.01089	571.833	84.907
27	97.741	2550.169	97.486	66.825	272.340	210.322	0.01054	490.029	99.676
28	100.000	2214.027	94.555	62.158	294.531	204.849	0.01081	567.598	85.707
29	100.000	1000.000	97.540	26.854	269.078	227.647	0.02596	829.764	40.774
30	100.000	1000.000	97.895	26.973	270.154	227.444	0.02598	833.235	40.748

compromise among the six objectives. The performance objectives P, T, BSFC, CO, NO_x, and HC values obtained by the ERao-1 solution are 63.89 kW, 288.55 N m, 206.15 g/kWh, 0.01046%, 542.98 and 90.20 ppm, respectively. The SAP-Rao algorithm has achieved the least spacing value (0.10705), which is much better than that achieved by other algorithms compared. The ERao-1, SAP-Rao, and Jaya algorithms

Table 8.23 Best solutions obtained by various algorithms in MOO scenario of the turbocharged DI diesel engine case study

Algorithm	X_1 (%)	X_2 (rpm)	X_3 (%)	P (kW)	T (N.m)	BSFC (g/kWh)	CO (%)	NO_x (ppm)	HC (ppm)	Spacing
ABC	85.63	2208	97	63	**298**	**202.85**	0.012	560	88	–
Jaya	100	2257.30	94.806172	62.84	292.39	205.28	0.01065	558.13	87.42	0.16279
Rao-1	100	2325.90	95.28581	63.87	288.61	206.13	0.01047	543.21	90.16	0.32805
Rao-2	100	2264.31	94.85232	62.95	292.03	205.36	0.01063	556.61	87.70	0.15134
Rao-3	100	2250.98	94.76524	62.74	292.71	205.22	0.01067	559.49	**87.17**	0.13963
AMTPG-Jaya	100	2294.01	95.05487	63.40	290.44	205.71	0.01054	550.16	88.88	0.21219
SAP-Rao	100	2323.14	95.26661	63.83	288.78	206.09	0.01047	543.82	90.04	**0.10705**
ERao-1	100	2326.93	95.29346	63.89	288.55	206.15	0.01046	542.98	90.20	0.29966
ERao-2	100	2293.78	95.05325	63.40	290.45	205.71	0.01054	550.21	88.87	0.22398
ERao-3	100	2416.45	96.01875	**65.11**	282.78	207.58	**0.01034**	**523.28**	93.81	0.12800

Source ABC—Shirneshan et al. (2016)
Result in boldface indicates better values

Table 8.24 Ranks suggested by the MADM methods for different algorithm solutions presented in Table 8.23

Algorithm	SAW	WPM	TOPSIS	MTOPSIS	VIKOR	PROMETHEE	COPRAS	GRA	AR	CR	FR
ABC	10	10	10	10	9	6	10	2	8.31	9.83	10
Jaya	8	8	8	8	7	6	8	4	7.06	7.83	8
Rao-1	4	4	3	3	1	9	3	10	4.63	3.00	3
Rao-2	7	7	6	6	6	6	7	5	6.19	6.50	7
Rao-3	9	9	9	9	10	6	9	3	7.94	9.17	9
AMTPG-Jaya	6	6	4	4	5	10	6	7	6.00	5.17	6
SAP-Rao	3	3	2	2	1	2	2	9	3.00	2.17	2
ERao-1	2	2	1	1	1	6	1	8	2.69	1.33	1
ERao-2	5	5	5	5	1	1	5	6	4.13	4.33	4
ERao-3	1	1	7	7	8	6	4	1	4.31	4.67	5

AR—Average Rank; CR—Corrected Rank; FR—Final Rank

Table 8.25 Spearman's rank correlation coefficients between different pairs of MADM method's rankings presented in Table 8.24

Method	SAW	WPM	TOPSIS	MTOPSIS	VIKOR	PROMETHEE	COPRAS	GRA
SAW	1	1	0.73	0.73	0.61	0.11	0.93	−0.39
WPM	1	1	0.73	0.73	0.61	0.11	0.93	−0.39
TOPSIS	0.73	0.73	1	1	0.91	−0.01	0.92	−0.88
MTOPSIS	0.73	0.73	1	1	0.91	−0.01	0.92	−0.88
VIKOR	0.61	0.61	0.91	0.91	1	0.16	0.81	−0.89
PROMETHEE	0.11	0.11	−0.01	−0.01	0.16	1	0.11	0.11
COPRAS	0.93	0.93	0.92	0.92	0.81	0.11	1	−0.68
GRA	−0.39	−0.39	−0.88	−0.88	−0.89	0.11	−0.68	1

Table 8.26 Coverage (%) values of the Pareto-fronts obtained by the proposed algorithms in MOO for turbocharged DI diesel engine case study

Algorithm	Jaya	Rao-1	Rao-2	Rao-3	AMTPG-Jaya	SAP-Rao	ERao-1	ERao-2	ERao-3
Jaya	–	16.67	33.33	26.67	16.67	10	10	13.33	10
Rao-1	6.67	–	33.33	26.67	16.67	10	10	13.33	10
Rao-2	6.67	16.67	–	26.67	16.67	10	10	13.33	10
Rao-3	6.67	16.67	33.33	–	16.67	10	10	13.33	10
AMTPG-Jaya	6.67	16.67	33.33	26.67	–	10	10	13.33	10
SAP-Rao	6.67	16.67	33.33	26.67	16.67	–	10	13.33	10
ERao-1	6.67	16.67	33.33	26.67	16.67	10	–	13.33	10
ERao-2	6.67	16.67	33.33	26.67	16.67	10	10	–	16.67
ERao-3	6.67	16.67	33.33	26.67	16.67	10	10	20	–

have achieved better performance in terms of coverage values. In addition, the performances of the modified versions in this case study are better or competitive to those of the basic algorithms as well as the ABC algorithm.

8.3 Design Optimization of a Compression Ignition Biodiesel Engine with an EGR System

The description of the selected compression ignition biodiesel engine with an EGR system is presented in Sect. 2.3.3. In this case study, a compression ignition biodiesel engine with an exhaust gas recirculation system is considered for multi-objective optimization. This case study was presented by Jaliliantabar et al. (2019).

The design variables of this case study are the exhaust gas recirculation rate (ER), engine load percentage (EL), engine speed (ES) in rpm, and biodiesel percentage (BP). The ranges of design variables are as follows:$0 \le ER \le 30$, $25 \le EL \le 75$,

$1800 \leq ES \leq 2400$, and $0 \leq BP \leq 15$. The objective functions of this case study are maximization of power output (P), and minimization of the brake specific fuel consumption (BSFC) and emissions such as carbon monooxide emission (CO), nitrogen oxides emission (NO_x), hydrocarbon emission (HC), and smoke opacity (S_m). The regression models of the objectives of this case study are presented in Sect. 2.3.3.

Jaliliantabar et al. (2019) reported a single Pareto-optimal solution for this case study by implementing the NSGA-II algorithm. Now, the proposed algorithms, along with basic Jaya and Rao algorithms, are used to find the optimal parameters. The function evaluations taken by the NSGA-II algorithm was not specified. However, the proposed and other considered algorithms are executed for 10,000 function evaluations. The population sizes of all the algorithms are taken as 30. In the case of the SAP-Rao algorithm, 30 best-ranked solutions at termination are selected as the final population. The elite population size is taken as 20% of the population for the elitist Rao algorithms.

Furthermore, the solutions obtained by various algorithms are non-dominated. Hence, the average rank method was used to identify the best solution from the Pareto-front of the respective algorithm. Furthermore, the performances of the proposed algorithms are compared with the performances of the basic Jaya and Rao algorithms in terms of spacing and coverage performance metrics. The spacing performance metric is calculated for ten independent runs, and the statistical results are presented in terms of best, worst, mean, and standard deviation (SD).

Table 8.27 presents the optimal design parameters achieved by the SAP-Rao algorithm in single-objective optimization scenarios of each objective. In all the single objective optimization scenarios, the solutions obtained by the Jaya algorithm, Rao algorithms, and their modified versions are identical. Hence, the solutions achieved by the SAP-Rao algorithm in the respective scenarios are reported in Table 8.27.

The Pareto-optimal solutions achieved by the Jaya algorithm, Rao algorithms, and their modified versions in this case study are presented in Tables 8.28, 8.29, 8.30, 8.31, 8.32, 8.33, 8.34, 8.35 and 8.36. The power is varied between 0.86 and 3.53 kW, the BSFC is varied from 218 to 577 g/kWh, the CO emission is varied from 0.00044 to 2.16%, the NO_x emission is varied from 86 to 387 ppm, HC emission is varied from 16.01 to 250 ppm, and smoke emission is varied from 2.3 to 17.4 m^{-1}.

Table 8.27 Optimal solutions obtained by the SAP-Rao algorithm in single-objective optimization scenarios of the biodiesel engine with an EGR system case study

Objective	ER (%)	EL (%)	ES (rpm)	BP (%)	Objective value
P (kW)	30	75	2400	0	3.52314
CO (%)	30	25	1890.68	15	1.42286E-08
NO_x (ppm)	28.12839	25	2162.72	15	86.83693804
HC (ppm)	0	41.95999	1800	15	16.01092
BSFC (g/kWh)	5.18747	62.24392	1800	8.4683	218.6337404
S_m (1/m)	0	30.82292	1800	11.0903	2.301684075

Table 8.28 Pareto-optimal solutions obtained by the Jaya algorithm for the biodiesel engine with an EGR system case study

Solution	ER (%)	EL (%)	ES (rpm)	BP (%)	P (kW)	BSFC (g/kWh)	CO (%)	NO_x (ppm)	HC (ppm)	S_m (1/m)
1	0.000	42.020	1800.000	15.000	1.890	316.461	0.072	290.955	16.011	2.738
2	0.000	30.984	1800.000	10.894	1.361	409.994	0.117	218.026	22.028	2.302
3	30.000	25.000	2161.243	15.000	1.087	564.309	0.090	86.943	46.484	3.741
4	6.012	61.202	1800.000	8.078	2.548	218.904	0.399	357.422	54.371	6.732
5	29.378	25.988	1849.770	10.288	1.204	494.819	0.000	146.656	32.025	2.800
6	0.000	25.000	1992.674	2.002	0.831	548.314	0.162	130.140	45.304	2.773
7	0.000	25.000	2097.867	1.224	0.832	570.621	0.172	122.347	50.791	2.948
8	30.000	75.000	2400.000	0.000	3.523	424.208	2.160	256.076	249.948	17.390
9	0.000	71.693	1800.000	3.967	2.853	244.308	0.866	387.349	94.022	10.637
10	30.000	75.000	2400.000	1.848	3.503	411.455	2.077	251.066	239.231	16.854
11	30.000	75.000	2377.888	11.667	3.357	386.224	1.654	214.164	189.290	14.419
12	30.000	58.249	1948.955	4.356	2.445	292.100	0.484	278.509	89.589	7.341
13	19.402	72.140	1856.021	6.844	2.973	267.791	0.811	312.059	109.727	10.782
14	11.970	53.542	1800.000	2.226	2.170	247.489	0.358	330.792	60.766	5.974
15	29.323	75.000	2384.198	5.827	3.425	392.300	1.874	238.051	215.640	15.686
16	11.435	75.000	2053.031	12.919	3.009	343.749	1.011	271.798	126.986	11.725
17	30.000	75.000	2325.718	14.665	3.271	392.088	1.448	197.393	170.654	13.517
18	30.000	68.909	2396.850	3.834	3.227	367.370	1.643	259.566	195.277	13.767
19	0.441	25.000	1800.000	6.546	1.019	484.007	0.180	166.953	30.970	2.469
20	30.000	75.000	2400.000	6.429	3.449	390.737	1.888	236.584	215.312	15.676
21	20.082	75.000	1907.773	5.199	3.044	304.076	1.056	295.028	137.804	12.521

(continued)

Table 8.28 (continued)

Solution	ER (%)	EL (%)	ES (rpm)	BP (%)	P (kW)	BSFC (g/kWh)	CO (%)	NO_x (ppm)	HC (ppm)	S_m (1/m)
22	30.000	65.135	1995.096	8.373	2.757	298.936	0.637	260.534	102.257	8.822
23	0.000	27.056	2002.436	3.695	0.942	516.925	0.144	147.556	42.281	2.682
24	0.000	66.932	1800.000	1.124	2.631	243.933	0.788	384.518	89.300	9.623
25	30.000	42.333	2264.820	8.645	1.903	382.201	0.340	207.263	72.227	5.151
26	5.550	64.712	1800.000	3.088	2.598	230.785	0.643	367.217	80.551	8.640
27	0.000	27.946	1800.000	15.000	1.309	464.611	0.161	188.335	25.830	2.458
28	8.981	42.783	2145.635	5.634	1.700	377.501	0.293	231.746	64.452	4.534
29	28.210	70.095	1981.642	11.391	2.960	308.402	0.716	251.823	107.070	9.862
30	10.246	25.748	2175.782	13.267	0.979	550.222	0.161	106.736	49.393	3.328

Table 8.29 Pareto-optimal solutions obtained by the AMTPG-Jaya algorithm for the biodiesel engine with an EGR system case study

Solution	ER (%)	EL (%)	ES (rpm)	BP (%)	*P* (kW)	BSFC (g/kWh)	CO (%)	NO_x (ppm)	HC (ppm)	S_m (1/m)
1	0.000	42.077	1800.000	15.000	1.893	316.017	0.072	291.288	16.012	2.742
2	0.000	31.343	1800.000	11.381	1.385	406.998	0.114	220.556	21.560	2.303
3	30.000	25.000	2161.219	15.000	1.087	564.308	0.090	86.943	46.483	3.741
4	25.712	46.073	1878.288	13.441	2.070	312.306	0.000	245.005	32.131	3.689
5	0.000	71.687	1800.000	6.809	2.896	237.333	0.773	386.442	83.199	10.025
6	2.850	63.752	1800.000	8.443	2.638	219.206	0.466	370.217	56.936	7.287
7	0.775	25.000	2069.627	0.000	0.818	572.216	0.180	122.608	52.868	3.067
8	30.000	75.000	2400.000	0.733	3.515	418.849	2.126	254.146	245.625	17.173
9	30.000	75.000	2352.340	12.434	3.319	387.064	1.573	208.803	182.539	14.081
10	30.000	75.000	2396.374	7.234	3.434	388.721	1.850	233.315	210.956	15.463
11	30.000	75.000	2400.000	5.906	3.456	392.315	1.908	238.384	217.849	15.800
12	0.000	25.227	1800.000	0.000	0.892	501.229	0.221	167.032	39.512	2.836
13	26.003	33.233	2400.000	15.000	1.538	458.455	0.327	160.011	60.135	4.676
14	30.000	75.000	2400.000	0.878	3.513	417.832	2.120	253.754	244.778	17.131
15	30.000	25.000	1938.071	15.000	1.170	543.703	0.006	106.402	35.074	3.128
16	23.486	66.170	2389.103	9.420	2.976	342.463	1.296	255.044	155.411	11.487
17	30.000	75.000	2319.562	15.000	3.262	393.063	1.425	195.585	168.616	13.420
18	11.127	59.763	2028.955	12.736	2.433	299.121	0.442	278.581	72.853	6.803
19	18.198	59.960	2400.000	10.249	2.675	331.936	1.024	265.692	126.695	9.321
20	22.859	74.583	1800.000	1.062	3.023	281.142	1.079	328.817	139.279	12.824
21	6.456	65.674	2306.992	9.682	2.719	346.879	1.090	289.051	130.349	10.155

(continued)

Table 8.29 (continued)

Solution	ER (%)	EL (%)	ES (rpm)	BP (%)	P (kW)	BSFC (g/kWh)	CO (%)	NO_x (ppm)	HC (ppm)	S_m (1/m)
22	0.000	25.000	2054.034	5.037	0.856	550.044	0.147	124.568	43.617	2.661
23	0.000	60.729	1800.000	13.008	2.585	227.793	0.304	368.882	36.798	5.833
24	19.874	69.307	1824.075	11.590	2.959	247.671	0.521	312.469	77.326	8.795
25	30.000	43.537	2400.000	7.386	2.107	370.547	0.566	230.943	88.534	6.166
26	18.525	66.823	1914.548	7.469	2.747	270.389	0.656	300.149	97.052	9.140
27	30.000	75.000	2158.985	9.830	3.182	368.471	1.267	219.017	162.497	13.097
28	17.200	43.910	2156.799	11.233	1.815	367.350	0.239	214.474	59.897	4.534
29	0.000	28.852	1800.000	14.854	1.345	452.015	0.148	196.317	24.531	2.420
30	18.661	75.000	1873.335	14.021	3.150	294.184	0.717	287.019	97.452	10.624

Table 8.30 Pareto-optimal solutions obtained by the Rao-1 algorithm for the biodiesel engine with an EGR system case study

Solution	ER (%)	EL (%)	ES (rpm)	BP (%)	*P* (kW)	BSFC (g/kWh)	CO (%)	NO_x (ppm)	HC (ppm)	S_m (1/m)
1	0.000	30.567	1800.000	10.864	1.343	414.947	0.121	214.639	22.402	2.302
2	0.000	41.993	1800.000	15.000	1.889	316.677	0.072	290.793	16.011	2.735
3	23.749	28.556	1800.000	13.944	1.409	454.528	0.001	168.852	26.127	2.731
4	29.586	25.000	2154.583	15.000	1.084	563.987	0.088	86.926	46.412	3.721
5	6.484	62.534	1800.000	8.510	2.608	218.703	0.422	357.893	56.775	7.036
6	0.450	69.368	1800.000	4.932	2.784	233.771	0.747	385.098	82.687	9.613
7	30.000	75.000	2400.000	0.000	3.523	424.208	2.160	256.076	249.948	17.390
8	0.522	25.000	2263.460	0.000	0.931	591.255	0.230	127.383	59.038	3.359
9	1.341	25.000	2018.976	3.297	0.847	547.345	0.152	126.502	45.627	2.786
10	29.975	74.559	2400.000	13.267	3.346	385.895	1.631	210.914	184.387	14.148
11	30.000	66.744	2301.641	15.000	2.925	354.537	1.010	212.792	129.877	10.348
12	30.000	75.000	2400.000	4.225	3.475	398.763	1.976	243.916	226.344	16.215
13	30.000	69.881	2228.089	7.384	3.034	353.317	1.231	236.237	159.731	12.114
14	26.151	72.618	2391.127	14.595	3.201	377.472	1.473	218.468	168.634	13.095
15	26.862	75.000	2152.880	5.462	3.152	371.760	1.420	239.525	180.473	14.022
16	5.801	60.774	1896.858	7.080	2.448	256.468	0.480	330.285	69.241	7.110
17	0.000	71.894	1818.750	14.382	2.991	256.837	0.597	372.005	65.132	8.952
18	25.365	64.342	2400.000	3.408	3.001	352.834	1.421	272.931	173.811	12.125
19	30.000	75.000	2400.000	0.782	3.515	418.503	2.124	254.014	245.338	17.159
20	29.643	44.639	2370.336	15.000	2.055	377.221	0.446	200.928	72.700	5.664
21	30.000	75.000	2400.000	3.468	3.484	402.353	2.007	246.280	230.340	16.412

(continued)

Table 8.30 (continued)

Solution	ER (%)	EL (%)	ES (rpm)	BP (%)	P (kW)	BSFC (g/kWh)	CO (%)	NO_x (ppm)	HC (ppm)	S_m (1/m)
22	0.000	46.229	1862.286	9.992	1.911	292.669	0.142	301.795	28.103	3.485
23	30.000	55.591	2302.769	10.798	2.498	334.089	0.694	232.463	102.550	7.682
24	28.036	56.853	2293.258	15.000	2.498	342.603	0.641	218.799	95.060	7.528
25	8.646	65.839	2066.941	5.881	2.623	312.512	0.852	297.845	114.695	9.680
26	30.000	25.000	2286.419	9.064	1.130	546.286	0.182	114.576	56.307	4.076
27	30.000	60.002	2309.842	11.373	2.684	333.450	0.845	232.827	116.221	8.812
28	30.000	30.523	2400.000	5.795	1.523	476.051	0.352	177.085	70.267	4.836
29	6.032	58.228	1800.000	15.000	2.546	237.983	0.201	340.836	33.380	5.274
30	23.802	52.854	2155.279	12.720	2.242	332.286	0.381	228.110	74.524	6.029

Table 8.31 Pareto-optimal solutions obtained by the Rao-2 algorithm for the biodiesel engine with an EGR system case study

Solution	ER (%)	EL (%)	ES (rpm)	BP (%)	P (kW)	BSFC (g/kWh)	CO (%)	NO_x (ppm)	HC (ppm)	S_m (1/m)
1	27.643	25.000	2169.690	15.000	1.068	563.530	0.106	86.863	48.052	3.768
2	4.999	61.961	1800.000	7.719	2.567	218.853	0.430	361.847	56.424	6.964
3	0.000	30.090	1800.000	11.065	1.326	421.273	0.125	210.554	22.778	2.304
4	30.000	25.000	1800.000	12.931	1.268	504.354	0.000	144.332	27.355	2.768
5	0.000	41.933	1800.000	15.000	1.887	317.144	0.072	290.444	16.011	2.731
6	0.000	25.000	2126.198	11.206	0.889	560.184	0.161	116.718	42.786	2.692
7	0.000	68.145	1800.000	4.758	2.734	231.291	0.710	385.811	78.768	9.217
8	30.000	75.000	2400.000	1.249	3.509	415.308	2.103	252.740	242.634	17.024
9	0.000	25.000	2333.125	15.000	0.924	580.589	0.275	121.555	49.643	3.215
10	30.000	75.000	2400.000	5.315	3.463	394.344	1.932	240.374	220.778	15.943
11	23.923	65.938	2400.000	0.000	3.089	378.219	1.622	281.555	197.310	13.504
12	11.060	34.513	2219.750	5.979	1.385	447.748	0.231	183.601	60.093	3.883
13	16.417	75.000	2365.919	6.766	3.262	389.166	1.768	261.837	200.992	15.063
14	19.874	75.000	1969.442	0.000	2.991	349.807	1.358	289.249	174.867	14.266
15	28.062	25.000	1800.000	15.000	1.300	511.886	0.015	137.623	27.951	2.886
16	30.000	63.485	2400.000	3.418	3.011	351.522	1.391	268.179	171.632	11.893
17	28.680	58.028	2400.000	0.000	2.798	365.038	1.257	279.040	163.704	10.858
18	0.000	73.337	1909.708	0.000	2.819	317.761	1.185	354.583	135.921	12.739
19	28.124	75.000	2163.837	0.000	3.172	402.749	1.676	249.998	212.051	15.574
20	7.138	64.799	1800.000	2.794	2.603	232.272	0.654	363.503	83.556	8.773
21	0.482	65.205	2125.464	10.792	2.568	326.257	0.773	308.574	96.607	8.596

(continued)

Table 8.31 (continued)

Solution	ER (%)	EL (%)	ES (rpm)	BP (%)	P (kW)	BSFC (g/kWh)	CO (%)	NO_x (ppm)	HC (ppm)	S_m (1/m)
22	0.000	68.258	1966.150	2.629	2.646	304.661	0.929	341.816	112.222	10.379
23	20.335	75.000	1813.400	1.057	3.018	286.152	1.117	328.547	142.179	13.079
24	14.712	59.237	1971.165	4.373	2.382	286.628	0.561	295.289	90.587	7.678
25	28.004	36.330	2332.027	6.442	1.687	426.060	0.349	195.064	72.706	4.923
26	19.540	62.520	2266.864	7.361	2.676	332.972	0.971	260.842	130.847	9.750
27	0.000	25.000	1800.000	6.506	1.017	484.278	0.182	167.441	30.684	2.451
28	8.723	36.454	2391.112	15.000	1.530	442.369	0.367	195.832	59.718	4.247
29	0.000	71.302	1941.942	1.358	2.753	312.919	1.076	347.664	126.050	11.715
30	0.000	45.481	1800.000	0.000	1.761	296.953	0.290	315.886	44.361	4.455

Table 8.32 Pareto-optimal solutions obtained by the Rao-3 algorithm for the biodiesel engine with an EGR system case study

Solution	ER (%)	EL (%)	ES (rpm)	BP (%)	P (kW)	BSFC (g/kWh)	CO (%)	NO_x (ppm)	HC (ppm)	S_m (1/m)
1	27.776	25.000	2160.506	15.000	1.069	563.399	0.100	86.843	47.590	3.741
2	5.221	62.031	1800.000	7.501	2.567	218.985	0.437	361.555	57.440	7.023
3	0.000	41.676	1800.000	15.000	1.877	319.184	0.072	288.920	16.015	2.711
4	30.000	25.000	1800.000	12.941	1.268	504.391	0.000	144.304	27.350	2.768
5	0.000	31.282	1800.000	11.376	1.383	407.715	0.115	220.063	21.614	2.302
6	0.000	25.000	2185.362	0.000	0.867	587.425	0.201	122.356	56.129	3.183
7	0.000	25.000	2058.976	0.000	0.813	571.259	0.179	123.675	51.654	3.007
8	30.000	74.976	2400.000	2.719	3.492	406.157	2.037	248.612	234.251	16.602
9	0.000	75.000	1800.000	6.769	3.016	251.354	0.902	385.441	95.371	11.262
10	27.008	32.422	2309.250	6.215	1.476	463.028	0.276	172.120	67.050	4.491
11	13.295	61.429	1800.000	15.000	2.702	236.356	0.233	325.905	42.575	6.098
12	30.000	56.070	2326.973	2.688	2.592	350.552	0.964	261.082	135.923	9.231
13	23.049	60.578	2016.259	4.514	2.492	301.027	0.638	275.878	104.078	8.333
14	28.420	30.193	2400.000	0.000	1.530	505.750	0.432	188.810	85.740	5.364
15	30.000	70.863	2358.541	15.000	3.136	371.147	1.307	208.894	155.082	12.168
16	0.000	75.000	1800.000	13.092	3.105	256.721	0.718	379.217	75.251	10.125
17	26.971	25.000	1961.108	15.000	1.133	545.020	0.031	102.578	38.167	3.217
18	0.000	25.000	1806.297	9.024	1.060	486.971	0.177	165.133	29.570	2.421
19	24.002	75.000	1800.000	7.229	3.148	262.848	0.838	314.591	111.122	11.424
20	30.000	75.000	2246.424	15.000	3.211	388.704	1.272	194.456	157.312	12.860
21	21.497	52.685	1920.748	3.904	2.158	289.192	0.344	283.065	72.207	6.014

(continued)

Table 8.32 (continued)

Solution	ER (%)	EL (%)	ES (rpm)	BP (%)	P (kW)	BSFC (g/kWh)	CO (%)	NO_x(ppm)	HC (ppm)	S_m (1/m)
22	30.000	38.197	2400.000	2.081	1.906	422.617	0.544	226.404	93.029	5.997
23	18.810	67.825	2362.261	15.000	2.904	358.201	1.178	241.484	140.267	10.991
24	11.084	46.727	1800.000	10.269	2.038	266.491	0.087	298.123	28.654	3.706
25	27.910	62.539	2400.000	10.446	2.874	333.629	1.134	247.722	139.246	10.282
26	17.697	75.000	2003.670	12.217	3.061	330.005	0.949	262.775	124.803	11.646
27	17.489	53.688	1924.644	15.000	2.306	292.288	0.155	265.757	45.362	4.968
28	30.000	25.000	2382.543	14.613	1.184	549.795	0.261	107.233	55.159	4.518
29	30.000	59.205	2195.955	15.000	2.567	338.083	0.556	215.769	90.315	7.505
30	0.000	37.189	1800.000	9.568	1.597	340.655	0.095	265.523	19.811	2.520

Table 8.33 Pareto-optimal solutions obtained by the ERao-1 algorithm for the biodiesel engine with an EGR system case study

Solution	ER (%)	EL (%)	ES (rpm)	BP (%)	P (kW)	BSFC (g/kWh)	CO (%)	NO_x (ppm)	HC (ppm)	S_m (1/m)
1	4.905	62.083	1800.000	8.889	2.591	218.709	0.405	361.159	53.314	6.818
2	30.000	25.000	1896.402	14.469	1.197	533.404	0.000	116.133	32.759	3.011
3	0.000	30.978	1800.000	11.099	1.365	410.586	0.117	217.822	21.961	2.302
4	28.350	25.000	2162.771	15.000	1.074	563.647	0.099	86.838	47.415	3.748
5	0.000	42.340	1800.000	15.000	1.903	313.967	0.073	292.829	16.018	2.763
6	0.000	70.533	1800.000	8.807	2.884	232.185	0.673	384.988	72.680	9.249
7	0.000	25.000	2146.485	15.000	0.898	576.037	0.193	110.696	44.714	2.880
8	30.000	75.000	2400.000	1.151	3.510	415.967	2.107	253.011	243.200	17.052
9	0.000	25.000	2098.864	6.808	0.868	553.977	0.148	121.431	43.597	2.657
10	30.000	75.000	2400.000	7.151	3.441	388.887	1.861	234.032	211.884	15.511
11	30.000	75.000	2340.388	12.402	3.307	386.760	1.548	208.125	180.848	13.992
12	30.000	75.000	2276.294	2.108	3.312	404.552	1.800	240.785	218.778	15.803
13	30.000	70.560	2172.665	9.690	3.017	347.898	1.086	228.807	145.007	11.532
14	30.000	75.000	2400.000	2.202	3.499	409.299	2.061	250.050	237.246	16.755
15	19.693	35.408	2353.998	10.646	1.573	432.536	0.322	183.774	64.227	4.508
16	30.000	75.000	2195.368	7.327	3.209	375.820	1.427	225.596	179.353	13.888
17	20.770	35.408	2228.764	11.353	1.494	435.696	0.193	169.828	56.185	4.017
18	30.000	75.000	1953.362	13.199	3.183	328.546	0.811	238.373	115.058	11.093
19	27.755	70.646	2259.762	8.277	3.066	356.821	1.296	235.811	163.673	12.444
20	30.000	75.000	1883.538	13.312	3.221	307.324	0.708	256.341	102.423	10.636
21	0.133	54.783	1800.000	6.122	2.247	231.881	0.311	357.363	40.846	5.282

(continued)

Table 8.33 (continued)

Solution	ER (%)	EL (%)	ES (rpm)	BP (%)	P (kW)	BSFC (g/kWh)	CO (%)	NO_x (ppm)	HC (ppm)	S_m (1/m)
22	0.000	45.509	1800.000	3.165	1.822	281.782	0.224	317.438	35.168	3.970
23	30.000	66.860	2175.816	10.769	2.871	335.732	0.899	230.254	126.177	10.095
24	19.801	65.965	1837.314	5.184	2.727	246.939	0.620	322.913	91.777	8.940
25	20.530	72.514	1817.279	13.278	3.112	260.956	0.573	305.449	81.686	9.557
26	24.569	66.921	2104.642	6.115	2.790	324.935	0.940	259.042	133.209	10.507
27	30.000	66.860	1925.840	11.852	2.875	286.844	0.506	262.805	85.217	8.399
28	6.151	25.000	1815.507	11.873	1.121	493.562	0.151	152.878	32.564	2.682
29	0.000	26.813	1800.000	10.080	1.167	461.400	0.158	182.901	26.604	2.359
30	30.000	75.000	2234.350	13.444	3.212	382.650	1.294	201.402	160.676	13.005

Table 8.34 Pareto-optimal solutions obtained by the ERao-2 algorithm for the biodiesel engine with an EGR system case study

Solution	ER (%)	EL (%)	ES (rpm)	BP (%)	*P* (kW)	BSFC (g/kWh)	CO (%)	NO_x (ppm)	HC (ppm)	S_m (1/m)
1	19.981	43.830	1800.000	11.854	1.992	292.973	0.000	268.552	25.473	3.315
2	27.626	25.000	2159.930	15.000	1.068	563.336	0.101	86.848	47.635	3.739
3	5.719	62.420	1800.000	8.241	2.596	218.669	0.428	360.111	56.652	7.027
4	0.000	42.023	1800.000	15.000	1.891	316.436	0.072	290.974	16.011	2.738
5	0.000	31.547	1800.000	11.482	1.396	404.875	0.113	222.114	21.359	2.304
6	0.000	25.000	2306.596	15.000	0.912	581.378	0.260	118.372	49.061	3.159
7	0.000	25.000	1955.282	1.917	0.838	540.271	0.165	134.890	43.821	2.748
8	5.493	69.067	1800.000	7.101	2.826	228.375	0.663	368.019	79.204	9.233
9	30.000	75.000	2400.000	1.179	3.510	415.782	2.106	252.935	243.041	17.044
10	18.931	63.018	2400.000	4.302	2.876	346.906	1.312	280.894	160.207	11.306
11	30.000	75.000	2369.484	15.000	3.308	393.961	1.535	198.727	176.159	13.814
12	30.000	75.000	2400.000	6.334	3.451	391.007	1.892	236.912	215.766	15.699
13	30.000	32.660	2400.000	9.460	1.596	450.843	0.341	177.085	65.335	4.768
14	14.168	75.000	1840.842	3.970	3.016	278.815	1.031	330.477	128.194	12.426
15	30.000	75.000	2400.000	11.933	3.381	386.382	1.696	215.295	191.565	14.550
16	30.000	75.000	2400.000	2.385	3.497	408.223	2.053	249.520	236.231	16.704
17	30.000	29.396	2400.000	9.681	1.442	484.241	0.307	157.702	61.944	4.592
18	24.177	25.000	1800.000	15.000	1.278	508.136	0.045	138.157	30.167	2.945
19	15.486	47.382	2343.443	8.475	2.073	356.789	0.559	244.543	86.678	6.089
20	10.053	75.000	1800.000	7.319	3.068	251.504	0.870	351.286	103.425	11.401
21	30.000	36.093	2400.000	15.000	1.706	432.594	0.349	172.468	61.777	4.903

(continued)

Table 8.34 (continued)

Solution	ER (%)	EL (%)	ES (rpm)	BP (%)	P (kW)	BSFC (g/kWh)	CO (%)	NO_x (ppm)	HC (ppm)	S_m (1/m)
22	30.000	66.023	1990.820	4.076	2.755	308.824	0.812	272.711	122.955	9.934
23	11.008	50.835	1800.000	10.932	2.214	246.659	0.124	313.702	31.758	4.269
24	19.628	73.073	2076.026	11.368	2.994	337.462	0.997	251.235	132.170	11.519
25	29.673	70.826	2388.052	13.896	3.174	366.970	1.397	217.392	162.462	12.527
26	5.529	49.194	2206.387	6.728	1.982	351.969	0.462	262.620	76.093	5.591
27	30.000	75.000	2255.346	11.547	3.237	380.905	1.396	208.966	170.978	13.485
28	28.511	25.000	2017.495	4.913	0.996	539.701	0.076	123.885	51.112	3.386
29	27.720	56.235	2143.901	5.786	2.400	325.688	0.609	250.299	102.727	7.613
30	30.000	75.000	1935.987	8.379	3.145	317.987	0.953	259.010	131.075	11.881

Table 8.35 Pareto-optimal solutions obtained by the ERao-3 algorithm for the biodiesel engine with an EGR system case study

Solution	ER (%)	EL (%)	ES (rpm)	BP (%)	P (kW)	BSFC (g/kWh)	CO (%)	NO_x (ppm)	HC (ppm)	S_m (1/m)
1	29.867	25.000	1800.000	15.000	1.311	513.937	0.000	137.682	26.716	2.848
2	0.000	30.989	1800.000	10.911	1.361	409.979	0.117	218.052	22.017	2.302
3	0.000	41.895	1800.000	15.000	1.885	317.444	0.072	290.219	16.011	2.728
4	27.357	25.000	2171.558	15.000	1.066	563.473	0.108	86.885	48.262	3.773
5	1.928	63.037	1800.000	8.534	2.609	219.055	0.446	371.823	53.969	7.043
6	30.000	75.000	2400.000	0.000	3.523	424.208	2.160	256.076	249.948	17.390
7	0.000	25.000	2139.850	15.000	0.900	575.441	0.191	110.805	44.503	2.871
8	0.000	25.000	2129.855	8.324	0.877	557.117	0.154	119.545	43.672	2.675
9	0.000	69.731	1800.000	7.005	2.828	230.796	0.697	385.914	75.945	9.305
10	30.000	71.998	2400.000	8.737	3.302	368.329	1.639	236.538	188.997	13.939
11	30.000	75.000	2102.253	8.411	3.156	359.287	1.216	229.231	159.267	12.985
12	30.000	66.596	2400.000	7.440	3.097	346.685	1.406	251.774	168.223	12.158
13	26.378	72.771	2400.000	3.961	3.351	385.223	1.847	256.839	213.765	15.279
14	9.552	75.000	2035.345	15.000	3.005	346.382	0.933	275.843	117.006	11.280
15	30.000	75.000	2400.000	3.924	3.479	400.140	1.988	244.866	227.921	16.293
16	30.000	75.000	2400.000	10.086	3.405	385.343	1.756	222.911	198.925	14.893
17	30.000	37.059	2353.974	15.000	1.710	429.216	0.302	170.096	60.094	4.749
18	20.508	69.508	2086.062	6.608	2.854	326.345	1.007	265.275	137.292	11.174
19	28.840	56.259	1940.448	0.000	2.301	312.255	0.566	287.243	101.783	7.701
20	22.916	75.000	2037.558	9.720	3.086	338.798	1.072	249.749	141.514	12.324
21	30.000	34.510	2253.641	15.000	1.537	456.784	0.167	148.476	51.689	4.130

(continued)

Table 8.35 (continued)

Solution	ER (%)	EL (%)	ES (rpm)	BP (%)	P (kW)	BSFC (g/kWh)	CO (%)	NO_x (ppm)	HC (ppm)	S_m (1/m)
22	30.000	75.000	2400.000	2.772	3.492	406.027	2.037	248.382	234.103	16.599
23	0.000	25.000	1800.000	13.410	1.153	497.183	0.195	163.656	29.460	2.522
24	0.000	55.588	1800.000	4.632	2.252	232.954	0.361	360.646	46.429	5.684
25	0.500	67.582	2098.699	15.000	2.658	337.176	0.743	304.802	91.731	8.726
26	29.881	73.546	1800.000	15.000	3.260	276.806	0.496	280.378	76.713	9.386
27	5.703	60.666	1854.740	7.149	2.469	241.250	0.445	341.493	63.029	6.914
28	18.037	50.083	1800.000	3.693	2.085	258.114	0.236	310.285	52.267	5.078
29	16.649	25.000	2323.298	15.000	1.028	561.634	0.244	100.656	55.869	4.041
30	10.647	65.579	2111.080	9.714	2.643	317.097	0.795	281.075	108.616	9.223

Table 8.36 Pareto-optimal solutions obtained by the SAP-Rao algorithm for the biodiesel engine with an EGR system case study

Solution	ER (%)	EL (%)	ES (rpm)	BP (%)	P (kW)	BSFC (g/kWh)	CO (%)	NO_x (ppm)	HC (ppm)	S_m (1/m)
1	0.000	41.608	1800.000	15.000	1.874	319.726	0.072	288.515	16.017	2.706
2	2.573	62.466	1800.000	9.000	2.598	218.952	0.418	368.509	51.897	6.838
3	30.000	25.000	2164.146	15.000	1.087	564.355	0.092	86.943	46.614	3.750
4	0.000	30.081	1800.000	10.765	1.320	420.639	0.125	210.699	22.879	2.304
5	0.039	71.261	1800.000	7.199	2.887	235.347	0.745	386.066	80.418	9.799
6	30.000	25.000	1895.402	15.000	1.207	535.929	0.000	114.501	32.575	3.033
7	0.000	25.000	2156.082	15.000	0.896	576.845	0.195	110.599	45.015	2.894
8	0.000	25.000	2006.552	5.339	0.866	540.802	0.144	128.730	41.406	2.600
9	30.000	75.000	2400.000	0.000	3.523	424.208	2.160	256.076	249.948	17.390
10	21.019	74.220	2400.000	8.614	3.298	380.535	1.746	250.382	197.299	14.726
11	30.000	74.574	2400.000	15.000	3.324	390.991	1.582	202.916	178.579	13.891
12	16.915	75.000	2400.000	5.040	3.336	395.523	1.903	268.393	214.100	15.729
13	30.000	75.000	2400.000	1.607	3.505	412.975	2.087	251.746	240.594	16.922
14	29.825	75.000	2400.000	4.381	3.472	398.013	1.969	243.709	225.472	16.174
15	19.350	65.053	2229.668	10.244	2.735	332.716	0.936	252.536	125.951	9.861
16	30.000	67.251	2349.973	12.677	3.008	349.942	1.183	225.179	145.346	11.154
17	29.077	51.512	2136.747	7.645	2.215	332.489	0.414	236.483	82.577	6.201
18	10.248	68.575	1800.000	9.327	2.865	227.390	0.570	351.145	73.983	8.769
19	30.000	75.000	2255.077	15.000	3.216	389.408	1.290	194.374	158.665	12.925
20	0.000	57.276	1800.000	9.685	2.402	225.008	0.290	363.642	36.694	5.371
21	2.573	48.956	1800.000	10.333	2.096	254.435	0.141	325.837	25.783	3.819

(continued)

Table 8.36 (continued)

Solution	ER (%)	EL (%)	ES (rpm)	BP (%)	P (kW)	BSFC (g/kWh)	CO (%)	NO_x (ppm)	HC (ppm)	S_m (1/m)
22	0.000	26.266	1800.000	8.369	1.110	466.624	0.165	178.753	27.945	2.380
23	23.501	67.410	1899.793	7.779	2.812	270.468	0.640	292.914	97.236	9.191
24	30.000	75.000	1893.304	4.722	3.113	311.412	1.039	280.125	140.763	12.457
25	0.000	52.570	1800.000	12.297	2.262	243.823	0.170	345.434	24.795	4.182
26	29.825	53.749	2305.836	6.440	2.445	339.513	0.740	245.849	110.963	7.828
27	18.792	32.541	2150.875	9.740	1.324	457.749	0.130	157.341	51.722	3.591
28	26.329	46.882	2168.578	13.360	2.017	358.695	0.255	207.096	62.543	5.007
29	14.377	40.087	2400.000	13.640	1.756	402.843	0.428	210.913	67.840	4.899
30	20.360	74.794	2400.000	4.152	3.371	396.828	1.937	263.050	220.051	15.967

Similar to the previous case studies, the best solutions from these Pareto-fronts are identified. Solution 2 of the Jaya algorithm, Solution 4 of the AMTPG-Jaya algorithm, Solution 1 of the Rao-1 algorithm, Solution 3 of the Rao-2 algorithm, Solution 24 of the Rao-3 algorithm, Solution 2 of the ERao-1 algorithm, Solution 1 of the ERao-2 algorithm, Solution 3 of the ERao-3 algorithm, and Solution 1 of the SAP-Rao algorithm are identified as the best solutions from the respective algorithm Pareto-front. Now, these best solutions are compared with those of the NSGA-II algorithm in Table 8.37.

In Table 8.37, the solutions with the best average rank from the Pareto-optimal solutions of the various algorithms are compared with that reported for the NSGA-II algorithm. The AMPG-Jaya algorithm solution achieved a higher power output, the Rao-3 solution has the least BSFC, the ERao-3 solution has the least HC emission, the Rao-3 solution has the least smoke opacity, and the ERao-1 solution has the least CO and NO_x emission values. Here, an observation can be made that these solutions are non-dominated. Hence, to identify the best solution among the solutions of different algorithms, which has the best compromise among the objectives, the MADM methods based average ranks are calculated and presented in Table 8.38.

The Spearman's correlation coefficients for different pairs of MADM methods rankings for different algorithm's solutions are shown in Table 8.39. The ranks given by the TOPSIS and MTOPSIS methods for each solution are identical. Furthermore, the Spearman's correlation for all the pairs of decision-making methods (except with the VIKOR method) is greater than 0.5. Spearman's correlation values for the pairs consisting of the VIKOR method are either negative or nearer to zero. Hence, the ranks suggested by the VIKOR method cannot be considered for calculating the average ranks. Now, the corrected average ranks are calculated, excluding the ranks suggested by the VIKOR method, and presented in Table 8.38 as the corrected ranks. The ERao-3 algorithm solution has the least average rank, which is 1.6. All the MADM methods, except the WPM and SAW, have ranked the ERao-2 solution as one. The WPM method ranked this solution as two, and the SAW method ranked it four. Hence, the ERao-2 solution has achieved the least rank.

The ERao-2 solution has the best compromise among the six objectives. Thus, it can be considered as the best solution among the compared solutions. The ERao-2 solution has better values in five objectives (power output, BSFC, CO, HC, and S_m) compared to those of the NSGA-II solution. Similarly, the ERao-2 solution has better values in four objectives (NO_x, CO, HC, and S_m) compared to those of the Rao-3 solution. It has better values in four objectives (power output, BSFC, CO, and NO_x) compared to those of the SAP-Rao and ERao-3 solutions. It has better values for the objectives BSFC, CO, HC, and S_m, when compared to those of the AMTPG-Jaya algorithm solution. Similarly, it has better values in three objectives (Power output, BSFC, and CO) compared to those of the Jaya, Rao-1, and Rao-2 solutions. Also, it has better values for the objectives power output, BSFC, and HC, when compared to those of the ERao-1 solution. Hence, the ERao-2 solution can be considered as the best solution.

The spacing values of the Pareto-fronts obtained by different algorithms in the MOO scenario are presented in Table 8.40. The ERao-2 algorithm has achieved better

Table 8.37 Best solutions obtained by various algorithms in MOO scenario of the biodiesel engine with an EGR system case study

Algorithm	ER (%)	EL (%)	ES (rpm)	BP (%)	P (kW)	BSFC (g/kWh)	CO (%)	NO_x (ppm)	HC (ppm)	S_m (1/m)
NSGA-II	6.6	40	2125	10.9	0.32	391.83	0.19	215.19	49.01	3.67
Jaya	0	30.9842	1800	10.893	1.3609	409.99	0.11727	218.026	22.02787	2.30211
Rao-1	0	30.5670	1800	10.864	1.3427	414.95	0.12068	214.639	22.40188	**2.30200**
Rao-2	0	30.0902	1800	11.064	1.3262	421.27	0.12475	210.554	22.77768	2.30385
Rao-3	11.084	46.7273	1800	10.269	2.0385	**266.49**	0.08665	298.123	28.65417	3.70621
AMTPG-Jaya	25.712	46.0734	1878.2	13.441	**2.0698**	312.31	0.00040	245.005	32.13142	3.68890
SAP-Rao	0	41.6085	1800	15	1.8739	319.73	0.07211	288.515	16.01710	2.70602
ERao-1	30	25.0000	1896.4	14.468	1.1969	533.40	**0.00004**	**116.133**	32.75884	3.01143
ERao-2	19.9815	43.8303	1800	11.853	1.9924	292.97	0.00018	268.552	25.47298	3.31472
ERao-3	0	41.8954	1800	15	1.8854	317.44	0.07232	290.219	**16.01113**	2.72791

Source NSGA-II:—Jaliliantabar et al. (2019)
Result in boldface indicates better values

Table 8.38 Ranks suggested by the MADM methods for different algorithm solutions presented in Table 8.37

Algorithm	SAW	WPM	TOPSIS	MTOPSIS	VIKOR	PROMETHEE	COPRAS	GRA	AR	CR	FR
NSGA-II	10	10	10	10	10	10	10	10	10	10.0	10
Jaya	5	7	7	7	1	4.5	7	7	5.69	6.4	7
Rao-1	8	8	8	8	1	6	8	8	6.88	7.7	8
Rao-2	9	9	9	9	1	8	9	9	7.88	8.9	9
Rao-3	6	6	6	6	9	8	6	6	6.63	6.3	6
AMTPG-Jaya	7	3	2	2	7	4.5	2	4	3.94	3.5	3
SAP-Rao	3	4	4	4	5	2.5	3	3	3.56	3.4	2
ERao-1	1	1	3	3	8	8	4	5	4.13	3.6	4
ERao-2	4	2	1	1	1	1	1	1	1.5	1.6	1
ERao-3	2	5	5	5	6	2.5	5	2	4.06	3.8	5

AR—Average Rank; CR—Corrected Rank; FR—Final Rank

Table 8.39 Spearman's rank correlation coefficients between different pairs of MADM method's rankings presented in Table 8.38

Method	SAW	WPM	TOPSIS	MTOPSIS	VIKOR	PROMETHEE	COPRAS	GRA
SAW	1	0.794	0.685	0.685	0.006	0.543	0.661	0.770
WPM	0.794	1	0.964	0.964	−0.056	0.531	0.927	0.830
TOPSIS	0.685	0.964	1	1	0.025	0.661	0.988	0.891
MTOPSIS	0.685	0.964	1	1	0.025	0.661	0.988	0.891
VIKOR	0.006	−0.056	0.025	0.025	1	0.526	0.063	0.113
PROMETHEE	0.543	0.531	0.661	0.661	0.526	1	0.729	0.846
COPRAS	0.661	0.927	0.988	0.988	0.063	0.729	1	0.915
GRA	0.770	0.830	0.891	0.891	0.113	0.846	0.915	1

spacing values in terms of mean value, and the Rao-3 algorithm has achieved better spacing values in terms of best, and the ERao-3 algorithm has the least standard deviation. The mean of the spacing values achieved by the ERao-2 algorithm is 11.8, 2.8, 3.7, 12.9, 6, 0.7, 1.7 and 4.7% lesser when compared to that of the Jaya, Rao-1, Rao-2, Rao-3, AMTPG-Jaya, SAP-Rao, ERao-1, and ERao-3 algorithms, respectively.

The coverage metric values of the Pareto-fronts obtained by the Jaya algorithm, Rao algorithms, and their modified versions are presented in Table 8.41. The ERao-1 algorithm has achieved better coverage values than the other algorithms compared. Only two solutions of the ERao-1 algorithm are dominated by other algorithm Pareto-fronts. From the coverage values, it can be observed that four solutions from the Pareto-front of the SAP-Rao are dominated, and six solutions from the Pareto-fronts of the Jaya, Rao-1, Rao-3, AMTPG-Jaya, and ERao-3 algorithms are dominated. Similarly, seven solutions of the Pareto-front achieved by the ERao-2 algorithm are dominated, and 13 solutions are dominated from the Pareto-front achieved by the Rao-2 algorithm. The performances of the proposed modified algorithms are better or competitive to that of the basic algorithms in terms of coverage and spacing values.

From the computational results of the biodiesel engine with an EGR system case study, it can be observed that the solutions obtained by the proposed algorithms are better than those reported in the literature. The solution of the ERao-2 algorithm has a better compromise among the six objectives. By the solution of the ERao-1 solution, the output power is increased by 522.6% than that of the NSGA-II algorithm solution. In addition, the objectives BSFC, CO, HC, and S_m values are reduced by 25.2, 99, 48 and 9.7%, respectively, when compared to those of the NSGA-II solution. The ERao-2 algorithm has achieved the best mean spacing value (0.13532), which is much better than that achieved by other algorithms compared. The ERao-1 algorithm has achieved better performance in terms of coverage values. In addition, the performances of the modified versions in this case study are better or competitive to those of the basic algorithms as well as the NSGA-II algorithm.

Table 8.40 Spacing values of the Pareto-fronts obtained by the proposed algorithms in MOO for the biodiesel engine with an EGR system case study

Algorithm	Jaya	Rao-1	Rao-2	Rao-3	AMTPG-Jaya	SAP-Rao	ERao-1	ERao-2	ERao-3
Best	0.11925	0.11453	0.09774	**0.08456**	0.10115	0.09628	0.09932	0.10129	0.11349
Worst	0.19209	0.18317	0.20105	0.21666	**0.16852**	0.17344	0.18537	0.17749	0.17338
Mean	0.15338	0.13915	0.14050	0.15537	0.14391	0.13626	0.13773	**0.13532**	0.14202
SD	0.02864	0.02076	0.03055	0.03471	0.02102	0.02328	0.02893	0.02066	**0.01989**

Result in boldface indicates better values

Table 8.41 Coverage (%) values of the Pareto-fronts obtained by the proposed algorithms in MOO for the biodiesel engine with an EGR system case study

Algorithm	Jaya	Rao-1	Rao-2	Rao-3	AMTPG-Jaya	SAP-Rao	ERao-1	ERao-2	ERao-3
Jaya	–	13.33	26.67	6.67	13.33	10.00	6.67	6.67	13.33
Rao-1	6.67	–	20.00	10.00	3.33	13.33	3.33	3.33	6.67
Rao-2	3.33	3.33	–	6.67	3.33	0.00	0.00	0.00	0.00
Rao-3	10.00	16.67	36.67	–	6.67	3.33	3.33	3.33	10.00
AMTPG-Jaya	20.00	13.33	26.67	6.67	–	6.67	3.33	6.67	3.33
SAP-Rao	16.67	20.00	33.33	6.67	6.67	–	3.33	10.00	6.67
ERao-1	16.67	16.67	43.33	20.00	20.00	10.00	–	23.33	20.00
ERao-2	6.67	10.00	26.67	3.33	3.33	6.67	3.33	–	3.33
ERao-3	10.00	13.33	26.67	16.67	16.67	13.33	3.33	16.67	–

8.4 Process Optimization of a Microalgae-Based Biomass Cultivation Process

The description of the selected microalgae-based biomass cultivation process is presented in Sect. 2.3.4. In this case study, a multi-objective optimization case study of fertilizer-assisted cultivation of the Nannochloropsis species biomass production for biofuel feedstock is considered. This case study was presented by Banerjee et al. (2016).

The objective functions of this case study include the maximization of biomass production (BMP), eicosapentaenoic acid (EPA), and lipid productions (TLP). The design variables of this case study are cultivation light intensity (X_1 in μmol/m^2/s), temperature (X_2in °C), and concentrations of NaCl (X_3 in M), $NaHCO_3$(X_4 in g/L) and NPK-10:26:26 fertilizers (X_5 in g/L). The ranges of design variables are as follows: $25 \leq X_1 \leq 125$, $17 \leq X_2 \leq 29$, $0.25 \leq X_3 \leq 1.25$, $0.05 \leq X_4 \leq 1.85$ and $0.15 \leq X_5 \leq 1.1$. The regression models of the objectives of this case study are presented in Sect. 2.3.4.

Banerjee et al. (2016) presented the optimum design parameters for this case study using the NSGA-II algorithm. Now, the proposed algorithms, along with basic Jaya and Rao algorithms, are used to find the optimal parameters through multi-objective optimization. Furthermore, the Pareto-optimal solutions obtained by the NSGA-II algorithm were not reported in the literature. However, the performances of the proposed algorithms are compared with those achieved by the basic Jaya and Rao algorithms. Furthermore, the solutions obtained by various algorithms are non-dominated. Hence, the average rank method was used to identify the best solution from the Pareto-front of the respective algorithm. Furthermore, the performances of the proposed algorithms are compared with the performances of the basic Jaya and Rao algorithms in terms of hypervolume, spacing andcoverage performance metrics. The hypervolume and spacing performance metrics are calculated for ten independent runs and the statistical results are presented in terms of best, worst, mean and standard deviation (SD).

The Pareto-optimal solutions achieved by the Jaya algorithm, Rao algorithms and their modified versions in this case study are presented in Tables 8.42, 8.43, 8.44, 8.45, 8.46, 8.47, 8.48, 8.49 and 8.50 Here, the biomass productivity is varied between 0.731 and 0.7471 g/L, the lipid generation is varied from 45.11 to 47.12%, and the EPA generation is varied from 21.3 to 23.51%. Similar to the previous case studies, the best solutions from these Pareto-fronts are identified. Solution 5 of the

Table 8.42 Pareto-optimal solutions obtained by the Jaya algorithm for the case study of a microalgae-based biomass cultivation process

Solution	X_1	X_2	X_3	X_4	X_5	BMP (g/L)	TLP (%)	EPA (%)
1	125.0000	21.2426	0.2904	1.3984	0.6496	0.7410	47.1162	21.9670
2	125.0000	21.1474	0.3466	0.9651	0.8641	0.7471	46.2141	21.5826
3	101.4912	17.0000	0.2500	1.1607	0.6562	0.7320	45.1821	23.5014
4	125.0000	22.1296	0.3653	0.9700	0.8582	0.7471	46.2149	21.3031
5	125.0000	17.0000	0.2865	1.1115	0.7952	0.7452	46.5051	22.9416
6	123.6404	17.0000	0.2654	1.1757	0.7053	0.7429	46.6797	23.1403
7	125.0000	22.6981	0.3541	1.1111	0.7400	0.7455	46.8668	21.4290
8	125.0000	18.3874	0.2871	1.4147	0.6093	0.7392	46.9514	22.6363
9	124.9035	18.0441	0.2500	1.3972	0.5942	0.7384	46.8511	22.8469
10	125.0000	21.4339	0.3403	1.1318	0.7927	0.7464	46.7376	21.7001
11	125.0000	20.6925	0.2887	1.3413	0.6516	0.7417	47.1111	22.1250
12	125.0000	20.5394	0.3199	1.1934	0.8588	0.7462	46.4346	21.8569
13	105.2143	17.0000	0.2500	1.1571	0.6512	0.7336	45.4627	23.4904
14	123.9731	19.4937	0.3351	0.8201	0.8394	0.7460	46.0840	22.0355
15	112.7859	17.0000	0.2500	1.1481	0.6548	0.7371	45.9946	23.4198
16	110.9045	17.0000	0.2500	1.1481	0.6310	0.7356	45.8319	23.4397
17	125.0000	17.0000	0.3250	0.9773	0.8776	0.7457	45.9523	22.6162
18	125.0000	20.3025	0.2645	1.2128	0.6599	0.7429	47.0498	22.3302
19	122.2723	19.9065	0.2677	1.1183	0.7326	0.7440	46.7749	22.4524
20	125.0000	21.2644	0.3610	1.2117	0.6966	0.7441	47.0236	21.7617
21	104.3744	17.0000	0.2500	1.0788	0.6564	0.7336	45.3130	23.4909
22	121.2119	17.0000	0.2500	1.3724	0.7201	0.7404	46.6147	23.2149
23	125.0000	20.3670	0.3263	1.3712	0.7941	0.7444	46.7952	21.9599
24	109.5265	17.0000	0.2500	1.0581	0.7827	0.7381	45.5618	23.3661
25	125.0000	19.6649	0.2905	1.1817	0.8169	0.7459	46.6561	22.2529
26	117.8767	17.0000	0.2500	1.2384	0.6885	0.7396	46.4017	23.3266
27	122.4247	18.6161	0.3032	1.0617	0.8615	0.7452	46.1623	22.4289
28	108.5403	17.0000	0.2500	1.1589	0.7177	0.7365	45.7237	23.4485
29	107.1290	17.0000	0.2500	1.1390	0.6514	0.7346	45.5864	23.4786
30	122.4454	17.9353	0.2773	1.1639	0.7455	0.7437	46.6609	22.8750

Table 8.43 Pareto-optimal solutions obtained by the AMTPG-Jaya algorithm for the case study of a microalgae-based biomass cultivation process

Solution	X_1	X_2	X_3	X_4	X_5	BMP (g/L)	TLP (%)	EPA (%)
1	125.0000	20.9305	0.3461	0.9710	0.8606	0.7471	46.2436	21.6447
2	100.8574	17.0000	0.2500	1.1464	0.6540	0.7317	45.1126	23.5018
3	125.0000	21.6497	0.3065	1.3463	0.6535	0.7417	47.1185	21.8378
4	125.0000	22.7495	0.3339	1.0552	0.8188	0.7467	46.5341	21.3666
5	101.8510	17.0000	0.2500	1.1703	0.6298	0.7314	45.1770	23.4970
6	125.0000	20.4115	0.2616	1.1333	0.7823	0.7457	46.7872	22.2285
7	125.0000	18.4588	0.3364	1.1909	0.7254	0.7446	46.8865	22.4897
8	125.0000	18.9523	0.2500	1.2596	0.6403	0.7416	46.9716	22.6900
9	125.0000	21.7584	0.3514	0.9477	0.8351	0.7471	46.3570	21.4834
10	118.5807	17.0000	0.3122	1.2156	0.6259	0.7387	46.4317	23.0909
11	125.0000	21.3996	0.2723	1.1752	0.7002	0.7443	47.0370	22.0373
12	122.3157	17.3352	0.2746	1.1137	0.7762	0.7439	46.4697	22.9974
13	125.0000	22.3862	0.3158	1.2864	0.6755	0.7429	47.0906	21.6501
14	125.0000	19.3623	0.2550	1.2769	0.6481	0.7419	47.0188	22.5729
15	106.4583	17.0000	0.2517	1.1979	0.6561	0.7341	45.6046	23.4758
16	125.0000	18.1483	0.2500	1.2573	0.6094	0.7404	46.8531	22.8704
17	108.9231	17.0000	0.2500	1.1077	0.6429	0.7352	45.6681	23.4619
18	125.0000	20.3645	0.2911	1.0541	0.7973	0.7465	46.6750	22.1108
19	118.4000	17.0000	0.2521	1.0093	0.6940	0.7408	46.1989	23.2958
20	122.2519	17.0000	0.2500	1.1927	0.5830	0.7384	46.4557	23.2056
21	125.0000	17.5127	0.2803	1.1610	0.7740	0.7450	46.6681	22.8705
22	113.5406	17.0000	0.2500	1.0879	0.6365	0.7371	45.9504	23.4043
23	125.0000	17.1629	0.2736	1.2161	0.6848	0.7429	46.8135	23.0409
24	125.0000	20.3038	0.2875	1.1791	0.7476	0.7453	46.9330	22.2065
25	105.1946	17.0000	0.2500	1.1291	0.6427	0.7335	45.4177	23.4895
26	118.5759	17.0000	0.2500	1.0554	0.6631	0.7400	46.2575	23.3150
27	110.1718	17.0000	0.2500	1.1488	0.6639	0.7362	45.8274	23.4514
28	104.1054	17.0000	0.2500	1.0773	0.6453	0.7332	45.2738	23.4914
29	109.7695	17.0000	0.2500	1.1035	0.6581	0.7360	45.7436	23.4543
30	114.5881	17.0000	0.2500	1.0809	0.6642	0.7384	46.0452	23.3896

Jaya algorithm, Solution 26 of the AMTPG-Jaya algorithm, Solution 19 of the Rao-1 algorithm, Solution 6 of the Rao-2 algorithm, Solution 19 of the Rao-3 algorithm, Solution 17 of the ERao-1 algorithm, Solution 13 of the ERao-2 algorithm, Solution 9 of the ERao-3 algorithm, and Solution 22 of the SAP-Rao algorithm are identified as the best solutions from the respective algorithm's Pareto-front. Now, these best solutions are compared in Table 8.51.

Table 8.44 Pareto-optimal solutions obtained by the Rao-1 algorithm for the case study of a microalgae-based biomass cultivation process

Solution	X_1	X_2	X_3	X_4	X_5	BMP (g/L)	TLP (%)	EPA (%)
1	125.0000	21.3965	0.3447	0.9615	0.8585	0.7471	46.2462	21.5433
2	101.3886	17.0000	0.2500	1.1460	0.6485	0.7318	45.1473	23.5013
3	101.5146	17.0000	0.2500	1.0443	0.6691	0.7327	45.0537	23.4922
4	125.0000	21.7300	0.2819	1.3317	0.6505	0.7417	47.1169	21.9068
5	125.0000	23.7469	0.3355	1.0469	0.7852	0.7462	46.6322	21.1890
6	125.0000	19.5936	0.2757	1.2321	0.6863	0.7435	47.0285	22.4510
7	125.0000	18.3735	0.2873	1.0538	0.7738	0.7457	46.6537	22.6301
8	125.0000	22.3511	0.3263	0.9864	0.8407	0.7470	46.3769	21.4306
9	125.0000	19.9073	0.2993	0.9427	0.7063	0.7451	46.7661	22.2696
10	125.0000	22.3531	0.3048	1.1725	0.6914	0.7442	47.0431	21.7111
11	104.4553	17.0000	0.2500	1.0903	0.6587	0.7337	45.3368	23.4915
12	123.0035	17.7472	0.3020	1.0385	0.7592	0.7446	46.5297	22.7932
13	125.0000	21.2037	0.2983	1.0498	0.6925	0.7448	46.9529	22.0000
14	106.2541	17.0000	0.2500	1.1160	0.6620	0.7345	45.5081	23.4837
15	122.0862	17.2891	0.2812	1.3828	0.6521	0.7395	46.7486	23.0240
16	117.0355	17.6248	0.2535	1.0969	0.7834	0.7416	46.1553	23.1015
17	125.0000	20.5471	0.2901	1.3631	0.6643	0.7419	47.1092	22.1461
18	110.5626	17.1974	0.2500	1.2364	0.6519	0.7356	45.9469	23.3933
19	123.0035	17.7207	0.2811	1.3101	0.7150	0.7423	46.8012	22.9055
20	119.0762	17.2408	0.2513	1.1893	0.7403	0.7415	46.4141	23.2144
21	115.9004	17.0000	0.2500	1.1715	0.6617	0.7385	46.2226	23.3715
22	108.7157	17.0387	0.2500	1.1121	0.6472	0.7353	45.6708	23.4551
23	125.0000	20.9354	0.3379	1.1936	0.8046	0.7461	46.7194	21.8043
24	114.9694	17.0489	0.2674	1.3713	0.6746	0.7371	46.3161	23.2786
25	125.0000	21.7902	0.2907	1.2692	0.6622	0.7427	47.1078	21.8814
26	112.4077	17.0000	0.2500	1.0611	0.6725	0.7377	45.8838	23.4177
27	123.0035	17.9714	0.2853	1.3207	0.6403	0.7404	46.8440	22.8395
28	110.4306	17.0000	0.2500	1.0534	0.6572	0.7364	45.7250	23.4421
29	108.6494	17.0000	0.2500	1.3592	0.6174	0.7326	45.7997	23.4232
30	116.6414	17.0000	0.2500	1.1144	0.7404	0.7407	46.1766	23.3225

In Table 8.51, the solutions with the best average rank from the Pareto-optimal solutions of the various algorithms are presented. Figure 8.1 presents the Pareto-fronts achieved by the AMTPG-Jaya, SAP-Rao, and elitist Rao algorithms, including the optimal solutions reported for the Jaya, and Rao algorithms. The Jaya algorithm solution has achieved higher biomass productivity, the SAP-Rao algorithm solution

Table 8.45 Pareto-optimal solutions obtained by the Rao-2 algorithm for the case study of a microalgae-based biomass cultivation process

Solution	X_1	X_2	X_3	X_4	X_5	BMP (g/L)	TLP (%)	EPA (%)
1	100.9575	17.0000	0.2500	1.1470	0.6611	0.7319	45.1302	23.5016
2	125.0000	21.1658	0.2500	1.3561	0.6599	0.7414	47.0986	22.1371
3	125.0000	21.3381	0.2500	0.9529	0.8439	0.7463	46.3402	21.9128
4	125.0000	22.5894	0.2500	1.0018	0.8117	0.7461	46.5431	21.6920
5	99.9722	17.0000	0.2500	1.0560	0.6961	0.7325	44.9585	23.4852
6	117.5120	17.0000	0.2500	0.9924	0.8103	0.7419	45.8898	23.1973
7	125.0000	17.4875	0.2500	1.1197	0.7359	0.7443	46.6991	23.0205
8	125.0000	20.6525	0.2500	1.2662	0.6756	0.7428	47.0700	22.2835
9	125.0000	19.5495	0.2500	1.0000	0.7955	0.7459	46.5736	22.4390
10	125.0000	18.7056	0.2500	1.0121	0.7483	0.7452	46.6751	22.7033
11	125.0000	20.5581	0.2500	0.9452	0.8728	0.7462	46.1487	22.0303
12	125.0000	17.7822	0.2500	1.1885	0.7088	0.7436	46.8240	22.9679
13	105.9998	17.0000	0.2500	1.0690	0.7069	0.7355	45.4465	23.4671
14	125.0000	17.7965	0.2500	1.5457	0.6838	0.7391	46.8894	22.8596
15	110.3063	17.0000	0.2500	0.9452	0.6817	0.7370	45.5722	23.4151
16	124.3719	18.5339	0.2500	1.4333	0.6774	0.7405	46.9533	22.7538
17	125.0000	22.6996	0.2500	1.1838	0.6874	0.7435	47.0301	21.8120
18	112.3285	17.0000	0.2500	1.1401	0.7182	0.7383	45.9618	23.4069
19	124.6895	19.5487	0.2500	1.3191	0.6400	0.7411	47.0176	22.5414
20	123.6743	19.3461	0.2500	1.2427	0.7091	0.7431	46.9152	22.6208
21	115.3399	17.0000	0.2500	1.0645	0.7396	0.7402	46.0437	23.3396
22	102.1732	17.0000	0.2500	1.1860	0.6673	0.7324	45.2740	23.4991
23	101.2771	17.0000	0.2500	1.0668	0.7004	0.7332	45.0796	23.4851
24	120.0439	17.0000	0.2500	1.1604	0.6470	0.7399	46.4430	23.2899
25	107.3820	17.0000	0.2500	1.2320	0.6664	0.7345	45.7070	23.4718
26	103.8756	17.0000	0.2500	1.0130	0.6754	0.7340	45.2006	23.4801
27	107.5547	17.0000	0.2500	1.3472	0.6633	0.7335	45.7836	23.4473
28	107.5884	17.0000	0.2500	1.1456	0.7252	0.7363	45.6361	23.4510
29	112.0074	17.0000	0.2500	1.0836	0.7513	0.7388	45.8333	23.3794
30	116.3680	17.0000	0.2500	1.1667	0.7057	0.7397	46.2504	23.3525

has a higher lipid production percentage, and the ERao-3 solution has a higher EPA generation percentage.

These solutions are non-dominated, and to identify the best solution among the solutions of different algorithms, which has the best compromise among the objectives, the MADM methods based average ranks are calculated and presented in Table 8.52. The Spearman's correlation coefficients for different pairs of rankings

Table 8.46 Pareto-optimal solutions obtained by the Rao-3 algorithm for the case study of a microalgae-based biomass cultivation process

Solution	X_1	X_2	X_3	X_4	X_5	BMP (g/L)	TLP (%)	EPA (%)
1	125.0000	21.6905	0.3045	1.3420	0.6383	0.7413	47.1150	21.8360
2	125.0000	22.0396	0.3182	0.9025	0.8602	0.7470	46.1803	21.4638
3	103.0106	17.0000	0.2500	1.0242	0.6506	0.7329	45.1210	23.4878
4	102.6524	17.0000	0.2501	1.1259	0.6522	0.7326	45.2311	23.4990
5	125.0000	18.8661	0.2523	1.3691	0.7036	0.7423	46.9894	22.6657
6	125.0000	19.3395	0.3043	0.9607	0.7962	0.7464	46.5362	22.2890
7	125.0000	18.4287	0.2934	0.8792	0.8130	0.7460	46.2879	22.4858
8	125.0000	17.0000	0.2935	1.5403	0.7020	0.7398	46.8312	22.9011
9	105.4449	17.0000	0.2500	1.1974	0.6482	0.7334	45.5157	23.4871
10	125.0000	20.7472	0.3478	1.1550	0.6864	0.7443	47.0135	21.9386
11	125.0000	17.8788	0.2822	1.2620	0.6408	0.7417	46.9069	22.8334
12	106.6410	17.0000	0.2500	1.0147	0.7429	0.7364	45.3872	23.4264
13	123.9464	19.3048	0.2879	0.7791	0.7126	0.7440	46.3887	22.4068
14	125.0000	22.4777	0.3026	1.1030	0.7472	0.7457	46.8873	21.6475
15	109.4060	17.0785	0.2500	1.1384	0.7705	0.7377	45.6840	23.3725
16	118.0435	17.0000	0.2500	1.5760	0.6919	0.7357	46.4390	23.2000
17	125.0000	21.3115	0.3503	0.8689	0.8250	0.7469	46.3148	21.5851
18	112.8090	17.0000	0.2673	1.1828	0.7581	0.7391	45.9900	23.3062
19	123.8676	17.0000	0.2810	1.1183	0.6252	0.7411	46.6109	23.0823
20	122.0134	20.0628	0.3297	1.0675	0.7788	0.7451	46.6034	22.1496
21	112.4611	17.0000	0.2574	1.2345	0.6728	0.7370	46.0735	23.3923
22	123.3290	17.8696	0.2500	1.1353	0.6400	0.7414	46.6968	23.0034
23	114.3495	17.0000	0.2614	1.2214	0.6931	0.7384	46.1919	23.3479
24	107.1340	17.0000	0.2500	1.2454	0.6851	0.7347	45.7072	23.4688
25	125.0000	21.4680	0.2512	1.3048	0.6902	0.7427	47.0803	22.0712
26	114.8987	17.0012	0.2772	0.9941	0.7247	0.7402	45.9779	23.2480
27	125.0000	22.1379	0.2973	1.1090	0.7155	0.7451	46.9736	21.7744
28	109.3780	17.0000	0.2529	1.3061	0.6512	0.7345	45.8867	23.4313
29	112.6610	17.0000	0.2500	1.1346	0.6030	0.7355	45.8713	23.4054
30	125.0000	17.0000	0.2603	1.2146	0.6180	0.7407	46.7216	23.1188

for different algorithm's solutions are shown in Table 8.53. The ranks given by the TOPSIS and MTOPSIS methods and WPM and COPRAS methods for each solution are identical. Furthermore, the Spearman's correlation for all the pairs of decision-making methods formed by the SAW, WPM, TOPSIS, MTOPSIS, and COPRAS methods is greater than 0.5. However, the Spearman's correlation values for the VIKOR-TOPSIS, VIKOR—MTOPSIS, and VIKOR-GRA pairs are less than 0.5.

Table 8.47 Pareto-optimal solutions obtained by the ERao-1 algorithm for the case study of a microalgae-based biomass cultivation process

Solution	X_1	X_2	X_3	X_4	X_5	BMP (g/L)	TLP (%)	EPA (%)
1	101.0886	17.0000	0.2500	1.1427	0.6609	0.7320	45.1358	23.5015
2	125.0000	21.6117	0.3026	1.3323	0.6486	0.7417	47.1185	21.8648
3	125.0000	21.8584	0.3390	0.9676	0.8721	0.7471	46.1630	21.4243
4	100.7535	17.0000	0.2500	1.0690	0.6613	0.7321	45.0177	23.4969
5	103.6297	17.0000	0.2500	1.2669	0.6208	0.7313	45.3838	23.4789
6	120.5450	17.0400	0.2500	1.2386	0.6670	0.7402	46.5551	23.2646
7	123.7574	18.2917	0.2879	1.0709	0.7452	0.7448	46.6815	22.7148
8	123.7711	17.5470	0.2664	1.4219	0.6242	0.7388	46.8230	22.9481
9	109.8401	17.0000	0.2500	1.0267	0.7326	0.7377	45.6455	23.4116
10	113.3839	17.0245	0.2500	1.1501	0.6640	0.7376	46.0478	23.4054
11	125.0000	21.5395	0.3339	1.0043	0.8367	0.7470	46.4278	21.6053
12	111.9448	17.0276	0.2500	1.1746	0.6514	0.7365	45.9646	23.4235
13	125.0000	17.9650	0.2500	1.1241	0.6490	0.7424	46.7913	22.9379
14	123.7574	18.6865	0.2902	1.0049	0.8489	0.7458	46.2428	22.4412
15	123.5544	20.8428	0.2748	1.0131	0.8328	0.7459	46.4100	22.0196
16	110.6783	17.0000	0.2500	1.1905	0.6346	0.7354	45.8649	23.4417
17	125.0000	17.4446	0.2613	1.1088	0.6946	0.7436	46.7426	23.0149
18	125.0000	22.3978	0.3058	1.1217	0.7313	0.7454	46.9423	21.6714
19	124.4933	19.4536	0.3118	1.0306	0.8536	0.7465	46.3169	22.1582
20	107.4926	17.0000	0.2500	1.2361	0.6553	0.7343	45.7079	23.4705
21	103.7171	17.0000	0.2500	1.1019	0.6422	0.7329	45.2701	23.4945
22	124.8521	20.5644	0.2609	1.2806	0.7142	0.7436	47.0336	22.2512
23	108.2576	17.0000	0.2500	1.2000	0.6812	0.7355	45.7513	23.4655
24	104.5102	17.0000	0.2500	1.1006	0.6422	0.7333	45.3300	23.4913
25	123.7710	17.0000	0.2500	1.3310	0.6698	0.7408	46.7662	23.1783
26	125.0000	21.0258	0.2500	1.3227	0.6405	0.7412	47.0886	22.1813
27	124.5728	18.7620	0.2883	1.3876	0.6310	0.7402	46.9914	22.5751
28	123.1681	18.2812	0.3253	0.9763	0.8137	0.7455	46.3222	22.4865
29	125.0000	22.1679	0.3024	1.2163	0.6975	0.7441	47.0588	21.7525
30	123.7574	18.9931	0.2641	1.0384	0.7248	0.7445	46.7260	22.6445

Similarly, the PROMETHEE method has the Spearman's correlation values less than 0.5 with the SAW, WPM, TOPSIS, MTOPSIS, and COPRAS methods. The GRA method has the Spearman's correlation values less than 0.5 with the SAW, WPM, TOPSIS, MTOPSIS, VIKOR, and COPRAS methods. Hence, the ranks suggested by the VIKOR, PROMETHEE, and GRA methods cannot be considered for calculating the average ranks. Now, the corrected average ranks are calculated, excluding the

Table 8.48 Pareto-optimal solutions obtained by the ERao-2 algorithm for the case study of a microalgae-based biomass cultivation process

Solution	X_1	X_2	X_3	X_4	X_5	BMP (g/L)	TLP (%)	EPA (%)
1	99.8541	17.0000	0.2500	1.1676	0.6668	0.7314	45.0679	23.5004
2	124.9695	21.1798	0.2925	1.3500	0.6577	0.7418	47.1179	21.9947
3	125.0000	21.8736	0.3401	0.9688	0.8579	0.7471	46.2554	21.4513
4	104.2653	17.0000	0.2500	1.1582	0.6817	0.7339	45.4181	23.4915
5	124.9639	19.5125	0.2534	1.3921	0.6572	0.7409	47.0494	22.5078
6	102.8818	17.0000	0.2500	1.0630	0.6607	0.7331	45.1803	23.4937
7	107.8746	17.0000	0.2500	1.1495	0.6679	0.7353	45.6691	23.4725
8	124.9974	18.4808	0.2541	1.1409	0.7702	0.7451	46.7301	22.7333
9	105.8389	17.0000	0.2593	1.1679	0.7481	0.7358	45.5131	23.4107
10	116.3560	17.0000	0.2500	1.2342	0.6079	0.7367	46.2159	23.3434
11	112.6548	17.0000	0.2510	1.0348	0.6975	0.7384	45.8727	23.3990
12	125.0000	22.4003	0.3084	1.2076	0.7335	0.7449	46.9780	21.6524
13	118.5760	17.1384	0.2500	1.2851	0.7153	0.7401	46.4671	23.2612
14	125.0000	18.5914	0.3554	1.4037	0.7159	0.7425	46.9500	22.3493
15	124.5756	20.1091	0.3304	1.1487	0.7706	0.7458	46.8114	22.0886
16	124.5918	20.1243	0.3389	1.0686	0.8096	0.7465	46.5918	21.9917
17	124.8914	17.0000	0.2640	1.1845	0.7537	0.7441	46.6696	23.0776
18	125.0000	20.5879	0.2611	1.4345	0.6234	0.7395	47.0728	22.1987
19	105.0548	17.0000	0.2500	1.2009	0.6342	0.7328	45.4665	23.4864
20	125.0000	17.9122	0.2500	1.0790	0.8148	0.7454	46.4608	22.8158
21	125.0000	21.3154	0.3281	1.1786	0.7929	0.7461	46.7719	21.7686
22	123.8124	20.7197	0.2762	1.2090	0.7436	0.7444	46.9106	22.1779
23	124.9894	21.4731	0.2500	1.1219	0.6670	0.7435	47.0087	22.1114
24	113.1203	17.0000	0.2500	1.1514	0.7283	0.7388	46.0129	23.3897
25	107.2724	17.0370	0.2548	1.1295	0.6768	0.7354	45.6236	23.4494
26	118.8557	17.0000	0.2783	1.0955	0.6839	0.7409	46.3656	23.2104
27	123.6308	17.0000	0.3196	1.3684	0.7027	0.7416	46.7941	22.9161
28	125.0000	21.4281	0.2693	1.0915	0.8163	0.7463	46.6253	21.9042
29	114.0206	17.0000	0.2500	1.1708	0.6503	0.7374	46.0922	23.4014
30	121.0026	17.1954	0.2556	1.3597	0.6159	0.7379	46.6027	23.1575

ranks suggested by the VIKOR, PROMETHEE, and GRA methods, and presented in Table 8.52 as the corrected ranks.

The SAP-Rao algorithm solution has the least average rank, which is 1.4. The SAP-Rao solution has the best compromise among the objectives. Thus, it can be considered as the best solution among the compared solutions. The ERao-2 algorithm

Table 8.49 Pareto-optimal solutions obtained by the ERao-3 algorithm for the case study of a microalgae-based biomass cultivation process

Solution	X_1	X_2	X_3	X_4	X_5	BMP (g/L)	TLP (%)	EPA (%)
1	100.5422	17.0000	0.2500	1.1464	0.6593	0.7317	45.0935	23.5017
2	125.0000	21.4992	0.3484	0.9525	0.8585	0.7471	46.2312	21.5037
3	125.0000	21.0129	0.3014	1.3673	0.6577	0.7417	47.1177	21.9966
4	100.7928	17.0000	0.2500	1.1453	0.6483	0.7315	45.0978	23.5015
5	110.5065	17.0000	0.2500	1.0698	0.6888	0.7372	45.7721	23.4375
6	105.0944	17.0000	0.2545	1.1319	0.6760	0.7343	45.4585	23.4731
7	107.4323	17.0000	0.2500	1.0831	0.6975	0.7359	45.5716	23.4640
8	103.5104	17.0000	0.2500	1.1444	0.6445	0.7327	45.3079	23.4968
9	116.5563	17.0000	0.2515	1.2811	0.7046	0.7390	46.3437	23.3340
10	113.3456	17.0000	0.2500	1.0952	0.6866	0.7383	45.9927	23.4052
11	101.3553	17.0000	0.2500	1.1320	0.7092	0.7331	45.1627	23.4860
12	124.9361	22.4904	0.2999	1.1160	0.7281	0.7453	46.9427	21.6748
13	124.9433	20.3368	0.3118	1.3365	0.7036	0.7433	47.0676	22.1202
14	110.0204	17.0000	0.2500	1.0135	0.7322	0.7378	45.6402	23.4073
15	125.0000	18.9546	0.3040	1.2732	0.6432	0.7420	47.0138	22.5004
16	125.0000	17.0000	0.2887	0.8857	0.7230	0.7442	46.3833	22.9537
17	118.7036	17.0000	0.2500	1.2418	0.7949	0.7414	46.2480	23.2216
18	125.0000	20.2792	0.3097	0.9884	0.8715	0.7469	46.2117	21.9056
19	125.0000	17.0000	0.2684	1.0401	0.8520	0.7453	46.1656	22.8961
20	124.9057	22.1211	0.3478	1.2476	0.7321	0.7446	46.9666	21.5794
21	119.4545	17.0000	0.2830	1.0168	0.8108	0.7431	46.0612	23.0467
22	125.0000	20.6079	0.2684	1.1476	0.7680	0.7456	46.8545	22.1777
23	122.8455	19.3137	0.2500	1.2251	0.6843	0.7423	46.8850	22.6624
24	120.4315	17.0000	0.2500	1.2350	0.6606	0.7400	46.5377	23.2773
25	124.8803	18.7456	0.3020	0.9784	0.8129	0.7463	46.4468	22.4181
26	122.4654	18.1019	0.2779	1.3559	0.7848	0.7427	46.6378	22.7615
27	124.9495	19.6096	0.3064	1.0638	0.7701	0.7461	46.7541	22.2774
28	124.8273	19.1598	0.2986	0.9547	0.8051	0.7463	46.4746	22.3399
29	119.2542	17.0000	0.2613	1.0370	0.8003	0.7427	46.0960	23.1496
30	125.0000	18.8517	0.2577	1.2427	0.7789	0.7447	46.7888	22.6154

has achieved the next least average rank of 1.8. Furthermore, the ERao-1 and ERao-3 algorithms have attained the same average rank of 3.6. However, the ERao-1 has achieved better ranks by three methods (SAW, WPM, and COPRAS), and the ERao-3 algorithm has achieved better ranks by two methods (TOPSIS and MTOPSIS). Hence, the ERao-1 and ERao-3 algorithms are given ranks 3 and 4, respectively.

Table 8.50 Pareto-optimal solutions obtained by the SAP-Rao algorithm for the case study of a microalgae-based biomass cultivation process

Solution	X_1	X_2	X_3	X_4	X_5	BMP (g/L)	TLP (%)	EPA (%)
1	125.0000	21.4132	0.3410	0.9599	0.8519	0.7471	46.2889	21.5671
2	125.0000	21.6982	0.2970	1.3766	0.6539	0.7414	47.1181	21.8468
3	99.8128	17.0000	0.2500	1.1397	0.6521	0.7312	45.0167	23.5012
4	100.2454	17.0000	0.2500	1.1734	0.6400	0.7309	45.0696	23.4998
5	125.0000	21.3542	0.2617	1.2456	0.7253	0.7442	47.0193	22.0563
6	112.2826	17.0000	0.2500	1.3431	0.7161	0.7368	46.0928	23.3855
7	111.0456	17.0000	0.2515	1.2092	0.7270	0.7375	45.9301	23.4105
8	125.0000	19.3442	0.2511	1.4296	0.6446	0.7401	47.0324	22.5381
9	125.0000	20.4652	0.3414	1.0417	0.8720	0.7469	46.2398	21.7596
10	104.4634	17.0000	0.2500	1.0954	0.6701	0.7340	45.3554	23.4907
11	123.7532	17.9644	0.2500	1.4901	0.6447	0.7385	46.8613	22.8797
12	125.0000	17.7572	0.2912	1.0573	0.7333	0.7449	46.7009	22.8059
13	116.5568	17.0000	0.2621	1.3677	0.7438	0.7390	46.3308	23.2519
14	106.7963	17.0000	0.2500	1.2368	0.6857	0.7347	45.6766	23.4720
15	120.8521	17.0000	0.2500	1.4818	0.6834	0.7381	46.6260	23.1932
16	125.0000	20.6454	0.2570	1.0827	0.6620	0.7436	46.9606	22.2819
17	125.0000	18.0298	0.2988	1.1255	0.7737	0.7455	46.6948	22.6792
18	125.0000	21.3728	0.3405	1.1213	0.8313	0.7467	46.5427	21.6454
19	125.0000	18.4498	0.2500	1.4425	0.6459	0.7398	46.9709	22.7520
20	125.0000	17.9062	0.2833	1.0539	0.8696	0.7459	46.1718	22.5878
21	125.0000	19.9202	0.3592	1.0741	0.8246	0.7467	46.5149	21.9303
22	125.0000	17.0000	0.2594	1.3109	0.6330	0.7405	46.8018	23.1113
23	125.0000	19.4712	0.2965	1.3944	0.6894	0.7422	47.0593	22.3636
24	108.3264	17.0000	0.2500	1.2407	0.6820	0.7353	45.7883	23.4608
25	125.0000	17.4676	0.2558	1.3759	0.7341	0.7424	46.8336	22.9746
26	108.3264	17.0000	0.2573	1.1870	0.7082	0.7362	45.7527	23.4296
27	122.5491	19.6445	0.2987	1.2075	0.6967	0.7431	46.9021	22.4206
28	125.0000	18.0456	0.3256	1.1110	0.6972	0.7442	46.8305	22.6427
29	118.1135	17.0000	0.2500	1.0320	0.6233	0.7387	46.1400	23.3151
30	101.6586	17.0000	0.2500	1.2087	0.6687	0.7321	45.2551	23.4980

The hypervolume and spacing values of the Pareto-fronts obtained by different algorithms in the MOO scenario are presented in Table 8.54. The Rao-1 algorithm has achieved better spacing values in terms of mean value, the SAP-Rao algorithm has achieved better spacing values in terms of best, and the Rao-3 algorithm has the least standard deviation. Similarly, the SAP-Rao algorithm has achieved better hypervolume values in terms of mean and best values, and the ERao-3 algorithm has

Table 8.51 Best solutions obtained by various algorithms in MOO scenario of the case study of a microalgae-based biomass cultivation process

Algorithm	X_1	X_2	X_3	X_4	X_5	BMP (g/L)	TLP (%)	EPA (%)
Jaya	125	17	0.2865	1.1115	0.7952	**0.7452**	46.5051	22.9416
Rao-1	123.003	17.7207	0.2811	1.3101	0.7150	0.7423	46.8012	22.9055
Rao-2	117.512	17	0.2500	0.9924	0.8103	0.7419	45.8898	23.1973
Rao-3	123.868	17	0.2810	1.1183	0.6252	0.7411	46.6109	23.0823
AMTPG-Jaya	118.576	17	0.2500	1.0554	0.6631	0.7400	46.2575	23.3150
SAP-Rao	125	17	0.2594	1.3109	0.6330	0.7405	**46.8018**	23.1113
ERao-1	125	17.4446	0.2613	1.1088	0.6946	0.7436	46.7426	23.0149
ERao-2	118.576	17.1384	0.2500	1.2851	0.7153	0.7401	46.4671	23.2612
ERao-3	116.556	17	0.2515	1.2811	0.7046	0.7390	46.3437	**23.3340**

Result in boldface indicates better values

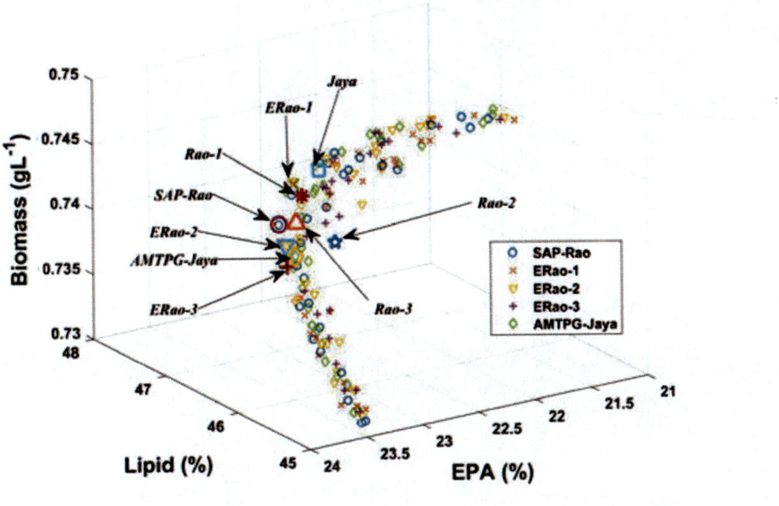

Fig. 8.1 Plot of Pareto-fronts of the proposed algorithms and solutions of the methods compared in the case study of a microalgae-based biomass cultivation process

the least standard deviation. The mean of the hypervolume values achieved by the SAP-Rao algorithm is 0.3, 0.08, 0.34, 0.6, and 0.4% higher when compared to that of the Jaya, AMTPG-Jaya, Rao-2, and Rao-3 algorithms, respectively.

The coverage metric values of the Pareto-fronts obtained by the Jaya algorithm, Rao algorithms, and their modified versions are presented in Table 8.55. The ERao-3 algorithm has achieved better coverage values than the other algorithms compared. Only four solutions of the ERao-3 algorithm are dominated by other algorithm Pareto-fronts. From the coverage values, it can be observed that five solutions from the Pareto-fronts of the SAP-Rao and ERao-1 are dominated, and six solutions from the

Table 8.52 Ranks suggested by the MADM methods for different algorithm solutions presented in Table 8.51

Algorithm	SAW	WPM	TOPSIS	MTOPSIS	VIKOR	PROMETHEE	COPRAS	GRA	AR	CR	FR
Jaya	8	8	8	8	5	3.5	8	1	6.187	8	8
Rao-1	7	7	7	7	7	3.5	7	4	6.187	7	7
Rao-2	9	9	9	9	9	7.5	9	9	8.812	9	9
Rao-3	6	6	6	6	2	5	6	8	5.625	6	6
AMTPG-Jaya	5	5	4	4	6	9	5	6	5.5	4.6	5
SAP-Rao	1	1	2	2	3	1.5	1	3	1.812	1.4	1
ERao-1	2	3	5	5	1	1.5	3	2	2.812	3.6	3
ERao-2	3	2	1	1	4	6	2	7	3.25	1.8	2
ERao-3	4	4	3	3	8	7.5	4	5	4.812	3.6	4

AR—Average Rank; CR—Corrected Rank; FR—Final Rank

Table 8.53 Spearman's rank correlation coefficients between different pairs of MADM method's rankings presented in Table 8.52

Method	SAW	WPM	TOPSIS	MTOPSIS	VIKOR	PROMETHEE	COPRAS	GRA
SAW	1	0.9833	0.8667	0.8667	0.6000	0.3798	0.9833	0.3000
WPM	0.9833	1	0.9333	0.9333	0.5500	0.3038	1	0.2167
TOPSIS	0.8667	0.9333	1	1	0.3333	0	0.9333	0.0333
MTOPSIS	0.8667	0.9333	1	1	0.3333	0	0.9333	0.0333
VIKOR	0.6000	0.5500	0.3333	0.3333	1	0.6498	0.5500	0.3333
PROMETHEE	0.3798	0.3038	0	0	0.6498	1	0.3038	0.6920
COPRAS	0.9833	1	0.9333	0.9333	0.5500	0.3038	1	0.2167
GRA	0.3000	0.2167	0.0333	0.0333	0.3333	0.6920	0.2167	1

Table 8.54 Hypervolume and spacing values of the Pareto-fronts obtained by the proposed algorithms in MOO of a microalgae-based biomass cultivation process

Algorithm		Jaya	Rao1	Rao2	Rao3	AMTPG-Jaya	SAP-Rao	ERao-1	ERao-2	ERao-3
Spacing	B	0.0557	0.0565	0.0589	0.0655	0.0577	**0.0549**	0.0554	0.0560	0.0612
	W	0.1022	**0.0968**	0.1176	0.1019	0.1078	0.0977	0.1042	0.0982	0.1106
	M	0.0750	**0.0728**	0.0788	0.0806	0.0747	0.0750	0.0755	0.0811	0.0832
	SD	0.0146	0.0157	0.0215	**0.0096**	0.0163	0.0132	0.0140	0.0136	0.0153
Hyper volume	B	687.29	687.38	687.23	687.45	687.09	**687.67**	687.23	687.10	687.29
	W	669.85	669.42	669.11	670.37	685.23	686.58	686.35	685.57	**686.75**
	M	685.25	684.82	683.19	684.58	686.54	**687.11**	686.91	686.77	686.93
	SD	5.41	5.46	6.99	5.06	0.71	0.33	0.24	0.45	**0.15**

Result in boldface indicates better values

Table 8.55 Coverage (%) values of the Pareto-fronts obtained by the proposed algorithms in MOO of a microalgae-based biomass cultivation process

Algorithm	Jaya	Rao-1	Rao-2	Rao-3	AMTPG-Jaya	SAP-Rao	ERao-1	ERao-2	ERao-3
Jaya	–	20	13.3	30	13.3	13.3	6.7	10	10
Rao-1	20	–	16.7	33.3	3.3	13.3	13.3	20	10
Rao-2	16.7	10	–	30	13.3	16.7	10	16.7	10
Rao-3	16.7	3.3	13.3	–	6.7	0	13.3	3.3	6.7
AMTPG-Jaya	30	23.3	26.7	36.7	–	6.7	10	20	13.3
SAP-Rao	20	10	16.7	23.3	6.7	–	13.3	10	6.7
ERao-1	6.7	20	13.3	33.3	13.3	13.3	–	20	10
ERao-2	26.7	10	13.3	23.3	20	3.3	16.7	–	13.3
ERao-3	30	13.3	13.3	23.3	10	16.7	6.7	6.7	–

Pareto-fronts of the AMTPG-Jaya, and ERao-2 algorithms are dominated. Similarly, seven solutions of the Pareto-front achieved by the Rao-1 algorithm are dominated; eight solutions are dominated from the Pareto-front achieved by the Rao-2 algorithm; nine solutions are dominated from the Pareto-front achieved by the Jaya algorithm; 11 solutions are dominated from the Pareto-front achieved by the Rao-3 algorithm. The performances of the proposed modified algorithms are better or competitive to that of the basic algorithms in terms of the coverage and spacing values.

From the computational results of the case study of a microalgae-based biomass cultivation process, it can be observed that the SAP-Rao algorithm has achieved the best solution. The Rao-1 algorithm has achieved the best mean spacing value (0.0728). The SAP-Rao algorithm has achieved the best mean hypervolume value (687.11). The ERao-3 algorithm has achieved better performance in terms of coverage values. In addition, the performances of the improved versions in this case study are better or competitive to those of the basic algorithms. The next chapter presents the application of the Jaya algorithm and Rao algorithms along with their modified versions to the hydropower generation system and geothermal energy system case studies.

References

Banerjee, A., Guria, C., & Maiti, S. K. (2016). Fertilizer assisted optimal cultivation of microalgae using response surface method and genetic algorithm for biofuel feedstock. *Energy, 115*, 1272–1290. https://doi.org/10.1016/j.energy.2016.09.066

Dhingra, S., Bhushan, G., & Dubey, K. K. (2014). Multi-objective optimization of combustion, performance and emission parameters in a jatropha biodiesel engine using non-dominated sorting genetic algorithm-II. *Frontiers of Mechanical Engineering, 9*(1), 81–94. https://doi.org/10.1007/s11465-014-0287-9

Jaliliantabar, F., Ghobadian, B., Najafi, G., Mamat, R., & Carlucci, A. P. (2019). Multi-objective NSGA-II optimization of a compression ignition engine parameters using biodiesel fuel and exhaust gas recirculation. *Energy, 187*, 115970. https://doi.org/10.1016/j.energy.2019.115970

Shirneshan, A., Samani, B. H., & Ghobadian, B. (2016). Optimization of biodiesel percentage in fuel mixture and engine operating conditions for diesel engine performance and emission characteristics by artificial bees colony algorithm. *Fuel, 184*, 518–526. https://doi.org/10.1016/j.fuel.2016.06.117

Chapter 9
Optimization of Hydroenergy and Geothermal Energy Systems

Abstract This chapter presents the applications of optimization algorithms to the problems of hydro- and geothermal energy systems. A single-objective optimization case study of the Nigerian Jebba hydropower plant performance and a multi-objective optimization case study of a ground source heat pump-radiant ceiling air conditioning system are considered for optimization to see if there can be any improvement in the performances of the selected systems. The optimization is carried out using the Jaya algorithm, Rao algorithms, and their modified versions. Furthermore, to identify the best solutions from the Pareto-fronts, the average rank method described in Chap. 5 is used. Computational results revealed that the performances of the modified Jaya and Rao algorithms are superior to those of the other algorithms. Also, the performances of the selected systems are improved by the solutions of the proposed algorithms.

9.1 Optimization of a Hydropower Generation System

Optimization of the Nigerian Jebba hydropower plant performance characteristics to enhance its electricity generation case study is considered to demonstrate the application of advanced optimization algorithms. This case study was presented by Onokwai et. al. (2020) and a brief description of this case study is presented in Sect. 2.4.1. This is a single-objective optimization case study, and the objective is to improve the estimated power generation (EP_w) of the Jebba hydropower plant which is given in Eq. 2.65. The decision variables considered are the discharge (D), pressure drop between the head and tail-water (P_d), the stator temperature (T_{st}), the water pressure (P_{water}) for cooling the generator, and the oil pressure (P_{oil}) for cooling and lubrication of the bearings. The lower and upper boundaries of the decision variables are as follows:

$$7139.25\text{m}^3/\text{s} \leq D \leq 8783.3667\text{m}^3/\text{s} \tag{9.1}$$

$$26.1257\text{m} \leq P_d \leq 28.5867\text{m} \tag{9.2}$$

V. R. Ravipudi et al., *Design Optimization of Renewable Energy Systems Using Advanced Optimization Algorithms*, Green Energy and Technology,
https://doi.org/10.1007/978-3-030-95589-2_9

$$53^\circ\mathrm{C} \le T_{\mathrm{st}} \le 58^\circ\mathrm{C} \tag{9.3}$$

$$29\mathrm{N/m}^2 \le P_{\mathrm{water}} \le 33\mathrm{N/m}^2 \tag{9.4}$$

$$31\mathrm{N/m}^2 \le P_{\mathrm{oil}} \le 37\mathrm{N/m}^2 \tag{9.5}$$

Onokwai et al. (2020) had reported an optimal solution using the response surface methodology (RSM) for this case study. Now, the Jaya algorithm, Rao algorithms, and their improved versions are employed to find the optimal solution for this case study. The proposed algorithms are executed using only 10,000 function evaluations as the termination criterion. In all the computational experiments, the population size is maintained as 25 for all the algorithms. The elite size for the elitist Rao algorithms is taken as 20%. Table 9.1 presents the single-objective optimization results of this case study. The Jaya algorithm, Rao algorithms, and their improved versions have obtained identical solution in this case study. Hence, only one solution is presented here. Also, the solutions obtained by the proposed algorithms are superior to those obtained by the RSM. The maximum estimated power generation obtained by the proposed algorithms are 3166.874 MW. The proposed algorithms solutions have resulted 18.5% increment in the total estimated power generation.

Here, it can be observed that the values achieved by the proposed algorithms for the decision variables discharge, stator temperature, water, and oil pressure are at the boundaries of the variables. To investigate why this has happened, decision variables versus estimated power generation plots are drawn and presented in Fig. 9.1. From the plots, it can be observed that the increase in discharge, water pressure for cooling the generator, and stator temperature will result in increase of estimated power generation. Similarly, as the pressure drop and oil pressure for lubrication increases, the net power generation is decreased.

Furthermore, the larger the discharge, the more energy will be made available for conversion to electric power as shown in Fig. 9.1. Also, from the hydrological data obtained from the Jebba hydropower plant, lower turbine discharge values rendered the turbine less efficient such that at very low reservoir levels. This is as a result of the fact that Jebba hydropower plant operates via fixed turbine blades which cannot be adjusted to run at low reservoir level (Onokwai et al., 2020). In addition, thermal instability of generator windings occurs when the temperature fluctuated.

Table 9.1 Optimum solutions obtained using the RSM and proposed algorithms in the optimization of the Jebba hydropower plant system case study

Method	D (m^3/s)	P_d(m)	T_{st}(°C)	P_{water}($\mathrm{N/m}^2$)	P_{oil}($\mathrm{N/m}^2$)	$EP_w(MW)$
RSM	8783.37	28.59	58	31	31	2671.4
Proposed algorithms	8783.367	26.2361	58	33	31	**3166.874**

Result in boldface indicates a better performing algorithm

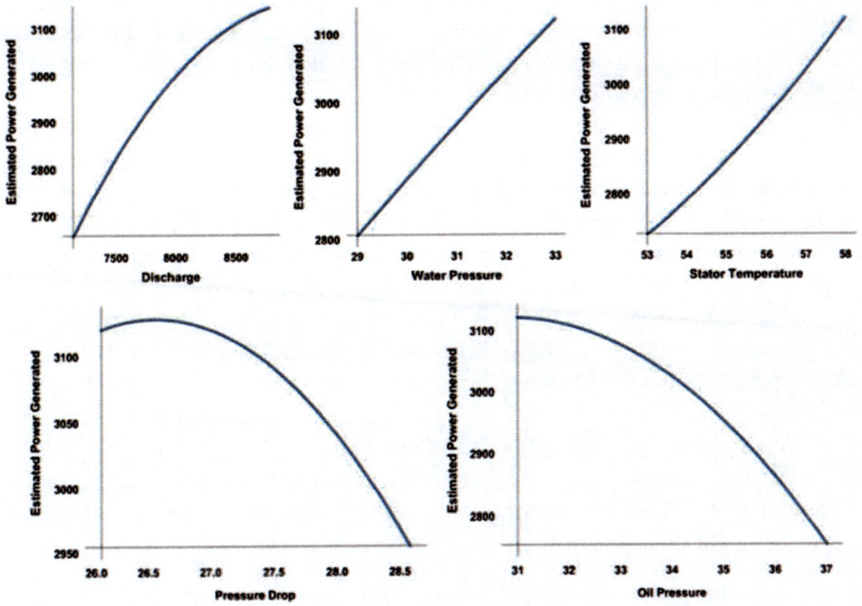

Fig. 9.1 Plot of decision variables versus estimated power generation

Maintaining the windings temperature at 58 °C or lower would enable the generator run at its maximum capacity without risk of insulation failure. The hotter the windings temperature, the less load the generator would sustain (Onokwai et al., 2020). The following section presents the application of the Jaya and Rao algorithms along with their modified versions to a ground source heat pump system case study.

9.2 Optimization of a Ground Source Heat Pump-Radiant Ceiling Air Conditioning System

The description of the ground source heat pump-radiant ceiling (GSHP-RC) air conditioning system is presented in Sect. 2.4.2. The detailed system description and technical specifications of the GSHP-RC system considered in this case study were presented by Xie et al. (2020). This case study consists of three objectives, which are seasonal performance factor (SPF), thermal comfort predicted mean value (PMV), and the operating cost (OC). In the optimization of this system, SPF is a maximization function and the absolute value of PMV and operating costs is minimization functions. The decision variables of these functions are the water supply temperature of radiant ceiling (a), the indoor set temperature (b), and the water supply temperature of the heat pump (c). The lower and upper boundaries of these variables are $16 < a < 20$, $24 < b < 28$, and $7 < c < 11$.

Xie et al. (2020) optimized this multi-objective optimization case study using the NSGA-II algorithm and reported two Pareto-optimal solutions. These two solutions were selected from a Pareto-front based on the proximity to a reference solution using utility method for two different weight combinations of the objectives. The number of function evaluations taken by the NSGA-II was not specified. However, the Jaya algorithm, Rao algorithms, and their modified versions are tested by taking 10,000 function evaluations as the termination criterion. The best Pareto-optimal solutions are reported using the average rank based on multiple decision-making methods. In all the computational experiments, the population size is maintained as 20 for all the algorithms. The elite size for the elitist Rao algorithms is taken as 20%.

The Pareto-optimal solutions achieved by the Jaya algorithm, Rao algorithms, and their improved versions in this case study are presented in Tables 9.2, 9.3, 9.4, 9.5, 9.6, 9.7, 9.8, 9.9 and 9.10 Here, an observation can be made that the seasonal performance factor is varied between 3.65 and 3.75, the thermal comfort predicted mean value is varied between 1e-5 and 0.75, and the operating cost is varied from 2200 to 2500. The best solutions from the Pareto-fronts obtained by the proposed algorithms are identified based on the average rank method. Solution 4 of the Jaya algorithm,

Table 9.2 Pareto-optimal solutions obtained by the Jaya algorithm for the GSHP-RC system case study

Solution	a	b	c	SPF	PMV	OC
1	16.1691	25.9539	9.337048	3.718262	3.57E-06	24,945.33
2	16	28	9.746768	3.762294	0.71256	23,098.13
3	16	25.94701	9.194334	3.719283	0.007977	24,963.48
4	16.56782	25.9273	11	3.695793	0.00543	23,835.93
5	16.24631	28	11	3.740916	0.713688	22,087.41
6	16.38885	28	10.77912	3.744283	0.714341	22,324.12
7	16	27.67985	9.843536	3.755992	0.591089	23,197.08
8	16.31228	27.9283	10.47108	3.75012	0.686818	22,623.24
9	16	26.80147	9.260826	3.736557	0.275096	24,173.19
10	16.00461	27.63492	10.70988	3.744806	0.574356	22,513.75
11	16	27.39271	9.285821	3.748246	0.485004	23,741.47
12	16.48853	26.21918	11	3.702682	0.094432	23,500.48
13	16.40366	25.74062	10.095	3.710895	0.057692	24,748.86
14	16.04723	26.39836	9.646177	3.730488	0.139874	24,274.95
15	16	26.98305	10.25315	3.73975	0.338343	23,332.9
16	16.09935	26.65432	11	3.717215	0.227029	22,952.95
17	16.529	27.35783	11	3.724056	0.479647	22,550.85
18	16.40899	27.12582	11	3.721459	0.395881	22,684.19
19	16.34138	26.4258	10.61972	3.717977	0.157036	23,597.33
20	16.55944	26.76018	9.849759	3.729401	0.273479	24,010.42

Table 9.3 Pareto-optimal solutions obtained by the Rao-1 algorithm for the GSHP-RC system case study

Solution	a	b	c	SPF	PMV	OC
1	16.04694	25.96695	10.01825	3.721415	4.62E-05	24,406.63
2	16	25.97329	9.496396	3.722447	0.000372	24,750.09
3	16	28	9.743059	3.762294	0.71256	23,100.75
4	16.1572	25.96175	11	3.702918	0.002064	23,607.25
5	16.12582	28	10.99627	3.742915	0.713136	22,069.82
6	16.17336	28	10.34625	3.755532	0.713354	22,665.24
7	16.08983	27.25775	11	3.729029	0.437454	22,496.87
8	16	26.86286	9.586978	3.740045	0.296358	23,917.96
9	16.05193	27.77799	9.699609	3.757165	0.628371	23,255.22
10	16.0604	27.67006	10.23338	3.752299	0.587993	22,915.4
11	16.02723	26.20045	10.94375	3.711107	0.074317	23,376.86
12	16.10831	26.44729	10.70825	3.720252	0.157783	23,407.14
13	16.12806	27.54888	10.6124	3.743183	0.54379	22,679.87
14	16.16195	26.59041	9.99303	3.731804	0.206993	23,901.2
15	16.00082	26.66473	9.864731	3.736305	0.228203	23,878.66
16	16.2986	26.75558	11	3.716039	0.266037	22,939.54
17	16.03245	27.33034	10.16909	3.746825	0.462784	23,169.09
18	16.1184	26.83472	9.969582	3.737336	0.289137	23,704.57
19	16.07005	27.72215	10.65147	3.746676	0.607547	22,530.4
20	16.0433	27.49273	10.28793	3.748514	0.522165	22,972.61

Table 9.4 Pareto-optimal solutions obtained by the Rao-2 algorithm for the GSHP-RC system case study

Solution	a	b	c	SPF	PMV	OC
1	16	28	9.742384	3.762294	0.71256	23,101.22
2	16	25.97139	10.43529	3.717768	0.000232	24,044.45
3	16	25.97362	11	3.70562	0.000476	23,526.32
4	16	27.94971	11	3.743818	0.693258	22,070.29
5	16	28	9.939083	3.761866	0.71256	22,958.33
6	16	25.96738	9.692646	3.722976	0.00151	24,623.5
7	16	26.1877	10.90721	3.712246	0.069336	23,412.64
8	16	27.65644	10.45286	3.750067	0.582338	22,727.75
9	16	27.5293	10.74163	3.742145	0.53513	22,545.8
10	16	26.32809	9.428732	3.728885	0.115307	24,463.06

(continued)

Table 9.4 (continued)

Solution	a	b	c	SPF	PMV	OC
11	16	27.76579	10.86784	3.743749	0.623365	22,292.95
12	16	26.48194	10.78965	3.720812	0.166433	23,262.13
13	16	27.39644	10.2889	3.747323	0.486364	23,021.67
14	16	27.08576	10.60852	3.73632	0.374593	22,956.1
15	16	27.71986	10.19788	3.754584	0.606085	22,903.68
16	16	27.22655	10.60736	3.739064	0.424851	22,859.97
17	16	26.65326	10.17318	3.734208	0.224277	23,651.8
18	16	26.92921	10.50032	3.735239	0.319476	23,164.81
19	16	26.7355	9.655706	3.737768	0.252388	23,970.08
20	16	26.54602	10.07694	3.732951	0.187957	23,815.94

Table 9.5 Pareto-optimal solutions obtained by the Rao-3 algorithm for the GSHP-RC system case study

Solution	a	b	c	SPF	PMV	OC
1	16.47459	25.91993	11	3.697117	8.03E-05	23,798.19
2	16.00771	25.9718	9.256978	3.720357	0.000161	24,905.23
3	16	28	9.750632	3.762293	0.71256	23,095.4
4	16.21532	28	10.39278	3.754226	0.713546	22,633.62
5	16.0022	28	11	3.744755	0.71257	22,045.85
6	16	27.91716	9.363954	3.759106	0.680806	23,394.16
7	16.0202	26.76323	9.572211	3.737748	0.262371	24,011.46
8	16.16614	26.70101	11	3.717067	0.244459	22,937.87
9	16.00432	27.11443	9.823609	3.745035	0.38485	23,570.24
10	16.33389	27.64992	11	3.732771	0.583127	22,304.72
11	16	27.38294	9.857441	3.75022	0.481441	23,365.09
12	16.30292	27.67971	9.762669	3.751333	0.59383	23,320.72
13	16.003	27.4504	10.02407	3.750745	0.506137	23,198.84
14	16.00398	26.92806	9.704893	3.741495	0.319155	23,789.27
15	16.00888	26.48933	9.395937	3.731623	0.169143	24,346.81
16	16.51486	25.76827	10.48706	3.70492	0.045152	24,452.34
17	16.06765	26.62883	10.00623	3.733955	0.217622	23,825.32
18	16.66508	25.89608	10.17595	3.709084	0.000771	24,648.12
19	16.13211	27.39453	11	3.731008	0.48743	22,417.02
20	16.03355	26.90615	9.608356	3.740422	0.312108	23,881.15

Table 9.6 Pareto-optimal solutions obtained by the AMTPG-Jaya algorithm for the GSHP-RC system case study

Solution	a	b	c	SPF	PMV	OC
1	16.05922	25.96562	11	3.704534	5.02E-05	23,559.94
2	16.03247	25.97029	9.584395	3.722272	0.00052	24,708.77
3	16	28	9.743278	3.762294	0.71256	23,100.59
4	16.00539	28	11	3.744705	0.712585	22,046.36
5	16.01162	28	10.75841	3.750687	0.712613	22,276.3
6	16.00659	27.1389	11	3.728041	0.393598	22,555.87
7	16.0174	27.85123	11	3.741641	0.655812	22,123.73
8	16.0054	27.71231	9.744761	3.756648	0.603297	23,250.75
9	16.06314	25.88255	10.61108	3.712019	0.026135	24,006.94
10	16.00972	26.06049	9.76879	3.724643	0.028555	24,484.16
11	16.03848	27.31381	10.42345	3.743292	0.456885	22,971.94
12	16	27.49515	9.592036	3.752284	0.522539	23,481.56
13	16.0341	26.5311	9.578951	3.733067	0.183815	24,201.24
14	16.03099	26.65468	9.611074	3.73561	0.225505	24,075.89
15	16.01505	27.63702	9.755525	3.755039	0.575239	23,287.78
16	16.01168	26.29423	10.71704	3.718625	0.104503	23,497.79
17	16.00976	27.24628	9.657993	3.747492	0.432099	23,598.15
18	16.01796	26.78955	9.938515	3.738188	0.27138	23,730.55
19	16.03975	27.02236	10.99397	3.725434	0.35292	22,654.08
20	16.05525	26.4448	9.926435	3.730988	0.155517	24,036.94

Table 9.7 Pareto-optimal solutions obtained by the SAP-Rao algorithm for the GSHP-RC system case study

Solution	a	b	c	SPF	PMV	OC
1	16	28	9.743338	3.762294	0.71256	23,100.55
2	16.4169	25.92638	11	3.698149	8E-05	23,763.66
3	16	25.99068	9.078556	3.718573	0.005908	24,987.65
4	16.00436	28	11	3.744721	0.71258	22,046.19
5	16.1388	28	10.24448	3.757321	0.713196	22,743.03
6	16.0043	27.82129	10.57186	3.751158	0.64437	22,534.4
7	16.26081	26.32153	11	3.708243	0.120683	23,305.49
8	16.26984	26.02953	11	3.702456	0.027255	23,590.21
9	16	26.83961	9.46639	3.739019	0.28829	24,015.28
10	16.003	27.37273	11	3.732618	0.477765	22,399.04

(continued)

Table 9.7 (continued)

Solution	a	b	c	SPF	PMV	OC
11	16.0023	26.21195	9.534165	3.727214	0.0773	24,500.35
12	16.24963	26.61405	10.53811	3.72457	0.217108	23,467.77
13	16	26.1147	9.618033	3.725679	0.045686	24,533.56
14	16	26.56128	9.641764	3.734371	0.193102	24,121.19
15	16.38782	27.69624	10.15899	3.748402	0.600709	23,032.4
16	16.18683	26.42839	10.55398	3.721687	0.153659	23,592.02
17	16.00294	26.4988	10.15166	3.731376	0.172158	23,798.14
18	16.00408	27.6691	11	3.738329	0.587105	22,220.65
19	16	26.23018	10.92118	3.712705	0.083177	23,360.3
20	16	27.49377	11	3.735004	0.52203	22,322.87

Table 9.8 Pareto-optimal solutions obtained by the ERao-1 algorithm for the GSHP-RC system case study

Solution	a	b	c	SPF	PMV	OC
1	16.30606	25.93907	11	3.700138	4.66E-05	23,698.24
2	16	28	9.743159	3.762294	0.71256	23,100.68
3	16.03424	25.96858	10.58699	3.714594	3.69E-05	23,930.14
4	16	27.88057	11	3.742481	0.66685	22,105.38
5	16.00457	28	9.800415	3.762185	0.712581	23,060.65
6	16.04393	25.95323	9.52005	3.721491	0.0045	24,773.37
7	16.03211	26.45176	9.404069	3.730593	0.157211	24,382.42
8	16.00562	27.63591	9.658408	3.755089	0.574733	23,354.24
9	16	26.63605	9.80372	3.735887	0.218423	23,945.92
10	16.01316	27.78872	10.63283	3.749229	0.63212	22,499.29
11	16.00737	26.9612	10.49741	3.73579	0.330818	23,145.84
12	16.09543	26.08195	11	3.706213	0.038223	23,462.42
13	16	27.53361	10.62601	3.744638	0.536721	22,647.98
14	16	27.87609	10.67989	3.750172	0.665147	22,408.38
15	16	26.22183	10.1042	3.726474	0.080451	24,078.67
16	16.02753	27.43658	10.60461	3.742743	0.501388	22,732.83
17	16.02981	26.35499	10.48652	3.7239	0.125037	23,655.46
18	16.04679	26.85149	9.516769	3.738793	0.293404	23,988.1
19	16.01395	27.11582	11	3.727479	0.385512	22,574.03
20	16.03057	27.18223	9.350473	3.744304	0.409465	23,848.6

Table 9.9 Pareto-optimal solutions obtained by the ERao-2 algorithm for the GSHP-RC system case study

Solution	a	b	c	SPF	PMV	OC
1	16	25.9721	10.59807	3.714992	8.06E-06	23,902.01
2	16	25.97106	9.194449	3.719749	0.000338	24,939.89
3	16	28	9.738953	3.762294	0.71256	23,103.64
4	16.06596	25.96574	11	3.70443	0.000217	23,562.78
5	16.04473	28	10.46249	3.755852	0.712765	22,545.13
6	16.03005	28	11	3.744317	0.712698	22,050.3
7	16.04612	26.22184	9.747491	3.727196	0.081855	24,362.88
8	16	26.75796	9.558205	3.737909	0.260104	24,018.37
9	16	27.1879	9.664041	3.746528	0.410992	23,630.89
10	16.02597	26.93439	11	3.723783	0.321806	22,708.94
11	16	26.42954	10.86852	3.717902	0.148933	23,233.19
12	16.06419	27.89502	11	3.741751	0.672747	22,109.23
13	16.16997	27.37748	10.92519	3.732104	0.481764	22,510.1
14	16.02844	26.09683	11	3.707554	0.040837	23,419.93
15	16.05495	26.22524	10.92511	3.711642	0.083229	23,383.06
16	16.00725	26.34244	9.751825	3.730138	0.12025	24,235.5
17	16	27.5705	11	3.736488	0.55037	22,276.86
18	16	26.14559	10.02962	3.725536	0.055673	24,206.93
19	16	27.81759	9.614432	3.758586	0.642939	23,282.64
20	16	27.52896	10.19998	3.750872	0.535005	23,012.24

Solution 4 of the Rao-1 algorithm, Solution 3 of the Rao-2 algorithm, Solution 1 of the Rao-3 algorithm, Solution 1 of the AMTPG-Jaya algorithm, Solution 2 of the SAP-Rao algorithms, Solution 1 of the ERao-1 algorithm, Solution 4 of the ERao-2, and Solution 2 of the ERao-3 algorithm are identified as the best solutions from the respective algorithm Pareto-front. Now, these best solutions are compared with those of the NSGA-II algorithm in Table 9.11.

In Table 9.11, the solutions with the best average rank from the Pareto-optimal solutions of the proposed algorithms are compared with those reported by the NSGA-II algorithm. Figure 9.2 presents the Pareto-fronts achieved by the AMTPG-Jaya, SAP-Rao, and elitist Rao algorithms, including the optimal solutions reported for the NSGA-II, Jaya, and Rao algorithms. The conflicting nature of these objectives can be observed in Fig. 9.2. Any change in the design variables that will result in the increment of SPF also leads to the increment of the PMV and OC, which is not desirable. Also, the variation of the SPF is relatively less when compared to the other objectives. It can also be observed from the Pareto-optimal solutions obtained by the algorithms. It indicates that for a small change in SPF caused by the design variables, there is a relatively considerable change in the PMV and operating costs.

Table 9.10 Pareto-optimal solutions obtained by the ERao-3 algorithm for the GSHP-RC system case study

Solution	a	b	c	SPF	PMV	OC
1	16	28	9.743323	3.762294	0.71256	23,100.56
2	16.48512	25.91899	10.97858	3.697524	5.4E-06	23,825.1
3	16.11637	28	10.98185	3.743461	0.713093	22,082.21
4	16.25939	28	9.913288	3.757891	0.713748	23,021.83
5	16.17793	25.93108	9.313292	3.717463	0.006879	24,986.94
6	16.35046	26.40046	9.705634	3.725847	0.148988	24,351.24
7	16	26.54618	9.816065	3.734132	0.188011	24,011.28
8	16.42168	26.27675	9.52748	3.721838	0.110881	24,618.27
9	16.39246	26.79812	11	3.715384	0.282573	22,937.78
10	16.05327	26.90949	9.99245	3.739685	0.313677	23,608.91
11	16.20253	26.28523	11	3.708458	0.107164	23,315.25
12	16.09541	27.40063	10.99107	3.731951	0.48916	22,412.56
13	16.55551	26.63564	10.55846	3.719813	0.231811	23,548.06
14	16.03401	27.75923	10.17975	3.754989	0.621166	22,903.1
15	16.06842	26.06045	11	3.706222	0.030461	23,471.58
16	16.17544	26.48085	9.229082	3.72725	0.170733	24,518.54
17	16.02089	26.04747	10.83608	3.710985	0.024746	23,620.61
18	16.08193	25.99947	11	3.704831	0.011461	23,536.64
19	16.05195	27.30341	10.19902	3.745705	0.453324	23,167.82
20	16	27.21867	9.891739	3.746944	0.422022	23,447.59

The decision-making methods ranking of the Pareto-optimal solutions obtained by different algorithms is shown in Table 9.12. The Spearman's correlation coefficients for different pairs of rankings given by decision-making methods for different algorithm's solutions are shown in Table 9.13. The ranks given by the TOPSIS and MTOPSIS methods for each solution are identical. Furthermore, the Spearman's correlation for all the pairs of decision-making methods (except with the VIKOR, PROMETHEE, and GRA methods) is greater than 0.5. The Spearman's correlation value for the GRA-WPM pair is negative, and for the GRA method pairs with SAW, TOPSIS, MTOPSIS, and COPRAS, it is less than 0.5. Similarly, the PROMETHEE method has the Spearman's correlation values less than 0.5 with the SAW, WPM, and COPRAS methods. The VIKOR method has the Spearman's correlation values less than 0.5 with the SAW, WPM, and COPRAS methods. Hence, the ranks suggested by the VIKOR, PROMETHEE, and GRA methods cannot be considered for calculating the average ranks. Now, the corrected average ranks are calculated, excluding the ranks suggested by the VIKOR, PROMETHEE, and GRA methods and presented in Table 9.12 as corrected ranks.

Table 9.11 Best solutions and performance metrics obtained by various algorithms in MOO scenario of the GSHP-RC system case study

Algorithm	a	b	c	SPF	PMV	OC	Hypervolume	Spacing
NSGA-II solution 1	16.1	26.6	10.4	**3.741**	0.225	**23,525**	–	–
NSGA-II solution 2	16.2	26	9.7	3.734	0.0203	24,613	–	–
Jaya	16.56782	25.9273	11	3.695793	0.00543	23,835.93	18,072.62	0.1154082
Rao-1	16.1572	25.96175	11	3.702918	0.002064	23,607.25	18,303.26	0.0997092
Rao-2	16	25.97362	11	3.70562	0.000476	23,526.32	18,066.16	0.0837749
Rao-3	16.47459	25.91993	11	3.697117	8.03E-05	23,798.19	17,803.86	0.1268176
AMTPG-Jaya	16.05922	25.96562	11	3.704534	5.02E-05	23,559.94	18,028.67	0.0942512
SAP-Rao	16.4169	25.92638	11	3.698149	8E-05	23,763.66	17,998.07	**0.0699101**
ERao-1	16.30606	25.93907	11	3.700138	4.66E-05	23,698.24	18,098.15	0.1310586
ERao-2	16.06596	25.96574	11	3.70443	0.000217	23,562.78	**18,773.68**	0.1191293
ERao-3	16.48512	25.91899	10.97858	3.697524	**5.4E-06**	23,825.1	18,360.59	0.1408059

Source NSGA-II: Xie et al. (2020)
Result in boldface indicates better values

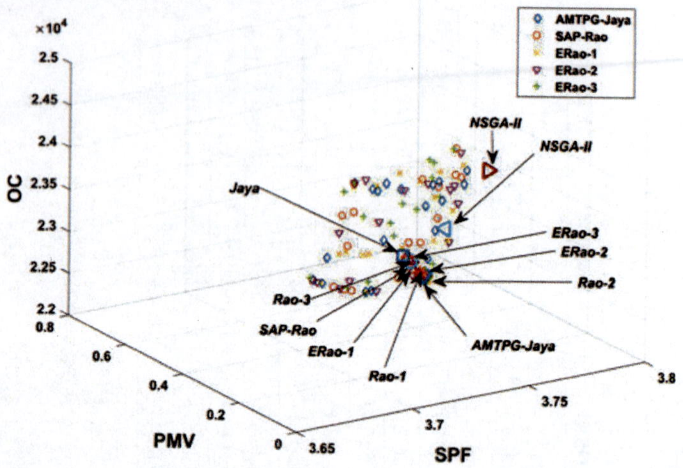

Fig. 9.2 Plot of Pareto-fronts of the proposed algorithms and solutions of the methods compared in the GSHP-RC system case study

The AMTPG-Jaya algorithm solution has the least average rank, which is 1.8. Thus, it can be regarded as the best solution. The ERao-1, ERao-2, ERao-3, and SAP-Rao algorithms' solutions have the next best average ranks, which are 2.8, 3.8, 4, and 4.6, respectively. The AMTPG-Jaya algorithm solution ranked one by three methods (TOPSIS, MTOPSIS, and COPRAS) and ranked three by two methods (SAW and WPM). Here, an observation can be made that the AMTPG-Jaya algorithm solution has a better value in any objective. The NSGA-II solution 1 has better seasonal performance factor and operating costs, and the ERao-3 algorithm solution has better predicted mean value. However, the AMTPG-Jaya algorithm's solution is ranked one because it has best compromise among the three objectives. If we compare the AMTPG-Jaya algorithm solution with NSGA-II solution 1, the AMTPG-Jaya algorithm's solution has 0.97% deterioration in SPF and 0.15% deterioration in operating cost, Whereas the PMV has improved 99% compared to NSAGA-II solution 1. This indicates that the relative improvement in the PMV is dominating the relative deterioration in the SPF and OC. Also, the solutions of the Jaya, Rao-1, Rao-3, SAP-Rao, and ERao-2 algorithms are dominated by the solution of the AMTPG-Jaya algorithm. Hence, the ERao-3 algorithm solution has achieved the least average rank and can be considered as the best solution among the compared.

Now, the performances of the Jaya algorithm, Rao algorithms, and their modified versions in the MOO scenario are evaluated based on the hypervolume and spacing indicators. These performance indicators are presented in Table 9.11. Here, an observation can be made that the ERao-2 algorithm Pareto-front has higher hypervolume and the SAP-Rao algorithm has lower spacing value. The SAP-Rao algorithm has a better spacing, which is 39, 29, 16, 44, 25, 46, 41, and 50% lesser value when compared to that of the Jaya, Rao-1, Rao-2, Rao-3, AMTPG-Jaya, ERao-1, ERao-2, and ERao-3 algorithms, respectively. The ERao-2 algorithm Pareto-front has a

Table 9.12 Ranks suggested by the MADM methods for different algorithm solutions presented in Table 9.11

Algorithm	SAW	WPM	TOPSIS	MTOPSIS	VIKOR	PROMETHEE	COPRAS	GRA	AR	CR	FR
NSGA-II solution 1	8	11	11	11	6	3	11	2	7.875	10.4	11
NSGA-II solution 2	11	10	10	10	10	10	10	10	10.125	10.2	10
Jaya	10	9	9	9	11	11	9	11	9.875	9.2	9
Rao-1	9	8	8	8	4	8	8	5	7.25	8.2	8
Rao-2	7	7	3	3	1	2	7	1	3.875	5.4	6
Rao-3	5	5	6	6	9	9	6	8	6.75	5.6	7
AMTPG-Jaya	3	3	1	1	2	1	1	3	1.875	1.8	1
SAP-Rao	4	4	5	5	7	6.5	5	7	5.4375	4.6	5
ERao-1	2	2	4	4	5	4	2	6	3.625	2.8	2
ERao-2	6	6	2	2	3	5	3	4	3.875	3.8	3
ERao-3	1	1	7	7	8	6.5	4	9	5.4375	4	4

AR—Average Rank; CR—Corrected Rank; FR—Final Rank

Table 9.13 Spearman's rank correlation coefficients between different pairs of MADM method's rankings presented in Table 9.12

Method	SAW	WPM	TOPSIS	MTOPSIS	VIKOR	PROMETHEE	COPRAS	GRA
SAW	1	0.945	0.591	0.591	0.273	0.478	0.836	0.127
WPM	0.945	1	0.645	0.645	0.209	0.296	0.891	−0.055
TOPSIS	0.591	0.645	1	1	0.700	0.592	0.864	0.445
MTOPSIS	0.591	0.645	1	1	0.700	0.592	0.864	0.445
VIKOR	0.273	0.209	0.700	0.700	1	0.843	0.455	0.900
PROMETHEE	0.478	0.296	0.592	0.592	0.843	1	0.460	0.875
COPRAS	0.836	0.891	0.864	0.864	0.455	0.460	1	0.164
GRA	0.127	−0.055	0.445	0.445	0.900	0.875	0.164	1

higher hypervolume value which is 3.9, 2.6, 3.9, 5.4, 4.1, 4.3, 3.7, and 2.2% higher than Jaya, Rao-1, Rao-2, Rao-3, AMTPG-Jaya, SAP-Rao, ERao-1, and ERao-3 algorithms, respectively. The ERao-3, Rao-1, and ERao-1 algorithms have achieved the next better hypervolume values, respectively. The performances of the improved algorithms are better or competitive to that of the basic algorithms in terms of the hypervolume and spacing values.

From the computational results of the GSHP-RC system case study, it can be observed that the performance of the considered system can be improved with the solutions achieved by the proposed algorithms. The solution of the AMTPG-Jaya algorithm has better compromise among the SPF, PMV, and OC. The SAP-Rao algorithm's performance in terms of spacing and the ERao-2 algorithm's performance in terms of hypervolume are much better than that achieved by other algorithms. In addition, the performances of the improved versions in this case study are better or competitive to those of the basic algorithms as well as the NSGA-II algorithm. The next chapter presents the conclusions.

References

Onokwai, A. O., Owamah, H. I., Ibiwoye, M. O., Ayuba, G. C., & Olayemi, O. A. (2020). Application of response surface methodology (RSM) for the optimization of energy generation from Jebba hydro-power plant, Nigeria. *ISH Journal of Hydraulic Engineering*. https://doi.org/10.1080/09715010.2020.1806120

Xie, Y., Hu, P., Zhu, N., Lei, F., Xing, L., & Xu, L. (2020). Collaborative optimization of ground source heat pump-radiant ceiling air conditioning system based on response surface method and NSGA-II. *Renewable Energy, 147*, 249–264. https://doi.org/10.1016/j.renene.2019.08.109

Appendices

Appendix A. Single-objective Optimization of Standard Benchmark Problems

Appendix A.1. Unconstrained Standard Benchmark Problems

Problem	Function	Formulation	D	Interval	C
F1	Sphere	$F_1(y) = \sum_{i=1}^{D} y_i^2$	30	[−100, 100]	US
F2	SumSquares	$F_2(y) = \sum_{i=1}^{D} i y_i^2$	30	[−10, 10]	US
F3	Beale	$F_3(y) = \sum_{i=1}^{D} (1.5 - y_1 + y_1 y_2)^2 + \left(2.25 - y_1 + y_1 y_2^2\right)^2 + \left(2.625 - y_1 + y_1 y_2^3\right)^2$	2	[−4.5, 4.5]	UN
F4	Easom	$F_4(y) = -\cos(y_1)\cos(y_2)\exp(-(y_1 - \pi)^2 - (y_2 - \pi)^2)$	2	[−100, 100]	UN
F5	Matyas	$F_5(y) = 0.26(y_1^2 + y_2^2) - 0.48 y_1 y_2$	2	[−10, 10]	UN
F6	Colville	$F_6(y) = 100(y_1^2 - y_2)^2 + (y_1 - 1)^2 + (y_3 - 1)^2 - 90(y_3^2 - y_4) + 10.1\left((y_2 - 1)^2 + (y_4 - 1)^2\right) + 19.8(y_2 - 1)(y_4 - 1)$	4	[−10, 10]	UN
F7	Trid 6	$F_7(y) = \sum_{i=1}^{D} (y_i - 1)^2 - \sum_{i=2}^{D} y_i y_{i-1}$	6	[−D^2, D^2]	UN
F8	Trid 10	$F_8(y) = \sum_{i=1}^{D} (y_i - 1)^2 - \sum_{i=2}^{D} y_i y_{i-1}$	10	[−D^2, D^2]	UN
F9	Zakharov	$F_9(y) = \sum_{i=1}^{D} y_i^2 + \left(\sum_{i=1}^{D} 0.5 i y_i\right)^2 + \left(\sum_{i=1}^{D} 0.5 i y_i\right)^4$	10	[−5, 10]	UN
F10	Schwefel 1.2	$F_{10}(y) = \sum_{i=1}^{D} \left(\sum_{j=1}^{i} y_j^2\right)^2$	30	[−100, 100]	UN
F11	Rosenbrock	$F_{11}(y) = \sum_{i=1}^{D} \left[100\left(y_i^2 - y_{i+1}\right)^2 + (1 - y_i)^2\right]$	30	[−30, 30]	UN
F12	Dixon-Price	$F_{12}(y) = (y_i - 1)^2 + \sum_{i=2}^{D} i\left(2y_i^2 - y_{i-1}\right)^2$	30	[−10, 10]	UN
F13	Shekel's Foxholes	$F_{23}(y) = \left[\frac{1}{500} + \sum_{j=1}^{25} \frac{1}{j + \sum_{i=1}^{2} (y_i - a_{ij})^6}\right]$	2	[−65.536, 65.536]	MS
F14	Branin	$F_{13}(y) = \left(y_2 - \frac{5.1}{4\pi^2} y_1^2 + \frac{5}{\pi} y_1 - 6\right)^2 + 10\left(1 - \frac{1}{8\pi}\right)\cos y_1 + 10$	2	[−5, 10] [0, 15]	MS

(continued)

V. R. Ravipudi et al., *Design Optimization of Renewable Energy Systems Using Advanced Optimization Algorithms*, Green Energy and Technology,
https://doi.org/10.1007/978-3-030-95589-2

(continued)

Problem	Function	Formulation	D	Interval	C
F15	Bohachevsky 1	$F_{14}(y) = y_1^2 + 2y_2^2 - 0.3\cos(3\pi y_1) - 0.4\cos(4\pi y_2) + 0.7$	2	[−100, 100]	MS
F16	Booth	$F_{17}(y) = (y_1 + 2y_2 - 7)^2 + (2y_1 + y_2 - 5)^2$	2	[−10, 10]	MS
F17	Michalewicz 2	$F_{18}(y) = -\sum_{i=1}^{D} \sin y_1 \left(\sin\left(iy_i^2/\pi\right)\right)^{20}$	2	[0, π]	MS
F18	Michalewicz 5	$F_{19}(y) = -\sum_{i=1}^{D} \sin y_1 \left(\sin\left(iy_i^2/\pi\right)\right)^{20}$	5	[0, π]	MS
F19	Bohachevsky 2	$F_{15}(y) = y_1^2 + 2y_2^2 - 0.3\cos(3\pi y_1)(4\pi y_2) + 0.3$	2	[−100, 100]	MN
F20	Bohachevsky 3	$F_{16}(y) = y_1^2 + 2y_2^2 - 0.3\cos(3\pi y_1 + 4\pi y_2) + 0.3$	2	[−100, 100]	MN
F21	GoldStein-Price	$F_{20}(y) = \left[1 + (y_1 + y_2 + 1)^2\left(19 - 14y_1 + 3y_1^2 - 14y_2 + 6y_1y_2 + 3y_2^2\right)\right] \left[30 + (2y_1 - 3y_2)^2\left(18 - 32y_1 + 12y_1^2 + 48y_2 - 36y_1y_2 + 27y_2^2\right)\right]$	2	[−2, 2]	MN
F22	Perm	$F_{21}(y) = \sum_{k=1}^{D}\left[\sum_{i=1}^{D}\left(i^k + \beta\right)\left((y_i/i)^k - 1\right)\right]^2$	4	[−D, D]	MN
F23	Hartmann 3	$F_{24}(y) = -\sum_{i=1}^{4} c_i \exp\left[-\sum_{j=1}^{3} a_{ij}\left(y_j - p_{ij}\right)^2\right]$	3	[0, 1]	MN
F24	Ackley	$F_{22}(y) = -20\exp\left(-0.2\sqrt{\frac{1}{D}\sum_{i=1}^{D} y_i^2}\right) - \exp\left(\frac{1}{D}\sum_{i=1}^{D}\cos 2\pi y_i\right) + 20 + e$	30	[−32, 32]	MN
F25	Penalized 2	$F_{25}(y) = 0.1\left[\sin^2(\pi y_1) + \sum_{i=1}^{D-1} \begin{matrix} (y_i - 1)^2\left\{1 + \sin^2(3\pi y_{i+1})\right\} \\ +(y_D - 1)^2 + \left(1 + \sin^2(2\pi y_D)\right) \end{matrix}\right]$ $+\sum_{i=1}^{D} u(y_i, 5, 100, 4), \quad u(y_i, a, k, m) = \begin{Bmatrix} k(y_i - a)^m & y_i \succ a, \\ 0 & -a \le y_i \le a, \\ k(-y_i - a)^m & y_i \prec -a \end{Bmatrix}$	30	[−50, 50]	MN
F26	Langerman 2	$F_{26}(y) =$ $-\sum_{i=1}^{D} c_i\left[\exp\left(-\frac{1}{\pi}\sum_{j=1}^{D}\left(x_j - a_{ij}\right)^2\right)\cos\left(\pi\sum_{j=1}^{D}\left(x_j - a_{ij}\right)^2\right)\right]$	2	[0, 10]	MN
F27	Langerman 5	$F_{27}(y) =$ $-\sum_{i=1}^{D} c_i\left[\exp\left(-\frac{1}{\pi}\sum_{j=1}^{D}\left(x_j - a_{ij}\right)^2\right)\cos\left(\pi\sum_{j=1}^{D}\left(x_j - a_{ij}\right)^2\right)\right]$	5	[0, 10]	MN
F28	Langerman 10	$F_{28}(y) =$ $-\sum_{i=1}^{D} c_i\left[\exp\left(-\frac{1}{\pi}\sum_{j=1}^{D}\left(x_j - a_{ij}\right)^2\right)\cos\left(\pi\sum_{j=1}^{D}\left(x_j - a_{ij}\right)^2\right)\right]$	10	[0, 10]	MN
F29	Fletcher Powell 5	$F_{29}(y) = \sum_{i=1}^{D}(A_i - B_i)^2$ $A_i = \sum_{j=1}^{D}\left(a_{ij}\mathrm{Sin}(\alpha_j) + b_{ij}\mathrm{Cos}(\alpha_j)\right), B_i =$ $\sum_{j=1}^{D}\left(a_{ij}\mathrm{Sin}(x_j) + b_{ij}\mathrm{Cos}(x_j)\right)$	5	[−π, π]	MN
F30	Fletcher Powell 1	$F_{30}(y) = \sum_{i=1}^{D}(A_i - B_i)^2$ $A_i = \sum_{j=1}^{D}\left(a_{ij}\mathrm{Sin}(\alpha_j) + b_{ij}\mathrm{Cos}(\alpha_j)\right), B_i =$ $\sum_{j=1}^{D}\left(a_{ij}\mathrm{Sin}(x_j) + b_{ij}\mathrm{Cos}(x_j)\right)$	10	[−π, π]	MN

D—Dimension, C—Characteristics, U—Unimodal, M—Multimodal, S—Separable, N—Non-separable

Appendix A.2. Unimodal and Multimodal Benchmark Problems Set-1

Problem	Formulation	Interval
P1	$G_1(x) = \sum_{i=1}^{n} x_i^2$	[−100, 100]
P2	$G_2(x) = \sum_{i=1}^{n} \lvert x_i\rvert + \Pi_{i=1}^{n}\lvert x_i\rvert$	[−10, 10]
P3	$G_3(x) = \sum_{i=1}^{n} \left(\sum_{j=1}^{i} x_j\right)^2$	[−100, 100]
P4	$G_4(x) = \max_i\{\lvert x_i\rvert,\ 1 \le i \le n\}$	[−100, 100]
P5	$G_5(x) = \sum_{i=1}^{n-1} \left[100\left(x_{i+1} - x_i^2\right)^2 + (x_i - 1)^2\right]$	[−30, 30]
P6	$G_6(x) = \sum_{i=1}^{n} (\lvert x_i + 0.5\rvert)^2$	[−100, 100]
P7	$G_7(x) = \sum_{i=1}^{n} i x_i^4 + \text{random}[0, 1)$	[−1.28, 1.28]
P8	$G_8(x) = \sum_{i=1}^{n} -\mathrm{x_i} \sin\left(\sqrt{\lvert x_i\rvert}\right)$	[−500, 500]
P9	$G_9(x) = \sum_{i=1}^{n} \left[x_i^2 - 10\cos(2\pi x_i) + 10\right]$	[−5.12, 5.12]
P10	$G_{10}(x) = -20\exp\left(-0.2\sqrt{\left(\frac{1}{n}\sum_{i=1}^{n} x_i^2\right)}\right) - \exp\left(\frac{1}{n}\sum_{i=1}^{n}\cos(2\pi x_i)\right) + 20 + e$	[−32, 32]
P11	$G_{11}(x) = \frac{1}{4000}\sum_{i=1}^{n} x_i^2 - \Pi_{i=1}^{n}\cos\left(\frac{x_i}{\sqrt{i}}\right) + 1$	[−600, 600]

(continued)

(continued)

Problem	Formulation	Interval
P12	$G_{12}(x) = \frac{\pi}{n}\left\{10\sin^2(\pi.y) + \sum_{i=1}^{n-1}(y_i - 1)^2\left[1 + 10\sin^2(\pi.y_{i+1})\right] + (y_n - 1)^2\right\} + \sum_{i=1}^{n} u(x_i, 10, 100, 4)$ $y_i = 1 + \left(\frac{x_i+1}{4}\right)$ $u(x_i, a.k, m) = \left\{\begin{array}{ll} k(x_i - a)^m & x_i > a \\ 0 & -a < x_i < a \\ k(-x_i - a)^m & x_i < -a \end{array}\right\}$	[−50, 50]
P13	$G_{13}(x) = 0.1\left\{\sin^2(3\pi x_1) + \sum_{i=1}^{n-1} \begin{array}{l} (x_i - 1)^2\left[1 + \sin^2(3\pi x_{i+1})\right] \\ +(x_n - 1)^2\left[1 + \sin^2(2\pi x_n)\right] \end{array}\right\} + \sum_{i=1}^{n} u(x_i, 5, 100, 4)$	[−50, 50]
P14	$G_{14}(x) = \left(\frac{1}{500} + \sum_{j=1}^{25}\frac{1}{j+\sum_{i=1}^{2}(x_i-a_{ij})^6}\right)^{-1}$	[−65, 65]
P15	$G_{15}(x) = \sum_{i=1}^{11}\left[a_i - \frac{x_1\left(b_i^2+b_i x_2\right)}{b_i^2+b_i x_3+x_4}\right]^2$	[−5, 5]

(continued)

(continued)

Problem	Formulation	Interval
P16	$G_{16}(x) = 4x_1^2 + 2.1x_1^4 + \frac{1}{3}x_1^6 + x_1x_2 - 4x_2^2 + 4x_2^4$	[−5, 5]
P17	$G_{17}(x) = \left(x_2 - \frac{5.1}{4\pi^2}x_1^2 + \frac{5}{\pi}x_1 - 6\right)^2 + 10(1 - \frac{1}{8\pi})\cos x_1 + 10$	[−5, 5]
P18	$G_{18}(x) = \left[1 + (x_1 + x_2 + 1)^2\left(19 - 14x_1 + 3x_1^2 - 14x_2 + 6x_1x_2 + 3x_2^2\right)\right]$ $\times \left[30 + (2x_1 - 3x_2)^2 \times \left(18 - 32x_1 + 12x_1^2 + 48x_2 - 36x_1x_2 + 27x_2^2\right)\right]$	[−2, 2]
P19	$G_{19}(x) = -\sum_{i=1}^{4} C_i \exp\left(-\sum_{j=1}^{3} a_{ij}\left(x_j - P_{ij}\right)^2\right)$	[0, 1]
P20	$G_{20}(x) = -\sum_{i=1}^{4} C_i \exp\left(-\sum_{j=1}^{6} a_{ij}\left(x_j - P_{ij}\right)^2\right)$	[0, 1]
P21	$G_{21}(x) = -\sum_{i=1}^{5}\left[\sum_{j=1}^{4}\left(x_j - a_{ij}\right)^2 + c_i\right]^{=1}$	[0, 10]
P22	$G_{22}(x) = -\sum_{i=1}^{7}\left[\sum_{j=1}^{4}\left(x_j - a_{ij}\right)^2 + c_i\right]^{=1}$	[0, 10]
P23	$G_{23}(x) = -\sum_{i=1}^{10}\left[\sum_{j=1}^{4}\left(x_j - a_{ij}\right)^2 + c_i\right]^{=1}$	[0, 10]

Appendix A.3. Unimodal and Multimodal Benchmark Problems Set-2

Problem	Test function	Interval
P1	$TF_1(y) = \sum_{i=1}^{d} y_i^2$	$[-100, 100]^d$
P2	$TF_2(y) = \sum_{i=1}^{d} \text{abs}\lvert y_i \rvert + \prod_{i=1}^{d} \text{abs}\lvert y_i \rvert$	$[-10, 10]^d$
P3	$TF_3(y) = \sum_{m=1}^{d} \left(\sum_{n=1}^{m} y_n\right)^2$	$[-100, 100]^d$
P4	$TF_4(y) = \max\{\lvert y_i \rvert, 1 \le i \le d\}$	$[-100, 100]^d$
P5	$TF_5(y) = \sum_{m=1}^{d} \left[100\left(y_{m+1} - y_m^2\right) + (y_i - 1)^2\right]$	$[-30, 30]^d$
P6	$TF_6(y) = \sum_{i=1}^{d} \lfloor y_i + 0.5 \rfloor^2$	$[-100, 100]^d$
P7	$TF_7(y) = \sum_{i=1}^{d} i y_i^4 + \text{rand}[0, 1]$	$[-1.28, 1.28]^d$
P8	$TF_8(y) = \sum_{m=1}^{d} -y_m \sin\left(\sqrt{\lvert y_m \rvert}\right)$	$[-500, 500]^d$
P9	$TF_9(y) = \sum_{m=1}^{d} \left[y_m^2 - 10\cos(2\pi y_m) + 10\right]$	$[-5.12, 5.12]^d$
P10	$TF_{10}(y) =$ $20\left(-0.2\sqrt{\frac{1}{d}\sum_{m=1}^{d} y_m^2}\right) - \exp\left(\frac{1}{d}\sum_{m=1}^{d}\cos(2\pi y_m)\right) + 20 + e$	$[-32, 32]^d$

Appendix B. Multi-objective Optimization Standard Benchmark Problems

Multi-objective Optimization ZDT Test Problems

ZDT-1 problem:

Minimize: $f_1(x) = x_1$

Minimize: $f_2(x) = g(x) \times h(f_1(x), g(x))$

where $G(x) = 1 + \frac{9}{N-1} \sum_{i=2}^{N} x_i$

$$h(f_1(x), g(x)) = 1 - \sqrt{\frac{f_1(x)}{g(x)}}$$

$$0 \le x_i \le 1, 1 \le i \le 30$$

ZDT-2 problem:

Minimize: $f_1(x) = x_1$

Minimize: $f_2(x) = g(x) \times h(f_1(x), g(x))$

where $G(x) = 1 + \frac{9}{N-1} \sum_{i=2}^{N} x_i$

$$h(f_1(x), g(x)) = 1 - \left(\frac{f_1(x)}{g(x)}\right)^2$$

$$0 \leq x_i \leq 1, 1 \leq i \leq 30$$

ZDT-3 problem:

Minimize: $f_1(x) = x_1$

Minimize: $f_2(x) = g(x) \times h(f_1(x), g(x))$

where $G(x) = 1 + \frac{9}{29} \sum_{i=2}^{N} x_i$

$$h(f_1(x), g(x)) = 1 - \sqrt{\frac{f_1(x)}{g(x)}} - \left(\frac{f_1(x)}{g(x)}\right) \sin(10\pi f_1(x))$$

$$0 \leq x_i \leq 1, 1 \leq i \leq 30$$

ZDT-1 with linear Pareto-front (ZDT-1L) problem:

Minimize: $f_1(x) = x_1$

Minimize: $f_2(x) = g(x) \times h(f_1(x), g(x))$

where $G(x) = 1 + \frac{9}{N-1} \sum_{i=2}^{N} x_i$

$$h(f_1(x), g(x)) = 1 - \frac{f_1(x)}{g(x)}$$

$$0 \leq x_i \leq 1, 1 \leq i \leq 30$$

ZDT-2 with three-objective (ZDT2-3O) problem:

Minimize: $f_1(x) = x_1$

Minimize: $f_2(x) = x_2$

Minimize: $f_3(x) = g(x) \times h(f_1(x), g(x)) \times h(f_2(x), g(x))$

where $G(x) = 1 + \frac{9}{N-1} \sum_{i=3}^{N} x_i$

$$h(f_1(x), g(x)) = 1 - \left(\frac{f_1(x)}{g(x)}\right)^2$$

$$h(f_2(x), g(x)) = 1 - \left(\frac{f_2(x)}{g(x)}\right)^2$$

$$0 \le x_i \le 1, 1 \le i \le 30$$

Appendix C. Codes for Jaya and Rao Algorithms (and their improved versions) for Unconstrained Optimization Problems

The codes for unconstrained benchmark problems presented in Appendix A.1 for sample single-objective optimization functions are given below. The user has to create separate MATLAB files, but the files are to be saved in a single folder. These codes may be used for reference, and the user may define the objective function(s), design variables, and their ranges as per his or her own requirements. The algorithm codes are made as function file which needs to be called in the '*All_Algorithms_Main.m*' file.

C.1 All_Algorithms_Main.m: Main Program for Executing All Algorithms

This program is to initialize the optimization problem parameters and algorithm parameters. The variable TME defines the algorithm to be executed. The TME values for different algorithms are as follows: TME = −2 executes the AMTPG-Jaya algorithm, TME = −1 executes the MTPG-Jaya algorithm, TME = 0 executes the Jaya algorithm, TME = 1 executes the Rao-1 algorithm, TME = 2 executes the Rao-2 algorithm, TME = 3 executes the Rao-3 algorithm, TME = 4 executes the Rao-4 algorithm, TME = 5 executes the Elitist_Rao-1 algorithm, TME = 6 executes the Elitist_Rao-2 algorithm, TME = 7 executes the Elitist_Rao-3 algorithm, TME = 8 executes the Elitist_Rao-4 algorithm, and TME = 9 executes the SAP_Rao algorithm. The unconstrained benchmark problems presented in Appendix A.1 are coded in the '*Unconstrained_benchmark_func.m*' file. This program is only for demonstration purpose. The numbers assigned to the population size, generations, design variables, and maximum function evaluations in this program need not be taken as default values

```
%%%%Code Begins %%%%
%%%% All_Algorithms_Main.m%%%%
clear
clear global
clear global variables
clc
warning off
rehash
global func_num
global L
global U
global P
global D
global MFEs
global Fes
global Ter_Error
global TME
global convergence
global ES
global Max_N_Teams
global Min_N_Teams
convergence =[];
format short g
MFEs=100000; %% maximum number of function evaluations
for P=20 %% population size
runs=30; %% number of runs
%% TME is to select the algorithm that we want to run
%%% TME=-2 Executes the AMTPG-Jaya algorithm
%%% TME=-1 Executes the AMTPG-Jaya algorithm
%%% TME=0 Executes the Jaya algorithm
%%% TME=1 Executes the Rao-1 algorithm
%%% TME=2 Executes the Rao-2 algorithm
%%% TME=3 Executes the Rao-3 algorithm
%%% TME=4 Executes the Rao-4 algorithm
%%% TME=5 Executes the Elitist_Rao-1 algorithm
%%% TME=6 Executes the Elitist_Rao-2 algorithm
%%% TME=7 Executes the Elitist_Rao-3 algorithm
%%% TME=8 Executes the Elitist_Rao-4 algorithm
%%% TME=9 Executes the SAP_Rao algorithm
for TME=[-2,-1,0,1,2,3,4,5,6,7,8,9] %% Algorithm number
   %%% select the unconstrained benchmark function number given
   % in the function file named "Unconstrained_benchmark_func.m"
for func_num=1:30
% Problem boundaries & number of variables are presented in "Dand-
Bounds.m".
   [D,lb1,ub1,Ter_Error]=DandBounds(func_num);
   L=lb1*ones(1,D); %% Lower boundaries of variables
   U=ub1*ones(1,D); %% Upper boundaries of variables
      ObjectiveFunction=@Unconstrained_benchmark_func; %% Func-
tions file
```

```
   GO=zeros(runs,1); %% Global Optimum
   GV=zeros(runs,D); %% Global Optimum variables
        GNFE=zeros(runs,1);  %%  number  of  function  evalua-
tions to reach optimum
%   Avg_eva=nan((MFEs/P),runs);
   Avg_obj=nan(round(MFEs/P),runs);
   for run=1:runs
      rng('shuffle')
      if TME<=4 && TME>=0 %% Jaya and Rao algorithms
         [xf]=single_obj(ObjectiveFunction);
      elseif TME==-1 %% MTPG-Jaya algorithm
         [xf]=MTPG_single_obj(ObjectiveFunction);
      elseif TME==-2 %% AMTPG-Jaya algorithm
         Min_N_Teams=2;
         Max_N_Teams=6;
         [xf]=AMTPG_single_obj(ObjectiveFunction);
      elseif TME>=5 && TME<=8%% ElitistRao algorithms
         ES=round(0.1*P); %% Elite size
         [xf]=E_Rao_single_obj(ObjectiveFunction);
      elseif TME==9 %% SAP-Rao algorithm
         [xf]=SAP_Rao_single_obj(ObjectiveFunction);
      end
      [val,ind]=min(xf(:,D+1));
      A=xf(ind,1:D);
      B=val;
      GO(run,1)= B;
      GV(run,:)= A;
      GNFE(run,1)= Fes;
      [rows,~]=size(convergence);
%       Avg_eva(1:rows,run)=convergence(:,1);
      Avg_obj(1:rows,run)=convergence(:,2);
      convergence =[];
   end
%   AE=mean(Avg_eva(:,:),2);
   AO=mean(Avg_obj(:,:),2);
   bbest=min(GO);
   wbest=max(GO);
   mbest=mean(GO);
   stdbest=std(GO);
   mFes=mean(GNFE);
   stdFes=std(GNFE);
   disp(['--------[ Algorithm',' TME No.= ',num2str(TME), ' ]-----
-----'])
   disp(['--------[ ','Fun No.= ',num2str(func_num), ' ]---------
------'])
     fprintf(' best=%f  worst=%f \n mean=%f  std=%f \n mean Fes=
%f \n std Fes=%f \n',bbest,wbest,mbest,stdbest,mFes,stdFes);
   fprintf('\n ')
   beep;
end
end
end
%%%%Code Ends %%%%
```

C.2 *single_obj.m: Jaya and Rao Algorithm Programs*

This function file needs to be called in the *'All_Algorithms_Main.m'* program to execute the algorithms. As the flow of the Jaya and Rao algorithms is similar and they differ only in movement equation, a common function file is created for these algorithms with sub-function files for each algorithm. The sub-functions for the Jaya and Rao algorithms are as follows: *'updatepop'* for the Jaya algorithm, *'Rao1'* for the Rao-1 algorithm, *'Rao2'* for the Rao-2 algorithm, *'Rao3'* for the Rao-3 algorithm, and *'Rao4'* for the Rao-4 algorithm. The *'Trimr'* function is to check and update the boundary constraints of the variables.

```
%%%%Code Begins %%%%
%%%% single_obj.m%%%%
function [xf]=single_obj(ObjectiveFunction)
global L
global U
global P
global D
global MFEs
global TME
global Fes
global initial_flag
global Ter_Error
global convergence
x=repmat(L,P,1)+rand(P,D).*repmat((U-L),P,1);
initial_flag=0;
f=ObjectiveFunction(x);
xf=[x,f];
NFE=0;
Gfold(1,1)=min(f);
counter =1;
while NFE<MFEs
   nx=[];
   if TME==0
      [nx]=updatepop(xf); %% Jaya algorithm
   elseif TME==1
      [nx]=Rao1(xf); %% Rao-1 algorithm
   elseif TME==2
      [nx]=Rao2(xf); %% Rao-2 algorithm
   elseif TME==3
      [nx]=Rao3(xf); %% Rao-3 algorithm
   elseif TME==4
      [nx]=Rao4(xf); %% Rao-4 algorithm
   end
   [nx,~]=trim(nx,L,U);
   initial_flag=0;
   newf = ObjectiveFunction(nx);
   NFE = NFE + P;
   for i=1:P
      if(newf(i)<=f(i))
         x(i,:)=nx(i,:);
         f(i)=newf(i);
```

```
        end
    end
    xf=[x,f];
    Gfnew(1,1)=min(f);
    if Gfnew<Gfold
        Fes=NFE;
        Gfold=Gfnew;
    end
    if Gfold<=Ter_Error
        Fes=NFE;
        break
    end
    convergence(counter,1)=NFE;
    convergence(counter,2)=Gfold;
    counter =counter+1;
end
end

function [nx]=Rao1(f)
%% Rao-1 algorithm
[row,col]=size(f);
F=f(:,col);
[~,indx]=sort(F);
x=f(:,1:col-1);
xb=x(indx(1,1),:);
xw=x(indx(row,1),:);
bx=repmat(xb,row,1);
wx=repmat(xw,row,1);
r=rand(row,col-1);
nx=x+(r.*(bx-wx));
end

function [nx]=Rao2(f)
%% Rao-2 algorithm
[row,col]=size(f);
x=f(:,1:col-1);
F=f(:,col);
[~,indx]=sort(F);
Best=x(indx(1,1),:);
worst=x(indx(row,1),:);
xnew=zeros(row,col-1);
for i=1:row
    k=unidrnd(row);
    while i==k
        k=randi(row);
    end
    if F(i)<F(k)%% minimization objective
      xnew(i,:)=x(i,:)+rand(1,(col-1)).*(Best-worst)+rand(1,(col-
1)).*(abs(x(i,:))-abs(x(k,:)));
    else
      xnew(i,:)=x(i,:)+rand(1,(col-1)).*(Best-worst)+rand(1,(col-
1)).*(abs(x(k,:))-abs(x(i,:)));
    end
end
nx=xnew;
```

```
end

function [nx]=Rao3(f)
%% Rao-3 algorithm
[row,col]=size(f);
x=f(:,1:col-1);
F=f(:,col);
[~,indx]=sort(F);
Best=x(indx(1,1),:);
worst=x(indx(row,1),:);
xnew=zeros(row,col-1);
for i=1:row
   k=unidrnd(row);
   while i==k
      k=randi(row);
   end
   if F(i)<F(k)%% minimization objective
                         xnew(i,:)=x(i,:)+rand(1,(col-1)).*(Best-
abs(worst))+rand(1,(col-1)).*(abs(x(i,:))-x(k,:));
   else
                         xnew(i,:)=x(i,:)+rand(1,(col-1)).*(Best-
abs(worst))+rand(1,(col-1)).*(abs(x(k,:))-x(i,:));
   end
end
nx=xnew;
end

function [nx]=Rao4(f)
%% Rao-4 algorithm
[row,col]=size(f);
F=f(:,col);
[~,indx]=sort(F);
x=f(:,1:col-1);
xb=x(indx(1,1),:);
xw=x(indx(row,1),:);
bx=repmat(xb,row,1);
wx=repmat(xw,row,1);
r=rand(row,col-1);
r1=rand(row,col-1);
r2=rand(row,col-1);
r3=rand(row,col-1);
nx=x+(r.*(bx-wx))+(r1.*(wx-x)+r2.*(bx-x))*0.5-r3.*(bx-(x));
end

function [nx]=updatepop(f)
%% Jaya algorithm
[pop,c]=size(f);
F=f(:,c);
[~,indx]=sort(F);
x=f(:,1:c-1);
xb=x(indx(1,1),:);
xw=x(indx(pop,1),:);
bx=repmat(xb,pop,1);
wx=repmat(xw,pop,1);
r=rand(pop,c-1);
```

```
r1=rand(pop,c-1);
nx=x+r.*(bx-abs(x))-r1.*(wx-abs(x));
end

function [nx,violations]=trim(nx,L,U)
%% Boundary constraints violation check
[~,D]=size(nx);
violations=0;
for i = 1:D
   indexes = (nx(:,i) < L(i));
   violations = violations + abs(sum(L(i)-nx(indexes,i)));
   if isempty(indexes)==0
      nx(indexes,i) = L(i);
   end
   indexes = (nx(:,i) > U(i));
   violations = violations + abs(sum(nx(indexes,i)-U(i)));
   if isempty(indexes)==0
      nx(indexes,i) = U(i);
   end
end
end
%%%%Code Ends %%%%
```

C.3 E_Rao_single_obj.m: Elitist Rao Algorithm Programs

This function file needs to be called in the *'All_Algorithms_Main.m'* program to execute the algorithms. As the flow of the elitist Rao algorithms is similar and they differ only in movement equation, a common function file is created for these algorithms with sub-function files for movement equation of each algorithm. The sub-functions for the elitist Rao algorithms are as follows: '*Rao1*' for the elitist Rao-1 algorithm, '*Rao2*' for the elitist Rao-2 algorithm, '*Rao3*' for the elitist Rao-3 algorithm, and '*Rao4*' for the elitist Rao-4 algorithm. The '*Trimr*' function is to check and update the boundary constraints of the variables.

```
%%%%Code Begins %%%%
%%%% E_Rao_single_obj.m%%%%
function [xf]=E_Rao_single_obj(ObjectiveFunction)
global L
global U
global P
global D
global MFEs
global TME
global Fes
global initial_flag
global Ter_Error
global convergence
global ES
initial_flag=0;
```

```
x=repmat(L,P,1)+rand(P,D).*repmat((U-L),P,1);
f=nan(P,1);
for i =1:P
   f(i,1)=ObjectiveFunction(x(i,:));
end
xf=[x,f];
NFE=0;
Gfold(1,1)=min(f);
counter =1;
while NFE<MFEs
   xf=unique(xf, 'rows');
   xf=sortrows(xf,D+1); % sort the solutions in ascending order
   [r,~]= size(xf);
   if r<P
      for i=r+1:P
         for j = 1 : D
            xf(i,j) = L(j) + (U(j) - L(j))*rand(1);
         end
      end
      for i = r+1:P
         xf(i,D+1) = ObjectiveFunction(xf(i,1:D));
      end
   else
      xf(:,:)=xf(:,:);
   end
   %% Replace the worst solution with elite solutions
   if mod(counter,round(P))==0
      xf(P-ES+1:P,:)= xf(1:ES,:);
   end
   %%
   nx=[];
   if TME==5 %% Elitist Rao-1 algorithm
      [nx]=Rao1(xf);
   elseif TME==6 %% Elitist Rao-2 algorithm
      [nx]=Rao2(xf);
   elseif TME==7 %% Elitist Rao-3 algorithm
      [nx]=Rao3(xf);
   elseif TME==8 %% Elitist Rao-4 algorithm
      [nx]=Rao4(xf);
   end
   [nx,~]=trim(nx,L,U);
   newf=nan(P,1);
   for i =1:P
      newf(i,1)=ObjectiveFunction(nx(i,:));
   end
   NFE = NFE + P;
   for i=1:P
      if(newf(i)<=f(i))
         x(i,:)=nx(i,:);
         f(i)=newf(i);
      end
   end
   xf=[x,f];
   Gfnew(1,1)=min(f);
```

```
    if Gfnew<Gfold
      Fes=NFE;
      Gfold=Gfnew;
    end
    if Gfold<=Ter_Error
      Fes=NFE;
      break
    end
    convergence(counter,1)=NFE;
    convergence(counter,2)=Gfold;
    counter =counter+1;
end
end

function [nx]=Rao1(f)
%% Elitist Rao-1 algorithm
[row,col]=size(f);
F=f(:,col);
[~,indx]=sort(F);
x=f(:,1:col-1);
xb=x(indx(1,1),:);
xw=x(indx(row,1),:);
bx=repmat(xb,row,1);
wx=repmat(xw,row,1);
r=rand(row,col-1);
nx=x+(r.*(bx-wx));
end

function [nx]=Rao2(f)
%% Elitist Rao-2 algorithm
[row,col]=size(f);
x=f(:,1:col-1);
F=f(:,col);
[~,indx]=sort(F);
Best=x(indx(1,1),:);
worst=x(indx(row,1),:);
xnew=zeros(row,col-1);
for i=1:row
   k=unidrnd(row);
   while i==k
      k=randi(row);
   end
   if F(i)<F(k)%% minimization objective
     xnew(i,:)=x(i,:)+rand(1,(col-1)).*(Best-worst)+rand(1,(col-
1)).*(abs(x(i,:))-abs(x(k,:)));
   else
     xnew(i,:)=x(i,:)+rand(1,(col-1)).*(Best-worst)+rand(1,(col-
1)).*(abs(x(k,:))-abs(x(i,:)));
   end
end
nx=xnew;
end

function [nx]=Rao3(f)
%% Elitist Rao-3 algorithm
```

```
[row,col]=size(f);
x=f(:,1:col-1);
F=f(:,col);
[~,indx]=sort(F);
Best=x(indx(1,1),:);
worst=x(indx(row,1),:);
xnew=zeros(row,col-1);
for i=1:row
  k=unidrnd(row);
  while i==k
     k=randi(row);
  end
  if F(i)<F(k)%% minimization objective
                    xnew(i,:)=x(i,:)+rand(1,(col-1)).*(Best-
abs(worst))+rand(1,(col-1)).*(abs(x(i,:))-x(k,:));
  else
                    xnew(i,:)=x(i,:)+rand(1,(col-1)).*(Best-
abs(worst))+rand(1,(col-1)).*(abs(x(k,:))-x(i,:));
  end
end
nx=xnew;
end

function [nx]=Rao4(f)
%% Elitist Rao-4 algorithm
[row,col]=size(f);
F=f(:,col);
[~,indx]=sort(F);
x=f(:,1:col-1);
xb=x(indx(1,1),:);
xw=x(indx(row,1),:);
bx=repmat(xb,row,1);
wx=repmat(xw,row,1);
r=rand(row,col-1);
r1=rand(row,col-1);
r2=rand(row,col-1);
r3=rand(row,col-1);
nx=x+(r.*(bx-wx))+(r1.*(wx-x)+r2.*(bx-x))*0.5-r3.*(bx-(x));
end

function [nx,violations]=trim(nx,L,U)
%% Boundary constraints violation check
[~,D]=size(nx);
violations=0;
for i = 1:D
  indexes = (nx(:,i) < L(i));
  violations = violations + abs(sum(L(i)-nx(indexes,i)));
  if isempty(indexes)==0
     nx(indexes,i) = L(i);
  end
  indexes = (nx(:,i) > U(i));
  violations = violations + abs(sum(nx(indexes,i)-U(i)));
  if isempty(indexes)==0
     nx(indexes,i) = U(i);
  end
end
end
%%%%Code Ends %%%%
```

C.4 SAP_Rao_single_obj.m: Self-Adaptive Population Rao Algorithm Program

This function file needs to be called in the *'All_Algorithms_Main.m'* program to execute the SAP-Rao algorithm. This program is only for demonstration purpose.

```
%%%%Code Begins %%%%
%%%% SAP_Rao_single_obj.m%%%%
function [xf]=SAP_Rao_single_obj(ObjectiveFunction)
global L
global U
global P
global D
global MFEs
global Fes
global initial_flag
global Ter_Error
global convergence
%%
initial_flag=0;
x=repmat(L,P,1)+rand(P,D).*repmat((U-L),P,1);
f=nan(P,1);
for i =1:P
  f(i,1)=ObjectiveFunction(x(i,:));
end
xf=[x,f];
NFE=0;
Gfold(1,1)=min(f);
GOFV=min(f);
counter =1;
while NFE<MFEs
  nx=nan(P,D);
  sub_pop_size=floor(P/4);
  IND=randperm(P);
  from=1;
  for k=1:4
    if(k~=4)
      to=from+sub_pop_size-1;
      ind=IND(from:to);
      from=from+sub_pop_size;
    else
      ind=IND(from:end);
    end
    tempxf=xf(ind,:);
    tme=k;%% this is a local variable
    if tme==4
      [tempxnew]=Rao1(tempxf);
    elseif tme==1
      [tempxnew]=Rao2(tempxf);
    elseif tme==2
      [tempxnew]=Rao3(tempxf);
    elseif tme==3
      [tempxnew]=Rao4(tempxf);
```

```
      end
      nx(ind,:)=tempxnew;
    end
    [nx,~]=trim(nx,L,U);
    newf=nan(P,1);
    for i =1:P
      newf(i,1)=ObjectiveFunction(nx(i,:));
    end
    NFE = NFE + P;
    for i=1:P
      if(newf(i)<f(i))
        x(i,:)=nx(i,:);
        f(i)=newf(i);
      end
    end
    xf=[x,f];
    Gfnew(1,1)=min(f);
    %% %%%%%%%%%%%%%%%%
    xf=sortrows(xf,D+1);
    %%%%%%%%%%%%%%%%%%%%
    %%
    %   disp(mod(Gfold,Gfnew))
    if abs(mod(Gfold,Gfnew))<1e-1
      if P>20
        P=P-round(0.05*P);
        x=xf(1:P,1:D);
        f=xf(1:P,D+1);
        xf=xf(1:P,:);
      end
    else
      if P<1000
        P_old=P;
        P=P+round(0.05*P);
        P_diff=P-P_old;
        xfadded=xf(1:P_diff,:);
        x=[x;xfadded(:,1:D)];
        f=[f;xfadded(:,D+1)];
        xf=[x,f];
      end
    end
    if Gfnew<GOFV
      Fes=NFE;
      GOFV=Gfnew;
    end
    if GOFV<=Ter_Error
      Fes=NFE;
      break
    end
    Gfold=Gfnew;
    convergence(counter,1)=NFE;
    convergence(counter,2)=GOFV;
    convergence(counter,3)=P;
    counter =counter+1;
  end
```

```
xf=[x,f];
end
function [nx]=Rao1(f)
[row,col]=size(f);
F=f(:,col);
[~,indx]=sort(F);
x=f(:,1:col-1);
xb=x(indx(1,1),:);
xw=x(indx(row,1),:);
bx=repmat(xb,row,1);
wx=repmat(xw,row,1);
r=rand(row,col-1);
nx=x+(r.*(bx-wx));
end
function [nx]=Rao2(f)
[row,col]=size(f);
x=f(:,1:col-1);
F=f(:,col);
[~,indx]=sort(F);
Best=x(indx(1,1),:);
worst=x(indx(row,1),:);
xnew=zeros(row,col-1);
for i=1:row
   k=unidrnd(row);
   while i==k
      k=randi(row);
   end
   if F(i)<F(k)%% minimization objective
     xnew(i,:)=x(i,:)+rand(1,(col-1)).*(Best-worst)+rand(1,(col-
1)).*(abs(x(i,:))-abs(x(k,:)));
   else
     xnew(i,:)=x(i,:)+rand(1,(col-1)).*(Best-worst)+rand(1,(col-
1)).*(abs(x(k,:))-abs(x(i,:)));
   end
end
nx=xnew;
end
function [nx]=Rao3(f)
[row,col]=size(f);
x=f(:,1:col-1);
F=f(:,col);
[~,indx]=sort(F);
Best=x(indx(1,1),:);
worst=x(indx(row,1),:);
xnew=zeros(row,col-1);
for i=1:row
   k=unidrnd(row);
   while i==k
      k=randi(row);
   end
   if F(i)<F(k)%% minimization objective
                              xnew(i,:)=x(i,:)+rand(1,(col-1)).*(Best-
abs(worst))+rand(1,(col-1)).*(abs(x(i,:))-x(k,:));
   else
```

```
                         xnew(i,:)=x(i,:)+rand(1,(col-1)).*(Best-
abs(worst))+rand(1,(col-1)).*(abs(x(k,:))-x(i,:));
  end
end
nx=xnew;
end
function [nx]=Rao4(f)
[row,col]=size(f);
F=f(:,col);
[~,indx]=sort(F);
x=f(:,1:col-1);
xb=x(indx(1,1),:);
xw=x(indx(row,1),:);
bx=repmat(xb,row,1);
wx=repmat(xw,row,1);
r=rand(row,col-1);
r1=rand(row,col-1);
r2=rand(row,col-1);
r3=rand(row,col-1);
nx=x+(r.*(bx-wx))+(r1.*(wx-x)+r2.*(bx-x))*0.5-r3.*(bx-(x));
end
function [nx,violations]=trim(nx,L,U)
[~,D]=size(nx);
violations=0;
for i = 1:D
  indexes = (nx(:,i) < L(i));
  violations = violations + abs(sum(L(i)-nx(indexes,i)));
  if isempty(indexes)==0
    nx(indexes,i) = L(i);
  end
  indexes = (nx(:,i) > U(i));
  violations = violations + abs(sum(nx(indexes,i)-U(i)));
  if isempty(indexes)==0
    nx(indexes,i) = U(i);
  end
end
end
%%%%Code Ends %%%%
```

C.5 *MTPG_ single_obj.m: MTPG-Jaya Algorithm Program*

This function file needs to be called in the '*All_Algorithms_Main.m*' program to execute the MTPG-Jaya algorithm. This program is only for demonstration purpose. The numbers assigned to the population size, generations, design variables, and maximum function evaluations in this program need not be taken as default values.

```
%%%%Code Begins %%%%
%%%% MTPG_single_obj.m%%%%
function [xf]=MTPG_single_obj(ObjectiveFunction)
global L
```

```
global U
global P
global D
global MFEs
global Fes
global initial_flag
global Ter_Error
global convergence
%%%%%%%%%%%%%%%%%%%%%%%%%%%%%%%%%%%%%%%%%%%%%%
N_Teams = 4; %4
NE=5; %5
Pacc=0.35; %0.1
MaxStagnation = 5;  % The maximum iterations without improve-
ment to consider not stagnated
Stagnation=0;
Command_Center_Iter = 10;
x=repmat(L,P,1)+rand(P,D).*repmat((U-L),P,1);
initial_flag=0;
f=ObjectiveFunction(x);
NFE = 0;
sol_current = x ;
FV_current= f ;

%% Update Global Best
[FV_Global, index] = min(FV_current);
sol_Global = sol_current (index,:); % Global Best
Fes = NFE;
FV_Global_Previous=inf;
%%%%%%%%%%%%%%%%%%%%%%%%%%%%%%%%%%%%%%%%
%%% MAIN LOOP
%%%%%%%%%%%%%%%%%%%%%%%%%%%%%%%%%%%%%%%%
Teams_sols = cell(1, N_Teams); % matrix of solutions
Teams_FV = Inf * ones(P, N_Teams);  % matrix of objective func-
tion values
Teams_Rank = Teams_FV; % matrix of objective function values
Teams_viols = zeros(1, N_Teams);
FirmWare=randi(NE,[1,N_Teams]);
% FirmWare=1;
r1=rand;
r2=rand;
itr=0;
counter =1;
while NFE<MFEs
  itr=itr+1;
  sol_iteration = sol_current; % Local Best
  FV_iteration = FV_current;
  %% Find New solutions
  violations = zeros(1, N_Teams);
  for team=1:N_Teams
    f=[sol_iteration,FV_iteration];
    if FirmWare(team)==1
      [nx]=updatepop(f);
    elseif FirmWare(team)==2
      [nx,r1]=cjupdate(f,r1);
```

```
      elseif FirmWare(team)==3
         [nx]=randupdate(sol_iteration);
      elseif FirmWare(team)==4
         [nx,r2]=chaosupdate(sol_iteration,r2);
      elseif FirmWare(team)==5
         nx=quasi(L,U,sol_iteration);
      elseif FirmWare(team)==6
         [nx]=updatepopjh(f);
      end
nx,violations(team)]=boundcorrection(nx,L,U,violations(team));
      Teams_sols{team} = nx;
      initial_flag=0;
      Teams_FV(:,team) = ObjectiveFunction(Teams_sols{team});
      NFE = NFE + P;
   end
   %% %% COMMAND CENTER
   for i = 1:P
     [sorted, indices] = sort(Teams_FV(i, :)); % Sort the row posi-
tion idx
      FV_iteration(i) = sorted(1); % Get the minimum value
      [~,c]=size(Teams_sols{indices(1)}(i, :));
      if c<D
         disp('error')
      end
      sol_iteration (i,:) = Teams_sols{indices(1)}(i, :);
      Teams_Rank(i, :) = indices;
   end
      Teams_viols = Teams_viols + mean(Teams_Rank, 1) + viola-
tions; % average rank of each team
   improved = FV_iteration < FV_current;
   %% ACCEPT WORSE SOLUTIONS IF SEARCH STAGNATED ? Pacc% OF CHANCE
   if (Stagnation > MaxStagnation-1)
      %       fprintf(' Stagnated!!! Selecting non-improved solu-
tions.\n');
      indexes = find( (improved + (rand(P, 1) < Pacc)) > 0);
      Stagnation = 0;
   else
      indexes = find( improved );
   end;
   %% Update Current Solution with iteration solution
   FV_current(indexes) = FV_iteration(indexes);
   sol_current(indexes,:) = sol_iteration (indexes,:);
   %% Update Global Best
   [FV_Global_New,index] = min(FV_current);
   if FV_Global_New < FV_Global
      FV_Global=FV_Global_New;
      sol_Global=sol_current (index,:);
      Fes=NFE;% NFE_FinalBest = NFE;
   end
   %% % TIME TO UPDATE THE FIRMWARE
   if (rem (itr, Command_Center_Iter) == 0)
      [~, index]=sort(Teams_viols);
      worst_FW_Num=index(N_Teams);
      worst_FW=FirmWare(worst_FW_Num);
```

```
% best_FW_Num=index(1); % best_FW=FirmWare(best_FW_Num); % h=1;
      h=randi(NE);
      while h==worst_FW
         h=randi(NE);
      end
FirmWare(index(N_Teams))=h; % FirmWare(index(N_Teams))=best_FW;
      Teams_viols = zeros(1, N_Teams);
   end %% Command_Center_Iters
   %% Stagnation Check
 if ( abs(FV_Global - FV_Global_Previous) / abs(FV_Global_Previous)
< 1e-6 ) %% Needs better improvement
      Stagnation = Stagnation + 1;
   else
      Stagnation = 0;
   end
   %% update previous global by current global solution
   FV_Global_Previous=FV_Global;
   xf=[sol_Global,FV_Global];
   if FV_Global<=Ter_Error
      Fes=NFE;
      break
   end
   convergence(counter,1)=NFE;
   convergence(counter,2)=FV_Global;
   counter =counter+1;
end
end
function [nx,violations]=boundcorrection(nx,LB,UB,violations)
[~,D]=size(nx);
for i = 1:D
   indexes = find(nx(:,i) < LB(i));
   violations = violations + abs(sum(LB(i)-nx(indexes,i)));
   if isempty(indexes)==0
      nx(indexes,i) = LB(i);
   end
   indexes = find(nx(:,i) > UB(i));
   violations = violations + abs(sum(nx(indexes,i)-UB(i)));
   if isempty(indexes)==0
      nx(indexes,i) = UB(i);
   end
end
end
function [nx]=updatepopjh(f)
[pop,NT]=size(f);
x=f(:,1:NT-1);
[sorted, ~]=sortrows(f,NT);
b1=sorted(1,1:NT-1);
b2=sorted(2,1:NT-1);
b3=sorted(3,1:NT-1);
b=[b1; b2; b3];
w1=sorted(pop,1:NT-1);
w=repmat(w1,pop,1);
nx=zeros(pop,NT-1);
for j=1:NT-1
```

```
  rb=randi(3,[pop,1]);
  r=rand(pop,1);
          nx(:,j)=x(:,j)+r.*(b(rb,j)-abs(x(:,j)))-r.*(w(:,j)-
abs(x(:,j)));
end
end
function [nx]=randupdate(f)
[pop,NT]=size(f);
x=f(:,1:NT);
nx=zeros(pop,NT);
r1=rand;
for i=1:pop
   for j=1:NT
      r1=4*r1*(1-r1);
      nx(i,j)=x(i,j)+rand*(2*r1-1);
   end
end
end
function xn=quasi(umin,umax,x)
[pop,var]=size(x);
a1=(umin+umax)/2;
b1=repmat(umin+umax,pop,1);
a=repmat(a1,pop,1);
b=b1-x;
c=rand(pop,var);
xn= a+(b-a).*c;
end
function [nx,r1]=cjupdate(f,r1)
[pop,NT]=size(f);
x=f(:,1:NT-1);
[~, ind]=min(f(:,NT));
xb=x(ind,:);
bx=repmat(xb,pop,1);
[~, ind]=max(f(:,NT));
xw=x(ind,:);
wx=repmat(xw,pop,1);
r=zeros(pop,NT-1);
for i=1:pop
   for j=1:NT-1
      r(i,j)=4*r1*(1-r1);
      r1=r(i,j);
   end
end
nx=x+r.*(bx-abs(x))-r.*(wx-abs(x));
end
function [nx,r1]=chaosupdate(f,r1)
[pop,NT]=size(f);
x=f(:,1:NT);
nx=zeros(pop,NT);
for i=1:pop
   for j=1:NT
      r1=4*r1*(1-r1);
      nx(i,j)=x(i,j)+rand*(2*r1-1);
   end
```

```
end
end
function [nx]=updatepop(f)
[pop,NT]=size(f);
x=f(:,1:NT-1);
[~, ind]=min(f(:,NT));
xb=x(ind,:);
bx=repmat(xb,pop,1);
[~, ind]=max(f(:,NT));
xw=x(ind,:);
wx=repmat(xw,pop,1);
nx=x+rand(pop,NT-1).*(bx-abs(x))-rand(pop,NT-1).*(wx-abs(x));
end
%%%%Code Ends %%%%
```

C.6 AMTPG_single_obj.m: MTPG-Jaya Algorithm Program

This function file needs to be called in the *'All_Algorithms_Main.m'* program to execute the AMTPG-Jaya algorithm. This program is only for demonstration purpose. The numbers assigned to the population size, generations, design variables, and maximum function evaluations in this program need not be taken as default values.

```
%%%%Code Begins %%%%
%%%% AMTPG_single_obj.m%%%%
function [xf]=AMTPG_single_obj(ObjectiveFunction)
global L
global U
global P
global D
global MFEs
global Fes
global initial_flag
global Ter_Error
global convergence
global FirmWare
global NE
global N_Teams
%%%%%%%%%%%%%%%%%%%%%%%%%%%%%%%%%%%%%%%%%%%%%
N_Teams = 4; %4
NE=5; %5
Pacc=0.35; %0.1
MaxStagnation = 5;  % The maximum iterations without improve-
ment to consider not stagnated
Stagnation=0;
Command_Center_Iter = 10;
x=repmat(L,P,1)+rand(P,D).*repmat((U-L),P,1);
initial_flag=0;
f=ObjectiveFunction(x);
```

```
NFE = 0;
sol_current = x;
FV_current= f;

%% Update Global Best
[FV_Global, index] = min(FV_current);
sol_Global = sol_current (index,:); % Global Best
Fes = NFE;
FV_Global_Previous=inf;
FV_Global_Old=FV_Global;
%%%%%%%%%%%%%%%%%%%%%%%%%%%%%%%%%%%%%%%%%%
%%% MAIN LOOP
%%%%%%%%%%%%%%%%%%%%%%%%%%%%%%%%%%%%%%%%%%
Teams_sols = cell(1, N_Teams); % matrix of solutions
Teams_FV = Inf * ones(P, N_Teams);  % matrix of objective func-
tion values
Teams_Rank = Teams_FV; % matrix of objective function values
Teams_viols = zeros(1, N_Teams);
FirmWare=randi(NE,[1,N_Teams]);
% FirmWare=1;
r1=rand;
r2=rand;
itr=0;
counter =1;
while NFE<MFEs
   itr=itr+1;
   sol_iteration = sol_current; % Local Best
   FV_iteration = FV_current;
   %% Find New solutions
   violations = zeros(1, N_Teams);
   for team=1:N_Teams
      f=[sol_iteration,FV_iteration];
      if FirmWare(team)==1
         [nx]=updatepop(f);
      elseif FirmWare(team)==2
         [nx,r1]=cjupdate(f,r1);
      elseif FirmWare(team)==3
         [nx]=randupdate(sol_iteration);
      elseif FirmWare(team)==4
         [nx,r2]=chaosupdate(sol_iteration,r2);
      elseif FirmWare(team)==5
         nx=quasi(L,U,sol_iteration);
      elseif FirmWare(team)==6
         [nx]=updatepopjh(f);
      end
 [nx,violations(team)]=boundcorrection(nx,L,U,violations(team));
      Teams_sols{team} = nx;
      initial_flag=0;
      Teams_FV(:,team) = ObjectiveFunction(Teams_sols{team});
      NFE = NFE + P;
   end
   %% %% COMMAND CENTER
   for i = 1:P
     [sorted, indices] = sort(Teams_FV(i, :)); % Sort the row posi-
 tion idx
```

```
        FV_iteration(i) = sorted(1);  % Get the minimum value
        sol_iteration (i,:) = Teams_sols{indices(1)}(i, :);
        Teams_Rank(i, :) = indices;
    end
        Teams_viols  =  Teams_viols  +  mean(Teams_Rank,  1)  +  viola-
tions;  % average rank of each team
    improved = FV_iteration < FV_current;
    %% ACCEPT WORSE SOLUTIONS IF SEARCH STAGNATED ?  Pacc% OF CHANCE
    if (Stagnation > MaxStagnation-1)
      %        fprintf(' Stagnated!!! Selecting non-improved solu-
tions.\n');
        indexes = find( (improved + (rand(P, 1) < Pacc))  > 0);
        Stagnation = 0;
    else
        indexes = find( improved );
    end;
    %% Update Current Solution with iteration solution
    FV_current(indexes) = FV_iteration(indexes);
    sol_current(indexes,:) = sol_iteration (indexes,:);
    %% Update Global Best
    [FV_Global_New,index] = min(FV_current);
    if FV_Global_New < FV_Global
        FV_Global=FV_Global_New;
        sol_Global=sol_current (index,:);
        Fes=NFE;% NFE_FinalBest = NFE;
    end
    %% % TIME TO UPDATE THE FIRMWARE
    if (rem (itr,  Command_Center_Iter) == 0)
            [Teams_viols,Teams_sols,Teams_FV,Teams_Rank ]=command-
centre(Teams_viols,FV_Global_Old,FV_Global);
        FV_Global_Old=FV_Global;
    end %% Command_Center_Iters
    %% Stagnation Check
 if ( abs(FV_Global - FV_Global_Previous) / abs(FV_Global_Previous)
< 1e-6 )  %% Needs better improvement
        Stagnation = Stagnation + 1;
    else
        Stagnation = 0;
    end
    %% update previous global by current global solution
    FV_Global_Previous=FV_Global;
    xf=[sol_Global,FV_Global];
    if FV_Global<=Ter_Error
        Fes=NFE;
        break
    end
    convergence(counter,1)=NFE;
    convergence(counter,2)=FV_Global;
    counter =counter+1;
end
end
function [Teams_viols,Teams_sols,Teams_FV,Teams_Rank ]=command-
centre(Teams_viols,FV_Global_Old,FV_Global_New)
%N_Teams
```

```
%FirmWare
%Teams_viols,FV_Global_Old,CViol_Global_Old,
FV_Global_New,CViol_Global_New
global P
global FirmWare
global NE
global N_Teams
global Max_N_Teams
global Min_N_Teams
% global NFE
Old_N_Teams=N_Teams;
[~, index]=sort(Teams_viols);
worst_FW_Num=index(N_Teams);
worst_FW=FirmWare(worst_FW_Num);
best_FW_Num=index(1);
best_FW=FirmWare(best_FW_Num);
if NE==1
  h=1;
else
  h=randi(NE);
  while h==worst_FW
    h=randi(NE);
  end
end
FirmWare(index(N_Teams))=h;% number of teams not changed
% if NFE>4e5
%   N_Teams=2;
% else
  if FV_Global_New >= FV_Global_Old
    if N_Teams<Max_N_Teams
      N_Teams=N_Teams+1;
    end
  else
    if N_Teams>Min_N_Teams
      N_Teams=N_Teams-1;
    end
  end
% end
  if Old_N_Teams<N_Teams %% number of teams increased
    FirmWare(N_Teams)=best_FW;
  elseif Old_N_Teams>N_Teams %% number of teams decreased
    FirmWare(worst_FW_Num)=[];
  end
  Teams_viols = zeros(1, N_Teams);
  Teams_sols = cell(1, N_Teams); % matrix of solutions
   Teams_FV = Inf * ones(P, N_Teams);  % matrix of objective func-
tion values
  Teams_Rank = Teams_FV;  % matrix of objective function values
end
function [nx,violations]=boundcorrection(nx,LB,UB,violations)
[~,D]=size(nx);
for i = 1:D
  indexes = find(nx(:,i) < LB(i));
  violations = violations + abs(sum(LB(i)-nx(indexes,i)));
```

```
    if isempty(indexes)==0
      nx(indexes,i) = LB(i);
    end
    indexes = find(nx(:,i) > UB(i));
    violations = violations + abs(sum(nx(indexes,i)-UB(i)));
    if isempty(indexes)==0
      nx(indexes,i) = UB(i);
    end
end
end
function [nx]=updatepopjh(f)
[pop,NT]=size(f);
x=f(:,1:NT-1);
[sorted, ~]=sortrows(f,NT);
b1=sorted(1,1:NT-1);
b2=sorted(2,1:NT-1);
b3=sorted(3,1:NT-1);
b=[b1; b2; b3];
w1=sorted(pop,1:NT-1);
w=repmat(w1,pop,1);
nx=zeros(pop,NT-1);
for j=1:NT-1
   rb=randi(3,[pop,1]);
   r=rand(pop,1);
            nx(:,j)=x(:,j)+r.*(b(rb,j)-abs(x(:,j)))-r.*(w(:,j)-
abs(x(:,j)));
end
end
function [nx]=randupdate(f)
[pop,NT]=size(f);
x=f(:,1:NT);
nx=zeros(pop,NT);
r1=rand;
for i=1:pop
   for j=1:NT
      r1=4*r1*(1-r1);
      nx(i,j)=x(i,j)+rand*(2*r1-1);
   end
end
end
function xn=quasi(umin,umax,x)
[pop,var]=size(x);
a1=(umin+umax)/2;
b1=repmat(umin+umax,pop,1);
a=repmat(a1,pop,1);
b=b1-x;
c=rand(pop,var);
xn= a+(b-a).*c;
end
function [nx,r1]=cjupdate(f,r1)
[pop,NT]=size(f);
x=f(:,1:NT-1);
[~, ind]=min(f(:,NT));
xb=x(ind,:);
```

```
bx=repmat(xb,pop,1);
[~, ind]=max(f(:,NT));
xw=x(ind,:);
wx=repmat(xw,pop,1);
r=zeros(pop,NT-1);
for i=1:pop
   for j=1:NT-1
      r(i,j)=4*r1*(1-r1);
      r1=r(i,j);
   end
end
nx=x+r.*(bx-abs(x))-r.*(wx-abs(x));
end
function [nx,r1]=chaosupdate(f,r1)
[pop,NT]=size(f);
x=f(:,1:NT);
nx=zeros(pop,NT);
for i=1:pop
   for j=1:NT
      r1=4*r1*(1-r1);
      nx(i,j)=x(i,j)+rand*(2*r1-1);
   end
end
end
function [nx]=updatepop(f)
[pop,NT]=size(f);
x=f(:,1:NT-1);
[~, ind]=min(f(:,NT));
xb=x(ind,:);
bx=repmat(xb,pop,1);
[~, ind]=max(f(:,NT));
xw=x(ind,:);
wx=repmat(xw,pop,1);
nx=x+rand(pop,NT-1).*(bx-abs(x))-rand(pop,NT-1).*(wx-abs(x));
% nx=round(nx);
end
%%%%Code Ends %%%%
```

C.7 *Unconstrained_benchmark_func.m: 30 Unconstrained Benchmark Function Program*

This function file needs to be called in the *'All_Algorithms_Main.m'* program to execute any algorithm. This program is only for demonstration purpose. The numbers assigned in this program need not be taken as default values.

```
%%%%Code Begins %%%%
%%%% Unconstrained_benchmark_func.m%%%%
function f=Unconstrained_benchmark_func(x)
global func_num
```

```
global initial_flag
persistent fhd
% benchmark_func.m is the main function for 25 test func-
tions, all minimize
% problems
% e.g. f=benchmark_func(x,func_num)
% x is the variable, f is the function value
% func_num is the function num,

%     30 TEST FUCNTIONS
% Problem No. Problem Name Boundaries Dimension
% 1.   sphere                  Bounds[-100,100]   D=30
% 2.   SumSquares (sum2)    Bounds[-10,10]     D=30
% 3.   Beale                   Bounds[-4.5,4.5]   D=2,5
% 4.   Easom                  Bounds[-100,100]   D=2
% 5.   Matyas                 Bounds[-10,10]     D=2
% 6.   Colville                Bounds[-10,10]     D=4
% 7.   Trid6                   Bounds[-6,6]       D=6
% 8.   Trid10                  Bounds[-10,10]     D=10
% 9.   Zakharov (zakh)        Bounds[-5,10]      D=10
% 10.  Schwefel 1.2(schw12) Bounds[-100,100]   D=30
% 11.  Rosenbrock (rosen)    Bounds[-30,30]     D=30
% 12.  Dixon-Price (dp)       Bounds[-10,10]     D=30
% 13.  Foxholes               Bounds[-65.536,65.536] D=2
% 14.  Branin                 Bounds[0,15]       D=2
% 15.  Bohachevsky 1(bh1) Bounds[-100,100]   D=2
% 16.  Booth           Bounds[-10,10]     D=2
% 17.  Michalewicz 2(mich2)Bounds[0,pi]       D=2
% 18.  Michalewicz 5(mich5)Bounds[0,pi]       D=5
% 19.  Bohachevsky 2(bh2)  Bounds[-100,100]   D=2
% 20.  Bohachevsky 3(bh3)  Bounds[-100,100]   D=2
% 21.  GoldStein-Price(gold)Bounds[-2,2]      D=2
% 22.  Perm            Bounds[-4,4]       D=4
% 23.  Hartman 3         Bounds[0,1]        D=3
% 24.  Ackley           Bounds[-32,32]     D=30
% 25.  Penalized 2        Bounds[-50,50]     D=30
% 26.  Langarman 2        Bounds[0,10]       D=2
% 27.  Langarman 5        Bounds[0,10]       D=5
% 28.  Langarman 10       Bounds[0,10]       D=10
% 29.  Fletcher Powell 5  Bounds[-pi,pi]     D=5
% 30.  Fletcher Powell 1  Bounds[-pi,pi]     D=10

if initial_flag==0
  if func_num==1; fhd=str2func('sphere');
  elseif func_num==2; fhd=str2func('sum2');
  elseif func_num==3; fhd=str2func('beale');
  elseif func_num==4; fhd=str2func('easom');
  elseif func_num==5; fhd=str2func('matyas');
  elseif func_num==6; fhd=str2func('colville');
  elseif func_num==7; fhd=str2func('trid6');
  elseif func_num==8; fhd=str2func('trid10');
  elseif func_num==9; fhd=str2func('zakh');
  elseif func_num==10; fhd=str2func('schw12');
  elseif func_num==11; fhd=str2func('rosen');
  elseif func_num==12; fhd=str2func('dp');
```

```
    elseif func_num==13; fhd=str2func('Foxholes');
    elseif func_num==14; fhd=str2func('branin');
    elseif func_num==15; fhd=str2func('bh1');
    elseif func_num==16; fhd=str2func('booth');
    elseif func_num==17; fhd=str2func('mich2');
    elseif func_num==18; fhd=str2func('mich5');
    elseif func_num==19; fhd=str2func('bh2');
    elseif func_num==20; fhd=str2func('bh3');
    elseif func_num==21; fhd=str2func('gold');
    elseif func_num==22; fhd=str2func('perm');
    elseif func_num==23; fhd=str2func('hartman3');
    elseif func_num==24; fhd=str2func('ackley');
    elseif func_num==25; fhd=str2func('Penalized2');
    elseif func_num==26; fhd=str2func('Langarman2');
    elseif func_num==27; fhd=str2func('Langarman5');
    elseif func_num==28; fhd=str2func('Langarman10');
    elseif func_num==29; fhd=str2func('FletcherPowell5');
    elseif func_num==30; fhd=str2func('FletcherPowell10');
    end
end
[row,~]=size(x);
f=nan(row,1);
for i=1:row
    f(i)=feval(fhd,x(i,:));
end
function [f]=FletcherPowell10(x)
% Bounds[-pi,pi]      D=10
% FletcherPowell-10 function.
%
a=[-79 56 -62 -9 92 48 -22 -34 -39 -40;91 -9 -18 -59 99 -45 88 -14
-29 26;-38 8 -12 -73 40 26 -64 29 -82 -32;-78 -18 -49 65 66 -40 88
-95 -57 10;-1 -43 93 -18 -76 -68 -42 22 46 -14;34 -96 26 -56 -36 -85
-62 13 93 78;52 -46 -69 99 -47 -72 -11 55 -55 91;81 47 35 55 67 -13 33
14 83 -42;-50 66 -47 -75 89 -16 82 6 -85 -62];
b=[-65 -11 76 78 30 93 -86 -99 -37 52;59 67 49 -45 52 -33 -34 29
-39 -80;21 -23 -80 86 86 -30 39 -73 -91 5;-91 -75 20 -64 -15 17 -89 36
-49 -2;-79 99 -31 -8 -67 -72 -43 -55 76 -57;-89 -35 -55 75 15 -6 -53
-56 -96 87;-76 45 74 12 -12 -69 2 71 75 -60;-50 -88 93 68 10 -13 84
-21 65 14;-23 -95 99 62 -37 96 27 69 -64 -92;-5 -57 -30 -6 -96 75 25
-6 96 77];
alpha=[-2.7910 2.5623 -1.0429 0.5097 -2.8096 1.1883 2.0771 -2.9926
0.0715 0.4142];
xx=x;
s1=0;
for i=1:2
    sA=0;
    sB=0;
    for j=1:2
        sA=sA+(a(i,j)*sin(alpha(j)))+(b(i,j)*cos(alpha(j)));
        sB=sB+(a(i,j)*sin(xx(j)))+(b(i,j)*cos(xx(j)));
    end
    A(i)=sA;
    B(i)=sB;
    s1=s1+(A(i)-B(i))^2;
```

```
end
f=s1;

function [f]=FletcherPowell5(x)
% Bounds[-pi,pi]      D=5
% FletcherPowell-5 function.
%
a=[-79 56 -62 -9 92;91 -9 -18 -59 99;-38 8 -12 -73 40;-78 -18 -49
65 66;-1 -43 93 -18 -76];
b=[-65 -11 76 78 30;59 67 49 -45 52;21 -23 -80 86 86;-91 -75 20 -64
-15;-79 99 -31 -8 -67];
alpha=[-2.7910 2.5623 -1.0429 0.5097 -2.8096];
xx=x;
s1=0;
for i=1:2
   sA=0;
   sB=0;
   for j=1:2
      sA=sA+(a(i,j)*sin(alpha(j)))+(b(i,j)*cos(alpha(j)));
      sB=sB+(a(i,j)*sin(xx(j)))+(b(i,j)*cos(xx(j)));
   end
   A(i)=sA;
   B(i)=sB;
   s1=s1+(A(i)-B(i))^2;
end
f=s1;

function [f]=Langarman10(x)
% Bounds[0,10] D=10
% Langarman10 function.
%
d = size(x,2);
m=10;
a=[9.681 0.667 4.783 9.095 3.517 9.325 6.544 0.211 5.122 2.020;9.4
2.041 3.788 7.931 2.882 2.672 3.568 1.284 7.033 7.374;8.025 9.152
5.114 7.621 4.564 4.711 2.996 6.126 0.734 4.982;2.196 0.415 5.649
6.979 9.510
9.166 6.304 6.054 9.377 1.426;8.074 8.777 3.467 1.863 6.708 6.349
4.534
0.276 7.633 1.567;7.650 5.658 0.720 2.764 3.278 5.283 7.474 6.274
1.409
8.208;1.256 3.605 8.623 6.905 0.584 8.133 6.071 6.888 4.187
5.448;8.314
2.261 4.224 1.781 4.124 0.932 8.129 8.658 1.208 5.762;0.226 8.858
1.420
0.945 1.622 4.698 6.228 9.096 0.972 7.637;7.305 2.228 1.242 5.928
9.133
1.826 4.060 5.204 8.713 8.247];
C=[0.806 0.517 1.5 0.908 0.965 0.669 0.524 0.902 0.531 0.876];
xx=x;
s2=0;
for i=1:m
s1=0;
for j=1:d
s1=s1+(xx(j)-a(i,j))^2;
```

```
end
s2=s2+C(i)*(exp((-1/pi)*s1))*(cos(pi*s1));
end
f=-s2;

function [f]=Langarman5(x)
% Bounds[0,10] D=5
% Langarman5 function.
%
d = size(x,2);
m=5;
a=[9.681 0.667 4.783 9.095 3.517 9.325 6.544 0.211 5.122 2.020;9.4
2.041 3.788 7.931 2.882 2.672 3.568 1.284 7.033 7.374;8.025 9.152
5.114
7.621 4.564 4.711 2.996 6.126 0.734 4.982;2.196 0.415 5.649 6.979
9.510
9.166 6.304 6.054 9.377 1.426;8.074 8.777 3.467 1.863 6.708 6.349
4.534
0.276 7.633 1.567;7.650 5.658 0.720 2.764 3.278 5.283 7.474 6.274
1.409
8.208;1.256  3.605  8.623  6.905  0.584  8.133  6.071  6.888  4.187
5.448;8.314
2.261 4.224 1.781 4.124 0.932 8.129 8.658 1.208 5.762;0.226 8.858
1.420
0.945 1.622 4.698 6.228 9.096 0.972 7.637;7.305 2.228 1.242 5.928
9.133
1.826 4.060 5.204 8.713 8.247];
C=[0.806 0.517 1.5 0.908 0.965 0.669 0.524 0.902 0.531 0.876];
xx=x;
s2=0;
for i=1:m
s1=0;
for j=1:d
s1=s1+(xx(j)-a(i,j))^2;
end
s2=s2+C(i)*(exp((-1/pi)*s1))*(cos(pi*s1));
end
f=-s2;

function [f]=Langarman2(x)
% Bounds[0,10] D=2
% Langarman2 function.
%
d = size(x,2);
m=2;
a=[9.681 0.667 4.783 9.095 3.517 9.325 6.544 0.211 5.122 2.020;9.4
2.041 3.788 7.931 2.882 2.672 3.568 1.284 7.033 7.374;8.025 9.152
5.114
7.621 4.564 4.711 2.996 6.126 0.734 4.982;2.196 0.415 5.649 6.979
9.510
9.166 6.304 6.054 9.377 1.426;8.074 8.777 3.467 1.863 6.708 6.349
4.534
0.276 7.633 1.567;7.650 5.658 0.720 2.764 3.278 5.283 7.474 6.274
1.409
```

```
8.208;1.256 3.605 8.623 6.905 0.584 8.133 6.071 6.888 4.187
5.448;8.314
2.261 4.224 1.781 4.124 0.932 8.129 8.658 1.208 5.762;0.226 8.858
1.420
0.945 1.622 4.698 6.228 9.096 0.972 7.637;7.305 2.228 1.242 5.928
9.133
1.826 4.060 5.204 8.713 8.247];
C=[0.806 0.517 1.5 0.908 0.965 0.669 0.524 0.902 0.531 0.876];
xx=x;
s2=0;
for i=1:m
s1=0;
for j=1:d
s1=s1+(xx(j)-a(i,j))^2;
end
s2=s2+C(i)*(exp((-1/pi)*s1))*(cos(pi*s1));
end
f=-s2;

function [f]=Penalized2(x)
% 30 [-50, 50]
% Penalized2 function.
%
c = size(x,2);
a=5;k=100;m=4;
u=[];
xx=x;
for i=1:c
  if(xx(i)>a)
    u(i)=k*((xx(i)-a)^m);
  end
  if(xx(i)>=-a && xx(i)<=a)
    u(i)=0;
  end
  if(xx(i)<-a)
    u(i)=k*(-xx(i)-a)^m;
  end
end
s1=0;
for i=1:c-1
            s1=s1+((xx(i)-1)^2)*(1+(sin(3*pi*xx(i+1)))^2)+((xx(c)-
1)^2)*(1+(sin(2*pi*xx(c)))^2);
end
f=0.1*(((sin(pi*xx(1)))^2)+s1)+sum(u);

function y = ackley(x)
% 30 [-32, 32]
% Ackley function.
%
n = size(x,2);
a = 20; b = 0.2; c = 2*pi;
s1 = 0; s2 = 0;
for i=1:n;
  s1 = s1+x(i)^2;
  s2 = s2+cos(c*x(i));
```

```
end
y = -a*exp(-b*sqrt((1/n)*s1))-exp((1/n)*s2)+a+exp(1);

function [f]=hartman3(x)
% 3 [0, 1]
% hartman3 function
c = size(x,2);
C=[1.0 1.2 3.0 3.2];
P=(10^-4)*[3689 1170 2673;4699 4387 7470;1091 8732 5547;381 5743
8828];
a=[3 10 30;0.1 10 35;3.0 10 30;0.1 10 35];
s1=0;
for i=1:4
s2=0;
for j=1:c
s2=s2+a(i,j)*(x(j)-P(i,j))^2;
end
s1=s1+C(i)*exp(-s2);
end
f=-s1;

function y = perm(x)
% 4 [-D, D]
% Perm function
%
n = 2;
b = 10;
s_out = 0;
for k = 1:n;
   s_in = 0;
   for j = 1:n
      s_in = s_in+(j+b)*(x(j)^k-(1/j)^k);
   end
   s_out = s_out+s_in^2;
end
y = s_out;

function y = gold(x)
% 2 [-2, 2]
% Goldstein and Price function
%
a = 1+ ((x(1)+x(2)+1)^2) * (19-14*x(1)+3*x(1)^2 - 14*x(2) +
6*x(1)*x(2) + 3*x(2)^2);
b = 30+ (2*x(1)-3*x(2))^2 * (18 - 32*x(1) + 12*x(1)^2 + 48 * x(2) -
36*x(1)*x(2) + 27*x(2)^2);
y = a*b;

function y = bh3(x)
% 2 [-100 100]
% Bohachecsky function 3
%
y = x(1)^2+2*x(2)^2-0.3*cos(3*pi*x(1)+4*pi*x(2))+0.3;

function y = bh2(x)
% 2 [-100 100]
```

```
% Bohachecsky function 2
%
% z(i)=(x1^2)+(2*x2^2)-0.3*cos((3*pi*x1)*(4*pi*x2))+0.3;
y = x(1)^2+2*x(2)^2-0.3*cos(3*pi*x(1))*cos(4*pi*x(2))+0.3;

% function y = mich10(x)
% % 10 [0, pi]
% % Michalewicz function
% %
% n = size(x,2);
% m = 10;
% s = 0;
% for i = 1:n;
%    s = s+sin(x(i))*(sin(i*x(i)^2/pi))^(2*m);
% end
% y = -s;

function y = mich5(x)
% 5 [0, pi]
% Michalewicz function
%
n = size(x,2);
% m = 10;
s = 0;
for i = 1:n;
%    s = s+sin(x(i))*(sin(i*x(i)^2/pi))^(2*m);
   s = s+sin(x(i))*(sin(i*x(i)^2/pi))^20;
end
y = -s;

function y = mich2(x)
% 2 [0, pi]
% Michalewicz function
%
n = size(x,2);
m = 10;
s = 0;
for i = 1:n;
   s = s+sin(x(i))*(sin(i*x(i)^2/pi))^(2*m);
end
y = -s;

function y = booth(x)
% 2 [-10, 10]
% Booth function
%
y  = (x(1)+2*x(2)-7)^2+(2*x(1)+x(2)-5)^2;

function y=bh1(x)
% 2 [-100 100]
% Bohachecsky function 1
%
y=x(1)^2+2*x(2)^2-0.3*cos(3*pi*x(1))-0.4*cos(4*pi*x(2))+0.7;

function y = branin(x)
```

```
% 2 [0, 15]
% Branin function
%
% x1=x(1);
% x2=x(2);
%              z=(x2-(5.1/(4*pi^2))*(x1^2)+(5/pi)*x1-6)^2+10*(1-
(1/(8*pi)))*cos(x1)+10;
y       =      (x(2)-(5.1/(4*pi^2))*x(1)^2+5*x(1)/pi-6)^2+10*(1-
1/(8*pi))*cos(x(1))+10;

function [f]=Foxholes(x)
% 2 [-65.536, 65.536]
% Foxholes function.
%
a=[-32 -32;-16 -32;0 -32;16 -32;32 -32;-32 -16;-16 -16;0 -16;16
-16;32 -16;-32 0;-16 0;0 0;16 0;32 0;-32 16;-16 16;0 16;16 16;32
16;-32 32;-16 32;0 32;16 32;32 32];
s2=0;
for j=1:25
  s1=0;
  for i=1:2
    s1=s1+(x(i)-a(j,i))^6;
  end
  s2=s2+(1/(j+s1));
end
f=((1/500)+s2)^-1;

function y = dp(x)
% 30 [-10, 10]
% Dixon and Price function.
%
n =size(x,2);
s1 = 0;
for j = 2:n;
%   s1 = s1+j*( (2 * ( x(j)^2 ) - x(j-1) )^2 );
  s1=s1+j*(2*x(j)^2-x(j-1))^2;

end
y = s1+((x(1)-1)^2);

function y = rosen(x)
% 30 [-30, 30] UN
% Rosenbrock function
%
n =size(x,2);
sum = 0;
for j = 1:n-1;
  sum = sum+100 * (( x(j)^2 -x(j+1) )^2+( x(j)-1 )^2);
end
y = sum;

function y = schw12(x)
% 30 [-100, 100]
% Schwefel function
%
```

```
n =size(x,2);
sum=0;
for i=1:n
   sum2=0;
   for j=1:i
      sum2=sum2+(x(1,j))^2;
   end
   sum=sum+(sum2^2);
end
y =sum;

function y = zakh(x)
% 10 [-5, 10]
% Zakharov function
%
n = size(x,2);
s1 = 0;
s2 = 0;
for j = 1:n;
   s1 = s1+x(j)^2;
   s2 = s2+0.5*j*x(j);
end
y = s1+s2^2+s2^4;

function  y = trid10(x)
% 6 [-D2, D2] 10 [-D2, D2]
% Trid function
%
n = size(x,2);
s1 = 0;
s2 = 0;
for j = 1:n;
   s1 = s1+(x(j)-1)^2;
end
for j = 2:n;
   s2 = s2+x(j)*x(j-1);
end
y = s1-s2;

function  y = trid6(x)
% 6 [-D2, D2] 10 [-D2, D2]
% Trid function
%
n = size(x,2);
s1 = 0;
s2 = 0;
for j = 1:n;
   s1 = s1+(x(j)-1)^2;
end
for j = 2:n;
   s2 = s2+x(j)*x(j-1);
end
y = s1-s2;

function y = colville(x)
```

```
% 4 [-10, 10]
% Colville function
%
x1=x(1);
x2=x(2);
x3=x(3);
x4=x(4);
y=100*(((x1^2)-x2)^2)+((x1-1)^2)+((x3-1)^2)+90*(((x3^2)
-x4)^2)+10.1*(((x2-1)^2)+((x4-1)^2))+19.8*(x2-1)*(x4-1);

function y = matyas(x)
% 2 [-10, 10]
% Matyas function
%
y = 0.26*(x(1)^2+x(2)^2)-0.48*x(1)*x(2);

function y = easom(x)
% 2 [-100, 100]
% Easom function
%
y = -cos(x(1))*cos(x(2))*exp(-(x(1)-pi)^2-(x(2)-pi)^2);

function y = beale(x)
% 2, 5 [-4.5, 4.5]
% Beale function.
%
y   =    (1.5-x(1)*(1-x(2)))^2+(2.25-x(1)*(1-x(2)^2))^2+(2.625-
x(1)*(1-x(2)^3))^2;

function y = sum2(x)
% 30 [-10, 10]
% Sum Squares function
%
[~, n]=size(x);
s = 0;
for j = 1:n
  s=s+j*x(j)^2;
end
y = s;

function y = sphere(x)
% 30 [-100, 100]
% Sphere function
%
[~, n]=size(x);
s = 0;
for j = 1:n
  s = s+x(j)^2;
end
y = s;
%%%%Code Ends %%%%
```

C.8 DandBounds.m: 30 Unconstrained Benchmark Function Variable Dimensions and Their Boundary Programs

This function file needs to be called in the *'All_Algorithms_Main.m'* program to execute any algorithm. This program is only for demonstration purpose. The numbers assigned in this program need not be taken as default values.

```
%%%%Code Begins %%%%
%%%% DandBounds.m%%%%
function [D,L,U,Opt_Val]=DandBounds(func_num)
if func_num==1
   % 1.   sphere
   L=-100;U=100;D=30;Opt_Val=0;
elseif func_num==2
   % 2.   SumSquares (sum2)
   L=-10;U=10;D=30;Opt_Val=0;
elseif func_num==3
   % 3.   Beale
   L=-4.5;U=4.5;D=2;Opt_Val=0;
elseif func_num==4
   % 4.   Easom
   L=-100;U=100;D=2;Opt_Val=-1;
elseif func_num==5
   % 5.   Matyas
   L=-10;U=10;D=2;Opt_Val=0;

elseif func_num==6
   % 6.   Colville
   L=-10;U=10;D=4;Opt_Val=0;

elseif func_num==7
   % 7.   Trid6
   L=-36;U=36;D=6;Opt_Val=-49.9999999999984;

elseif func_num==8
   % 8.   Trid10
   L=-100;U=100;D=10 ;Opt_Val=-210;

elseif func_num==9
   % 9.   Zakharov (zakh)
   L=-5;U=10;D=10;Opt_Val=0;

elseif func_num==10
   % 10.  Schwefel 1.2(schw12)
   L=-100;U=100;D=30;Opt_Val=0;

elseif func_num==11
   % 11.  Rosenbrock (rosen)
   L=-30;U=30;D=30;Opt_Val=0;

elseif func_num==12
   % 12.  Dixon-Price (dp)
   L=-10;U=10;D=30;Opt_Val=0;
```

```
elseif func_num==13
   % 13.  Foxholes
   L=-65.536;U=65.536;D=2;Opt_Val=0.99800383;

elseif func_num==14
   % 14.  Branin
   L=0;U=15;D=2;Opt_Val=0.397887;

elseif func_num==15
   % 15.  Bohachevsky 1(bh1)
   L=-100;U=100;D=2;Opt_Val=0;

elseif func_num==16
   % 16.  Booth
   L=-10;U=10;D=2;Opt_Val=0;

elseif func_num==17
   % 17.  Michalewicz 2(mich2)
   L=0;U=pi;D=2;Opt_Val=-1.801303;

elseif func_num==18
   % 18.  Michalewicz 5(mich5)
   L=0;U=pi;D=5;Opt_Val=-4.6877;

elseif func_num==19
   % 19.  Bohachevsky 2(bh2)
   L=-100;U=100;D=2;Opt_Val=0;

elseif func_num==20
   % 20.  Bohachevsky 3(bh3)
   L=-100;U=100;D=2;Opt_Val=0;

elseif func_num==21
   % 21.  GoldStein-Price(gold)
   L=-2;U=2;D=2;Opt_Val=3;

elseif func_num==22
   % 22.  Perm
   L=-4;U=4;D=4;Opt_Val=0;

elseif func_num==23
   % 23.  Hartman 3
   L=0;U=1;D=3;Opt_Val=-3.862779;

elseif func_num==24
   % 24.  Ackley
   L=-32;U=32;D=30;Opt_Val=0;

elseif func_num==25
   % 25.  Penalized 2
   L=-50;U=50;D=30;Opt_Val=1e-16;

elseif func_num==26
   % 26.  Langarman 2
   L=0;U=10;D=2;Opt_Val=-1.08094;

elseif func_num==27
```

```
    % 27.  Langarman 5
    L=0;U=10;D=5;Opt_Val=-1.49;

elseif func_num==28
    % 28.  Langarman 10
    L=0;U=10;D=10;Opt_Val=-1.0528;

elseif func_num==29
    % 29.  Fletcher Powell 5
    L=-pi;U=pi;D=5;Opt_Val=0;

elseif func_num==30
    % 30.  Fletcher Powell 1
    L=-pi;U=pi;D=2;Opt_Val=0;
end
%%%%Code Ends %%%%
```

Index

A

Adaptive multi-team perturbation guiding Jaya, 61

B

Benchmark functions, 353, 375, 386
Bioenergy system, 5, 23, 37, 243

C

Codes, 352, 353, 355, 358, 362, 365, 370, 375, 386
Complex proportional assessment (COPRAS), 136, 138, 163, 164, 166, 168, 169, 177, 178, 180, 200–202, 208, 209, 217, 218, 223, 231, 232, 237, 245, 246, 271–273, 287, 288, 310, 311, 318, 319, 321, 324, 325, 338, 340–342
Compression ignition biodiesel engine, 26, 288
Compromise ranking method (VIKOR), 135, 138, 163, 164, 166, 169, 177, 178, 180, 201, 202, 208, 217, 218, 223, 224, 231, 232, 237, 245, 246, 271–273, 287, 288, 308, 310, 311, 319, 320, 324, 325, 338, 341, 342
Cost model, 13
Coverage, 53, 80, 170, 172, 174, 178, 200, 203, 204, 210, 212–214, 219, 226–229, 233, 240–242, 246, 271, 273, 274, 282, 284, 288, 289, 311, 313, 323, 327
Crowding distance, 66, 68, 70, 71, 79, 80, 216

E

Elitist Rao algorithm, 71–75, 82, 91, 93, 103, 106, 114, 122, 142, 152, 153, 156, 160, 166, 179, 180, 205, 206, 219, 221, 234, 236, 251, 276, 289, 316, 330, 332, 337, 358

G

Geothermal energy system, 4, 5, 29, 41, 43, 327, 329
Gray relational analysis (GRA), 39, 137, 138, 163, 164, 166, 168, 169, 178, 180, 201, 202, 208, 217, 218, 223, 224, 231, 232, 237, 238, 245, 246, 271–273, 278, 279, 287, 288, 310, 311, 319, 320, 324, 325, 338, 341, 342
Ground source heat pump-radiant ceiling air conditioning system, 11, 29, 329, 331

H

Hydro energy system, 3
Hydro power generation, 29, 39, 41, 327, 329
Hypervolume, 53, 81, 170, 171, 174, 178, 200, 203, 204, 210–212, 214, 219, 226, 227, 229, 233, 240, 241, 246, 313, 322, 326, 327, 339, 340, 343

I

Inverted generational distance, 53, 81

V. R. Ravipudi et al., *Design Optimization of Renewable Energy Systems Using Advanced Optimization Algorithms*, Green Energy and Technology,
https://doi.org/10.1007/978-3-030-95589-2

J

Jaya algorithm, 40–42, 45, 53–55, 57–59, 61–63, 66–68, 82, 83, 90, 91, 93, 98, 103, 104, 106, 110, 115, 122, 141–146, 149–151, 159, 160, 166, 167, 170–172, 179–181, 189, 200, 203, 205, 206, 210–214, 216, 219, 221, 222, 225–228, 234, 235, 238–242, 249, 251, 252, 254, 271, 273, 275–278, 282, 285, 289, 290, 292, 308, 311, 314–316, 323, 327, 329, 330, 332, 335, 337, 340, 343, 352, 353, 355, 365, 370

M

Microalgae-based biomass cultivation process, 28, 249, 313–323, 326, 327
Multi-attribute decision-making methods, 123, 131
Multi-objective AMTPG-Jaya algorithm, 66
Multi-objective elitist Rao algorithm, 74
Multi-objective Jaya algorithm, 66
Multi-objective optimization, 14, 21, 23, 26, 33, 35–37, 39, 42–44, 53, 66–72, 74, 75, 79–82, 110, 114, 116–120, 122, 123, 131, 138, 159, 160, 179, 234, 249, 251, 275, 288, 313, 329, 332, 350
Multi-objective Rao algorithm, 70, 74
Multi-objective SAP-Rao algorithm, 79
Multi-team perturbation guiding Jaya algorithm, 57, 59

N

Nuclear energy system, 6, 7

O

Ocean thermal energy system, 4

P

Power generation model, 13
Preference ranking organization method for enrichment evaluations (PROMETHEE), 132, 133, 138, 163, 164, 166, 168, 169, 177, 178, 180, 201, 202, 208, 217, 218, 223, 224, 231, 232, 237, 238, 245, 246, 271–273, 278, 279, 287, 288, 310, 311, 319, 324, 325, 338, 341, 342

R

Rao algorithm, 45, 53, 69–72, 74, 76, 77, 79, 82, 83, 90, 91, 93, 98, 103, 104, 106, 110, 114, 115, 123, 139, 141–147, 149–151, 154, 155, 159, 160, 162, 164, 166, 169–174, 179, 180, 191, 200, 203–206, 210–213, 216, 219, 221, 224–229, 234–236, 240, 241, 243, 249, 251, 268, 271, 275, 276, 281, 282, 285, 289, 306, 308, 311, 313–316, 320, 322, 323, 327, 329–332, 335, 337, 340, 343, 352, 353, 355, 358, 362

S

Self-adaptive population algorithm, 53, 74, 362
Simple additive weighing (SAW), 132, 138, 163, 164, 166, 168, 169, 177, 178, 180, 201, 202, 208, 217, 218, 223, 231, 232, 237, 245, 246, 271–273, 287, 288, 308, 310, 311, 318, 319, 321, 324, 325, 338, 340–342
Single-cylinder direct-injection diesel engine, 24, 39, 249, 252, 254, 256, 258, 260, 262, 264, 266, 268, 270, 273, 274
Single-objective optimization, 53, 54, 63, 66, 71, 73, 76, 82, 110, 160, 161, 179, 205, 219, 220, 233, 234, 251, 276, 289, 329, 330, 345, 352
Solar-assisted Brayton engine system, 161, 162, 167–176
Solar-assisted Carnot-like heat engine system, 11, 22, 23, 159, 230
Solar-assisted Stirling engine system, 11, 17–19, 21, 35, 159, 174, 176, 179, 181, 183, 185, 187, 189, 191, 193, 195, 197, 199, 200, 203–216, 219–230, 233
Solar energy system, 1, 2
Spacing, 53, 80, 81, 114–117, 119, 120, 170, 171, 174, 178, 200, 203, 204, 211, 212, 214, 219, 226, 227, 229, 233, 240, 241, 246, 251, 270, 271, 273, 281, 284–286, 289, 308, 311–313, 322, 326, 327, 339, 340, 343

T

Technique for order preference by similarity to ideal solution

(TOPSIS), 43, 133, 134, 138, 160, 163, 164, 166–170, 174, 175, 177–180, 199–202, 204–207, 209, 210, 215–218, 223, 231–233, 236–238, 240, 245, 271–273, 287, 288, 308, 310, 311, 318, 321, 324, 325, 338, 340–342
Turbocharged DI diesel engine, 25, 273, 277–286, 288

U

Unconstrained benchmark problems, 82, 83, 90, 91, 352
Unimodal and multimodal benchmark problems, 99, 104, 106, 346–350

W

Wake model, 11, 35
Weighted product method (WPM), 132, 138, 163, 164, 166, 168, 169, 177, 180, 200–202, 208, 209, 217, 218, 223, 231, 232, 245, 246, 271–273, 287, 288, 308, 310, 311, 318, 319, 321, 324, 325, 338, 340, 342
Wind energy system, 2, 3
Wind farm layout, 11, 14, 33, 34, 139, 141, 149, 151–156